Building Statistical Models in Python

Develop useful models for regression, classification, time series, and survival analysis

Huy Hoang Nguyen

Paul N Adams

Stuart J Miller

<packt>

BIRMINGHAM—MUMBAI

Building Statistical Models in Python

Copyright © 2023 Packt Publishing

All rights reserved. No part of this book may be reproduced, stored in a retrieval system, or transmitted in any form or by any means, without the prior written permission of the publisher, except in the case of brief quotations embedded in critical articles or reviews.

Every effort has been made in the preparation of this book to ensure the accuracy of the information presented. However, the information contained in this book is sold without warranty, either express or implied. Neither the author(s), nor Packt Publishing or its dealers and distributors, will be held liable for any damages caused or alleged to have been caused directly or indirectly by this book.

Packt Publishing has endeavored to provide trademark information about all of the companies and products mentioned in this book by the appropriate use of capitals. However, Packt Publishing cannot guarantee the accuracy of this information.

Group Product Manager: Ali Abidi

Publishing Product Manager: Sanjana Gupta

Senior Editor: Sushma Reddy

Technical Editor: Rahul Limbachiya

Copy Editor: Safis Editing

Book Project Manager: Kirti Pisat

Project Coordinator: Farheen Fathima

Proofreader: Safis Editing

Indexer: Hemangini Bari

Production Designer: Prashant Ghare

Marketing Coordinator: Nivedita Singh

First published: August 2023

Production reference: 3310823

Published by Packt Publishing Ltd.
Grosvenor House
11 St Paul's Square
Birmingham
B3 1RB, UK.

ISBN 978-1-80461-428-0

www.packtpub.com

To my parents, Thieu and Tang, for their enormous support and faith in me.

To my wife, Tam, for her endless love, dedication, and courage.

- Huy Hoang Nguyen

To my daughter, Lydie, for demonstrating how work and dedication regenerate inspiration and creativity. To my wife, Helene, for her love and support.

– Paul Adams

To my partner, Kate, who has always supported my endeavors.

– Stuart Miller

Contributors

About the authors

Huy Hoang Nguyen is a mathematician and data scientist with extensive experience in advanced mathematics, strategic leadership, and applied machine learning research. He holds a PhD in Mathematics, as well as two Master's degrees in Applied Mathematics and Data Science. His previous work focused on Partial Differential Equations, Functional Analysis, and their applications in Fluid Mechanics. After transitioning from academia to the healthcare industry, he has undertaken a variety of data science projects, ranging from traditional machine learning to deep learning.

Paul Adams is a Data Scientist with a background primarily in the healthcare industry. Paul applies statistics and machine learning in multiple areas of industry, focusing on projects in process engineering, process improvement, metrics and business rules development, anomaly detection, forecasting, clustering, and classification. Paul holds an MSc in Data Science from Southern Methodist University.

Stuart Miller is a Machine Learning Engineer with a wide range of experience. Stuart has applied machine learning methods to various projects in industries ranging from insurance to semiconductor manufacturing. Stuart holds degrees in data science, electrical engineering, and physics.

About the reviewers

Krishnan Raghavan is an IT Professional with over 20+ years of experience in software development and delivery excellence across multiple domains and technology ranging from C++ to Java, Python, Data Warehousing, and Big Data tools and technologies.

When not working, Krishnan likes to spend time with his wife and daughter, reading fiction and nonfiction as well as technical books. Krishnan tries to give back to the community by being part of the GDG Pune Volunteer Group, helping the team organize events. Currently, he is unsuccessfully trying to learn how to play the guitar.

You can connect with Krishnan at `mailtokrishnan@gmail.com` or via **LinkedIn**: `www.linkedin.com/in/krishnan-raghavan`.

I would like to thank my wife Anita and daughter Ananya for giving me the time and space to review this book.

Karthik Dulam is a Principal Data Scientist at EDB. He is passionate about all things data with a particular focus on data engineering, statistical modeling, and machine learning. He has a diverse background delivering machine learning solutions for the healthcare, IT, automotive, telecom, tax, and advisory industries. He actively engages with students as a guest speaker at esteemed universities delivering insightful talks on machine learning use cases.

I would like to thank my wife, Sruthi Anem, for her unwavering support and patience. I also want to thank my family, friends, and colleagues who have played an instrumental role in shaping the person I am today. Their unwavering support, encouragement, and belief in me have been a constant source of inspiration.

Table of Contents

Preface   xiii

Part 1: Introduction to Statistics

1

Sampling and Generalization   3

Software and environment setup	3	Sampling strategies – random, systematic, stratified, and clustering	11
Population versus sample	6	Probability sampling	11
Population inference from samples	8	Non-probability sampling	16
Randomized experiments	8	Summary	17
Observational study	9		

2

Distributions of Data   19

Technical requirements	19	Measuring shape	38
Understanding data types	20	The normal distribution and central limit theorem	42
Nominal data	20		
Ordinal data	21	The Central Limit Theorem	45
Interval data	21	Bootstrapping	45
Ratio data	22	Confidence intervals	46
Visualizing data types	22	Standard error	51
Measuring and describing distributions	26	Correlation coefficients (Pearson's correlation)	51
Measuring central tendency	26	Permutations	52
Measuring variability	33	Permutations and combinations	52
		Permutation testing	55

Transformations 57
Summary 59
References 59

3

Hypothesis Testing 61

The goal of hypothesis testing 61
Overview of a hypothesis test for the mean 62
Scope of inference 62
Hypothesis test steps 63

Type I and Type II errors 63
Type I errors 63
Type II errors 64

Basics of the z-test – the z-score, z-statistic, critical values, and p-values 65
The z-score and z-statistic 65
A z-test for means 72
z-test for proportions 78
Power analysis for a two-population pooled z-test 82

Summary 85

4

Parametric Tests 87

Assumptions of parametric tests 87
Normally distributed population data 88
Equal population variance 99

T-test – a parametric hypothesis test 102
T-test for means 103
Two-sample t-test – pooled t-test 108
Two-sample t-test – Welch's t-test 111
Paired t-test 112

Tests with more than two groups and ANOVA 114
Multiple tests for significance 114
ANOVA 117
Pearson's correlation coefficient 118
Power analysis examples 123

Summary 124
References 124

5

Non-Parametric Tests 125

When parametric test assumptions are violated 125
Permutation tests 126

The Rank-Sum test 128
The test statistic procedure 128
Normal approximation 129
Rank-Sum example 129

The Signed-Rank test 130

The Kruskal-Wallis test	132	Chi-square goodness-of-fit test power analysis	138
Chi-square distribution	133		
Chi-square goodness-of-fit	135	Spearman's rank correlation coefficient	139
Chi-square test of independence	136		
		Summary	141

Part 2: Regression Models

6

Simple Linear Regression   145

Simple linear regression using OLS	145	Homoscedasticity of the residuals	154
Coefficients of correlation and determination	148	Sample independence	155
		Testing for significance and validating models	155
Coefficients of correlation	148		
Coefficients of determination	151	Model validation	169
Required model assumptions	152	Summary	171
A linear relationship between the variables	152		
Normality of the residuals	153		

7

Multiple Linear Regression   173

Multiple linear regression	173	Ridge regression	189
Adding categorical variables	175	LASSO regression	192
Evaluating model fit	176	Elastic Net	194
Interpreting the results	181	Dimension reduction	196
Feature selection	184	PCA – a hands-on introduction	196
Statistical methods for feature selection	184	PCR – a hands-on salary prediction study	199
Performance-based methods for feature selection	186	Summary	202
Recursive feature elimination	187		
Shrinkage methods	188		

Part 3: Classification Models

8

Discrete Models — 205

Probit and logit models	205	The negative binomial regression model	217
Multinomial logit model	210		
Poisson model	213	Negative binomial distribution	217
The Poisson distribution	213	Summary	224
Modeling count data	215		

9

Discriminant Analysis — 225

Bayes' theorem	225	Linear Discriminant Analysis	229
Probability	225	Supervised dimension reduction	236
Conditional probability	227	Quadratic Discriminant Analysis	238
Discussing Bayes' Theorem	228	Summary	244

Part 4: Time Series Models

10

Introduction to Time Series — 247

What is a time series?	248	The white-noise model	263
Goals of time series analysis	249	Stationarity	265
Statistical measurements	249	Summary	270
Mean	249	References	270
Variance	251		
Autocorrelation	253		
Cross-correlation	257		

11

ARIMA Models — 271

Technical requirements	271	Models for non-stationary time series	295
Models for stationary time series	272	ARIMA models	296
Autoregressive (AR) models	272	Seasonal ARIMA models	304
Moving average (MA) models	283	More on model evaluation	311
Autoregressive moving average (ARMA) models	287	Summary	318
		References	319

12

Multivariate Time Series — 321

Multivariate time series	321	Step 2 – selecting the order of AR(p)	339
Time-series cross-correlation	322	Step 3 – assessing cross-correlation	340
ARIMAX	326	Step 4 – building the VAR(p,q) model	344
Preprocessing the exogenous variables	328	Step 5 – testing the forecast	346
Fitting the model	329	Step 6 – building the forecast	347
Assessing model performance	333	Summary	349
VAR modeling	335	References	349
Step 1 – visual inspection	338		

Part 5: Survival Analysis

13

Time-to-Event Variables – An Introduction — 353

What is censoring?	354	Survival data	356
Left censoring	354	Survival Function, Hazard and Hazard Ratio	357
Right censoring	354	Summary	360
Interval censoring	355		
Type I and Type II censoring	355		

14

Survival Models — 361

Technical requirements	361	Cox Proportional Hazards	
Kaplan-Meier model	362	regression model	372
Model definition	362	Step 1	374
Model example	364	Step 2	375
		Step 3	379
Exponential model	368	Step 4	380
Model example	370	Step 5	383
		Summary	384

Index — 385

Other Books You May Enjoy — 396

Preface

Statistics is a discipline of study used for applying analytical methods to answer questions and solve problems using data, in both academic and industry settings. Many methods have been around for centuries, while others are much more recent. Statistical analysis and results are fairly straightforward for presenting to both technical and non-technical audiences. Furthermore, producing results with statistical analysis does not necessarily require large amounts of data or compute resources and can be done fairly quickly, especially when using programming languages such as Python, which is moderately easy to work with and implement.

While artificial intelligence (AI) and advanced machine learning (ML) tools have become more prominent and popular over recent years with the increase of accessibility in compute power, performing statistical analysis as a precursor to developing larger-scale projects using AI and ML can enable a practitioner to assess feasibility and practicality before using larger compute resources and project architecture development for those types of projects.

This book provides a wide variety of tools that are commonly used to test hypotheses and provide basic predictive capabilities to analysts and data scientists alike. The reader will walk through the basic concepts and terminology required for understanding the statistical tools in this book prior to exploring the different tests and conditions under which they are applicable. Further, the reader will gain knowledge for assessing the performance of the tests. Throughout, examples will be provided in the Python programming language to get readers started understanding their data using the tools presented, which will be applicable to some of the most common questions faced in the data analytics industry. The topics we will walk through include:

- An introduction to statistics
- Regression models
- Classification models
- Time series models
- Survival analysis

Understanding the tools provided in these sections will provide the reader with a firm foundation from which further independent growth in the statistics domain can more easily be achieved.

Who this book is for

Professionals in most industries can benefit from the tools in this book. The tools provided are useful primarily at a higher level of inferential analysis, but can be applied to deeper levels depending on the industry in which the practitioner wishes to apply them. The target audiences of this book are:

- Industry professionals with limited statistical or programming knowledge who would like to learn to use data for testing hypotheses they have in their business domain
- Data analysts and scientists who wish to broaden their statistical knowledge and find a set of tools and their implementations for performing various data-oriented tasks

The ground-up approach of this book seeks to provide entry into the knowledge base for a wide audience and therefore should neither discourage novice-level practitioners nor exclude advanced-level practitioners from the benefits of the materials presented.

What this book covers

Chapter 1, Sampling and Generalization, describes the concepts of sampling and generalization. The discussion of sampling covers several common methods for sampling data from a population and discusses the implications for generalization. This chapter also discusses how to setup the software required for this book.

Chapter 2, Distributions of Data, provides a detailed introduction to types of data, common distributions used to describe data, and statistical measures. This chapter also covers common transformations used to change distributions.

Chapter 3, Hypothesis Testing, introduces the concept of statistical tests as a method for answering questions of interest. This chapter covers the steps to perform a test, the types of errors encountered in testing, and how to select power using the Z-test.

Chapter 4, Parametric Tests, further discusses statistical tests, providing detailed descriptions of common parametric statistical tests, the assumptions of parametric tests, and how to assess the validity of parametric tests. This chapter also introduces the concept of multiple tests and provides details on corrections for multiple tests.

Chapter 5, Non-parametric Tests, discuss how to perform statistical tests when the assumptions of parametric tests are violated with class of tests without assumptions called non-parametric tests.

Chapter 6, Simple Linear Regression, introduces the concept of a statistical model with the simple linear regression model. This chapter begins by discussing the theoretical foundations of simple linear regression and then discusses how to interpret the results of the model and assess the validity of the model.

Chapter 7, Multiple Linear Regression, builds on the previous chapter by extending the simple linear regression model into additional dimensions. This chapter also discusses issues that occur when modeling with multiple explanatory variables, including multicollinearity, feature selection, and dimension reduction.

Chapter 8, Discrete Models, introduces the concept of classification and develops a model for classifying variables into discrete levels of a categorical response variable. This chapter starts by developing the model binary classification and then extends the model to multivariate classification. Finally, the Poisson model and negative binomial models are covered.

Chapter 9, Discriminant Analysis, discusses several additional models for classification, including linear discriminant analysis and quadratic discriminant analysis. This chapter also introduces Bayes' Theorem.

Chapter 10, Introduction to Time Series, introduces time series data, discussing the time series concept of autocorrelation and the statistical measures for time series. This chapter also introduces the white noise model and stationarity.

Chapter 11, ARIMA Models, discusses models for univariate models. This chapter starts by discussing models for stationary time series and then extends the discussion to non-stationary time series. Finally, this chapter provides a detailed discussion on model evaluation.

Chapter 12, Multivariate Time Series, builds on the previous two chapters by introducing the concept of a multivariate time series and extends ARIMA models to multiple explanatory variables. This chapter also discusses time series cross-correlation.

Chapter 13, Survival Analysis, introduces survival data, also called time-to-event data. This chapter discusses the concept of censoring and the impact of censoring survival data. Finally, the chapter discusses the survival function, hazard, and hazard ratio.

Chapter 14, Survival Models, building on the previous chapter, provides an overview of several models for survival data, including the Kaplan-Meier model, the Exponential model, and the Cox Proportional Hazards model.

To get the most out of this book

You will need access to download and install open-source code packages implemented in the Python programming language and accessible through PyPi.org or the Anaconda Python distribution. While a background in statistics is helpful, but not necessary, this book assumes you have a decent background in basic algebra. Each unit of this book is independent of the other units, but the chapters within each unit build upon each other. Thus, we advise you to begin each unit with that unit's first chapter to understand the content.

Software/hardware covered in the book	Operating system requirements
Python version ≥ 3.8	Windows, macOS, or Linux
Statsmodels 0.13.2	
SciPy 1.8.1	
lifelines 0.27.4	
scikit-learn 1.1.1	
pmdarima 2.02	
Sktime 0.15.0	
Pandas 1.4.3	
Matplotlib 3.5.2	
Numpy 1.23.0	

If you are using the digital version of this book, we advise you to type the code yourself or access the code from the book's GitHub repository (a link is available in the next section). Doing so will help you avoid any potential errors related to the copying and pasting of code.

Download the example code files

You can download the example code files for this book from GitHub at `https://github.com/PacktPublishing/Building-Statistical-Models-in-Python`. If there's an update to the code, it will be updated in the GitHub repository.

We also have other code bundles from our rich catalog of books and videos available at `https://github.com/PacktPublishing/`. Check them out!

Conventions used

There are a number of text conventions used throughout this book.

`Code in text`: Indicates code words in text, database table names, folder names, filenames, file extensions, pathnames, dummy URLs, user input, and Twitter handles. Here is an example: "Mount the downloaded `WebStorm-10*.dmg` disk image file as another disk in your system."

A block of code is set as follows:

```
A = [3,5,4]
B = [43,41,56,78,54]
permutation_testing(A,B,n_iter=10000)
```

Any command-line input or output is written as follows:

```
pip install SomePackage
```

Bold: Indicates a new term, an important word, or words that you see onscreen. For instance, words in menus or dialog boxes appear in **bold**. Here is an example: "Select **System info** from the **Administration** panel."

> **Tips or important notes**
> Appear like this.

Get in touch

Feedback from our readers is always welcome.

General feedback: If you have questions about any aspect of this book, email us at customercare@packtpub.com and mention the book title in the subject of your message.

Errata: Although we have taken every care to ensure the accuracy of our content, mistakes do happen. If you have found a mistake in this book, we would be grateful if you would report this to us. Please visit www.packtpub.com/support/errata and fill in the form.

Piracy: If you come across any illegal copies of our works in any form on the internet, we would be grateful if you would provide us with the location address or website name. Please contact us at `copyright@packtpub.com` with a link to the material.

If you are interested in becoming an author: If there is a topic that you have expertise in and you are interested in either writing or contributing to a book, please visit `authors.packtpub.com`.

Share Your Thoughts

Once you've read *Building Statistical Models in Python*, we'd love to hear your thoughts! Please click here to go straight to the Amazon review page for this book and share your feedback.

Your review is important to us and the tech community and will help us make sure we're delivering excellent quality content.

Download a free PDF copy of this book

Thanks for purchasing this book!

Do you like to read on the go but are unable to carry your print books everywhere? Is your eBook purchase not compatible with the device of your choice?

Don't worry, now with every Packt book you get a DRM-free PDF version of that book at no cost.

Read anywhere, any place, on any device. Search, copy, and paste code from your favorite technical books directly into your application.

The perks don't stop there, you can get exclusive access to discounts, newsletters, and great free content in your inbox daily

Follow these simple steps to get the benefits:

1. Scan the QR code or visit the link below

https://packt.link/free-ebook/978-1-80461-428-0

2. Submit your proof of purchase
3. That's it! We'll send your free PDF and other benefits to your email directly

Part 1: Introduction to Statistics

This part will cover the statistical concepts that are foundational to statistical modeling. It includes the following chapters:

- *Chapter 1, Sampling and Generalization*
- *Chapter 2, Distributions of Data*
- *Chapter 3, Hypothesis Testing*
- *Chapter 4, Parametric Tests*
- *Chapter 5, Non-Parametric Tests*

1
Sampling and Generalization

In this chapter, we will describe the concept of populations and sampling from populations, including some common strategies for sampling. The discussion of sampling will lead to a section that will describe generalization. Generalization will be discussed as it relates to using samples to make conclusions about their respective populations. When modeling for statistical inference, it is necessary to ensure that samples can be generalized to populations. We will provide an in-depth overview of this bridge through the subjects in this chapter.

We will cover the following main topics:

- Software and environment setup
- Population versus sample
- Population inference from samples
- Sampling strategies – random, systematic, and stratified

Software and environment setup

Python is one of the most popular programming languages for data science and machine learning thanks to the large open source community that has driven the development of these libraries. Python's ease of use and flexible nature made it a prime candidate in the data science world, where experimentation and iteration are key features of the development cycle. While there are new languages in development for data science applications, such as **Julia**, Python currently remains the key language for data science due to its wide breadth of open source projects, supporting applications from statistical modeling to deep learning. We have chosen to use Python in this book due to its positioning as an important language for data science and its demand in the job market.

Python is available for all major operating systems: Microsoft Windows, macOS, and Linux. Additionally, the installer and documentation can be found at the official website: https://www.python.org/.

This book is written for Python version 3.8 (or higher). It is recommended that you use whatever recent version of Python that is available. It is not likely that the code found in this book will be compatible with Python 2.7, and most active libraries have already started dropping support for Python 2.7 since official support ended in 2020.

The libraries used in this book can be installed with the Python package manager, `pip`, which is part of the standard Python library in contemporary versions of Python. More information about `pip` can be found here: https://docs.python.org/3/installing/index.html. After `pip` is installed, packages can be installed using `pip` on the command line. Here is basic usage at a glance:

Install a new package using the latest version:

```
pip install SomePackage
```

Install the package with a specific version, version 2.1 in this example:

```
pip install SomePackage==2.1
```

A package that is already installed can be upgraded with the `--upgrade` flag:

```
pip install SomePackage -upgrade
```

In general, it is recommended to use Python virtual environments between projects and to keep project dependencies separate from system directories. Python provides a virtual environment utility, `venv`, which, like `pip`, is part of the standard library in contemporary versions of Python. Virtual environments allow you to create individual binaries of Python, where each binary of Python has its own set of installed dependencies. Using virtual environments can prevent package version issues and conflict when working on multiple Python projects. Details on setting up and using virtual environments can be found here: https://docs.python.org/3/library/venv.html.

While we recommend the use of Python and Python's virtual environments for environment setups, a highly recommended alternative is **Anaconda**. Anaconda is a free (enterprise-ready) analytics-focused distribution of Python by Anaconda Inc. (previously Continuum Analytics). Anaconda distributions come with many of the core data science packages, common IDEs (such as **Jupyter** and **Visual Studio Code**), and a graphical user interface for managing environments. Anaconda can be installed using the installer found at the Anaconda website here: https://www.anaconda.com/products/distribution.

Anaconda comes with its own package manager, `conda`, which can be used to install new packages similarly to `pip`.

Install a new package using the latest version:

```
conda install SomePackage
```

Upgrade a package that is already installed:

```
conda upgrade SomePackage
```

Throughout this book, we will make use of several core libraries in the Python data science ecosystem, such as NumPy for array manipulations, pandas for higher-level data manipulations, and matplotlib for data visualization. The package versions used for this book are contained in the following list. Please ensure that the versions installed in your environment are equal to or greater than the versions listed. This will help ensure that the code examples run correctly:

- statsmodels 0.13.2
- Matplotlib 3.5.2
- NumPy 1.23.0
- SciPy 1.8.1
- scikit-learn 1.1.1
- pandas 1.4.3

The packages used for the code in this book are shown here in *Figure 1.1*. The __version__ method can be used to print the package version in code.

```
import numpy as np
import scipy as sp
import pandas as pd
import statsmodels as sm
import matplotlib
import sklearn
```

```
print(f"numpy version: {np.__version__}")
print(f"scipy version: {sp.__version__}")
print(f"pandas version: {pd.__version__}")
print(f"statsmodels version: {sm.__version__}")
print(f"matplotlib version: {matplotlib.__version__}")
print(f"sklearn version: {sklearn.__version__}")
```

```
numpy version: 1.23.0
scipy version: 1.8.1
pandas version: 1.4.3
statsmodels version: 0.13.2
matplotlib version: 3.5.2
sklearn version: 1.1.1
```

Figure 1.1 – Package versions used in this book

Having set up the technical environment for the book, let's get into the statistics. In the next sections, we will discuss the concepts of population and sampling. We will demonstrate sampling strategies with code implementations.

Population versus sample

In general, the goal of statistical modeling is to answer a question about a group by making an inference about that group. The group we are making an inference on could be machines in a production factory, people voting in an election, or plants on different plots of land. The entire group, every individual item or entity, is referred to as the **population**. In most cases, the population of interest is so large that it is not practical or even possible to collect data on every entity in the population. For instance, using the voting example, it would probably not be possible to poll every person that voted in an election. Even if it was possible to reach all the voters for the election of interest, many voters may not consent to polling, which would prevent collection on the entire population. An additional consideration would be the expense of polling such a large group. These factors make it practically impossible to collect population statistics in our example of vote polling. These types of prohibitive factors exist in many cases where we may want to assess a population-level attribute. Fortunately, we do not need to collect data on the entire population of interest. Inferences about a population can be made using a subset of the population. This subset of the population is called a **sample**. This is the main idea of statistical modeling. A model will be created using a sample and inferences will be made about the population.

In order to make valid inferences about the population of interest using a sample, the sample must be *representative* of the population of interest, meaning that the sample should contain the variation found in the population. For example, if we were interested in making an inference about plants in a field, it is unlikely that samples from one corner of the field would be sufficient for inferences about the larger population. There would likely be variations in plant characteristics over the entire field. We could think of various reasons why there might be variation. For this example, we will consider some examples from *Figure 1.2*.

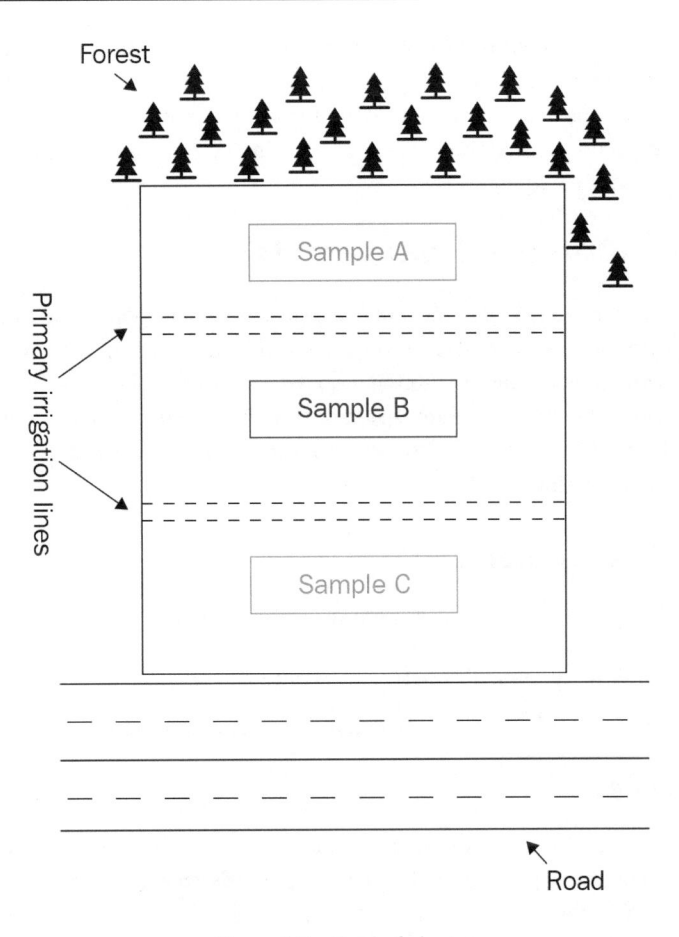

Figure 1.2 – Field of plants

The figure shows that **Sample A** is near a forest. This sample area may be affected by the presence of the forest; for example, some of the plants in that sample may receive less sunlight than plants in the other sample. **Sample B** is shown to be in between the main irrigation lines. It's conceivable that this sample receives more water on average than the other two samples, which may have an effect on the plants in this sample. The final **Sample C** is near a road. This sample may see other effects that are not seen in **Sample A** or **B**.

If samples were only taken from one of those sections, the inferences from those samples would be *biased* and would not provide valid references about the population. Thus, samples would need to be taken from across the entire field to create a sample that is more likely to be representative of the population of plants. When taking samples from populations, it is critical to ensure the sampling method is robust to possible issues, such as the influence of irrigation and shade in the previous example. Whenever taking a sample from a population, it's important to identify and mitigate possible influences of bias because biases in data will affect your model and skew your conclusions.

In the next section, various methods for sampling from a dataset will be discussed. An additional consideration is the sample size. The sample size impacts the type of statistical tools we can use, the distributional assumptions that can be made about the sample, and the confidence of inferences and predictions. The impact of sample size will be explored in depth in *Chapter 2, Distributions of Data* and *Chapter 3, Hypothesis Testing*.

Population inference from samples

When using a statistical model to make inferential conclusions about a population from a sample subset of that population, the study design must account for similar degrees of uncertainty in its variables as those in the population. This is the variation mentioned earlier in this chapter. To appropriately draw inferential conclusions about a population, any statistical model must be structured around a chance mechanism. Studies structured around these chance mechanisms are called **randomized experiments** and provide an understanding of both correlation and causation.

Randomized experiments

There are two primary characteristics of a randomized experiment:

- Random sampling, colloquially referred to as random selection
- Random assignment of treatments, which is the nature of the study

Random sampling

Random sampling (also called random selection) is designed with the intent of creating a sample representative of the overall population so that statistical models generalize the population well enough to assign cause-and-effect outcomes. In order for random sampling to be successful, the population of interest must be well defined. All samples taken from the population must have a chance of being selected. In considering the example of polling voters, all voters must be willing to be polled. Once all voters are entered into a lottery, random sampling can be used to subset voters for modeling. Sampling from only voters who are willing to be polled introduces sampling bias into statistical modeling, which can lead to skewed results. The sampling method in the scenario where only some voters are willing to participate is called **self-selection**. Any information obtained and modeled from self-selected samples – or any non-random samples – cannot be used for inference.

Random assignment of treatments

The random assignment of treatments refers to two motivators:

- The first motivator is to gain an understanding of specific input variables and their influence on the response – for example, understanding whether assigning treatment A to a specific individual may produce more favorable outcomes than a placebo.

- The second motivator is to remove the impact of external variables on the outcomes of a study. These external variables, called **confounding variables** (or **confounders**), are important to remove as they often prove difficult to control. They may have unpredictable values or even be unknown to the researcher. The consequence of including confounders is that the outcomes of a study may not be replicable, which can be costly. While confounders can influence outcomes, they can also influence input variables, as well as the relationships between those variables.

Referring back to the example in the earlier section, *Population versus sample*, consider a farmer who decides to start using pesticides on his crops and wants to test two different brands. The farmer knows there are three distinct areas of the land; plot A, plot B, and plot C. To determine the success of the pesticides and prevent damage to the crops, the farmer randomly chooses 60 plants from each plot (this is called **stratified random sampling** where random sampling is stratified across each plot) for testing. This selection is representative of the population of plants. From this selection, the farmer labels his plants (labeling doesn't need to be random). For each plot, the farmer shuffles the labels into a bag, to randomize them, and begins selecting 30 plants. The first 30 plants get one of two treatments and the other 30 are given the other treatment. This is a *random assignment of treatment*. Assuming the three separate plots represent a distinct set of confounding variables on crop yield, the farmer will have enough information to obtain an inference about the crop yield for each pesticide brand.

Observational study

The other type of statistical study often performed is an **observational study**, in which the researcher seeks to learn through observing data that already exists. An observational study can aid in the understanding of input variables and their relationships to both the target and each other, but cannot provide cause-and-effect understanding as a randomized experiment can. An observational study may have one of the two components of a randomized experiment – either random sampling or random assignment of treatment – but without both components, will not directly yield inference. There are many reasons why an observational study may be performed versus a randomized experiment, such as the following:

- A randomized experiment being too costly

- Ethical constraints for an experiment (for example, an experiment to determine the rate of birth defects caused by smoking while pregnant)

- Using data from prior randomized experiments, which thus removes the need for another experiment

One method for deriving some causality from an observational study is to perform random sampling and repeated analysis. Repeated random sampling and analysis can help minimize the impact of confounding variables over time. This concept plays a huge role in the usefulness of *big data* and *machine learning*, which has gained a lot of importance in many industries within this century. While almost any tool that can be used for observational analysis can also be used for a randomized experiment, this book focuses primarily on tools for observational analysis, as this is more common in most industries.

It can be said that statistics is a science for helping make the best decisions when there are quantifiable uncertainties. All statistical tests contain a null hypothesis and an alternative hypothesis. That is to say, an assumption that there is no statistically significant difference between data (the null hypothesis) or that there is a statistically significant difference between data (the alternative hypothesis). The term statistically significant difference implies the existence of a benchmark – or threshold – beyond which a measure takes place and indicates significance. This benchmark is called the **critical value**.

The measure that is applied against this critical value is called the **test statistic**. The critical value is a static value quantified based on behavior in the data, such as the average and variation, and is based on the hypothesis. If there are two possible routes by which a null hypothesis may be rejected – for example, we believe some output is either less than or more than the average – there will be two critical values (this test is called a **two-tailed** hypothesis test), but if there is only one argument against the null hypothesis, there will be only one critical value (this is called a **one-tailed** hypothesis test). Regardless of the number of critical values, there will always only be one test statistic measurement for each group within a given hypothesis test. If the test statistic exceeds the critical value, there is a statistically significant reason to support rejecting the null hypothesis and concluding there is a statistically significant difference in the data.

It is useful to understand that a hypothesis test can test the following:

- One variable against another (such as in a t-test)
- Multiple variables against one variable (for example, linear regression)
- Multiple variables against multiple variables (for example, MANOVA)

In the following figure, we can see visually the relationship between the test statistic and critical values in a two-tailed hypothesis test.

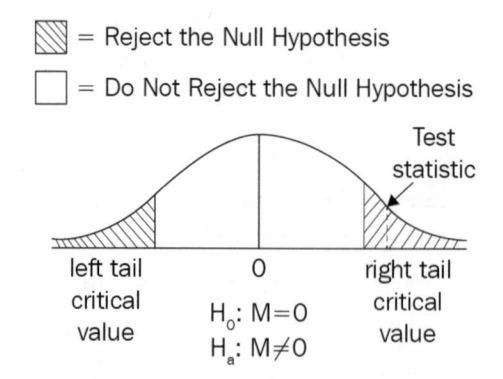

Figure 1.3 – Critical values versus a test statistic in a two-tailed hypothesis test

Based on the figure, we now have a visual idea of how a test statistic exceeding the critical value suggests rejecting the null hypothesis.

One concern with using only the approach of measuring test statistics against critical values in the hypothesis, however, is that test statistics can be impractically large. This is likely to occur when there may be a wide range of results that are not considered to fall within the bounds of a treatment effect. It is uncertain whether a result as extreme as or more extreme than the test statistic is possible. To prevent misleadingly rejecting the null hypothesis, a **p-value** is used. The p-value represents the probability that chance alone resulted in a value as extreme as the one observed (the one that suggests rejecting the null hypothesis). If a p-value is low, relative to the level of significance, the null hypothesis can be rejected. Common levels of significance are 0.01, 0.05, and 0.10. It is beneficial to confirm prior to making a decision on a hypothesis to assess both the critical value's relationship to the test statistic and the p-value. More will be discussed in *Chapter 3, Hypothesis Testing*, when we begin discussing hypothesis testing.

Sampling strategies – random, systematic, stratified, and clustering

In this section, we will discuss the different sampling methods used in research. Broadly speaking, in the real world, it is not easy or possible to get the whole population data for many reasons. For instance, the costs of gathering data are expensive in terms of money and time. Collecting all the data is impractical in many cases and ethical issues are also considered. Taking samples from the population can help us overcome these problems and is a more efficient way to collect data. By collecting an appropriate sample for a study, we can draw statistical conclusions or statistical inferences about the population properties. Inferential statistical analysis is a fundamental aspect of statistical thinking. Different sampling methods from probability strategies to non-probability strategies used in research and industry will be discussed in this section.

There are essentially two types of sampling methods:

- Probability sampling
- Non-probability sampling

Probability sampling

In *probability sampling*, a sample is chosen from a population based on the theory of probability, or it is chosen randomly using random selection. In *random selection*, the chance of each member in a population being selected is equal. For example, consider a game with 10 similar pieces of paper. We write numbers 1 through 10, with a separate piece of paper for each number. The numbers are then shuffled in a box. The game requires picking three of these ten pieces of paper randomly. Because the pieces of paper have been prepared using the same process, the chance of any piece of paper being selected (or the numbers one through ten) is equal for each piece. Collectively, the 10 pieces of paper are considered a population and the 3 selected pieces of paper constitute a random sample. This example is one approach to the probability sampling methods we will discuss in this chapter.

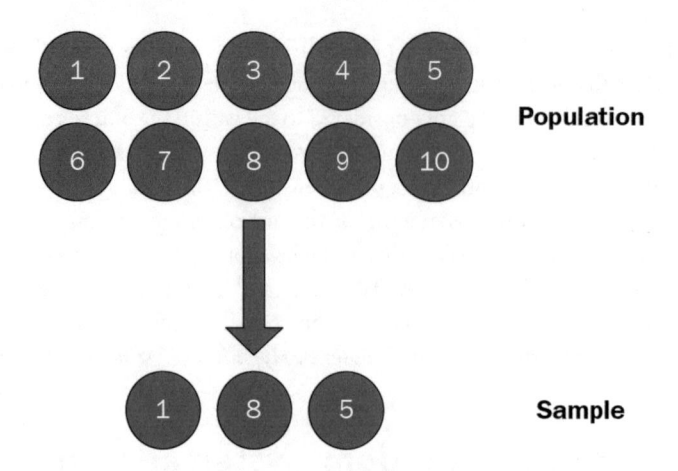

Figure 1.4 – A random sampling example

We can implement the sampling method described before (and shown in *Figure 1.4*) with `numpy`. We will use the `choice` method to select three samples from the given population. Notice that `replace==False` is used in the choice. This means that once a sample is chosen, it will not be considered again. Note that the random generator is used in the following code for reproducibility:

```
import numpy as np

# setup generator for reproducibility
random_generator = np.random.default_rng(2020)

population = np.arange(1, 10 + 1)
sample = random_generator.choice(
    population,     #sample from population
    size=3,         #number of samples to take
    replace=False   #only allow to sample individuals once
)
print(sample)
# array([1, 8, 5])
```

The purpose of random selection is to avoid a biased result when some units of a population have a lower or higher probability of being chosen in a sample than others. Nowadays, a random selection process can be done by using computer randomization programs.

Four main types of the probability sampling methods that will be discussed here are as follows:

- Simple random sampling
- Systematic sampling

- Stratified sampling
- Cluster sampling

Let's look at each one of them.

Simple random sampling

First, simple random sampling is a method to select a sample randomly from a population. Every member of the subset (or the sample) has an equal chance of being chosen through an unbiased selection method. This method is used when all members of a population have similar properties related to important variables (important features) and it is the most direct approach to probability sampling. The advantages of this method are to minimize bias and maximize representativeness. However, while this method helps limit a biased approach, there is a risk of errors with simple random sampling. This method also has some limitations. For instance, when the population is very large, there can be high costs and a lot of time required. Sampling errors need to be considered when a sample is not representative of the population and the study needs to perform this sampling process again. In addition, not every member of a population is willing to participate in the study voluntarily, which makes it a big challenge to obtain good information representative of a large population. The previous example of choosing 3 pieces of paper from 10 pieces of paper is a simple random sample.

Systematic sampling

Here, members of a population are selected at a random starting point with a fixed sampling interval. We first choose a fixed sampling interval by dividing the number of members in a population by the number of members in a sample that the study conducts. Then, a random starting point between the number one and the number of members in the sampling interval is selected. Finally, we choose subsequent members by repeating this sampling process until enough samples have been collected. This method is faster and preferable than simple random sampling when cost and time are the main factors to be considered in the study. On the other hand, while in simple random sampling, each member of a population has an equal chance of being selected, in systematic sampling, a sampling interval rule is used to choose a member from a population in a sample for a study. It can be said that systematic sampling is less random than simple random sampling. Similarly, as in simple random sampling, member properties of a population are similarly related to important variables/features. Let us discuss how we perform systematic sampling through the following example. In a class at one high school in Dallas, there are 50 students but only 10 books to give to these students. The sampling interval is fixed by dividing the number of students in the class by the number of books (50/10 = 5). We also need to generate a random number between one and 50 as a random starting point. For example, take the number 18. Hence, the 10 students selected to get the books will be as follows:

18, 23, 28, 33, 38, 43, 48, 3, 8, 13

The natural question arises as to whether the interval sampling is a fraction. For example, if we have 13 books, then the sampling interval will be 50/13 ~ 3.846. However, we cannot choose this fractional number as a sampling interval that represents the number of students. In this situation, we could choose number 3 or 4, alternatively, as the sampling intervals (we could also choose either 3 or 4 as the sampling interval). Let us assume that a random starting point generated is 17. Then, the 13 selected students are these:

17, 20, 24, 27, 31, 34, 38, 41, 45, 48, 2, 5, 9

Observing the preceding series of numbers, after reaching the number 48, since adding 4 will produce a number greater than the count of students (50 students), the sequence restarts at 2 (48 + 4 = 52, but since 50 is the maximum, we restart at 2). Therefore, the last three numbers in the sequence are 2, 5, and 9, with the sampling intervals 4, 3, and 4, respectively (passing the number 50 and back to the number 1 until we have 13 selected students for the systematic sample).

With systematic sampling, there is a biased risk when the list of members of a population is organized to match the sampling interval. For example, going back to the case of 50 students, researchers want to know how students feel about mathematics classes. However, if the best students in math correspond to numbers 2, 12, 22, 32, and 42, then the survey could be biased if conducted when the random starting point is 2 and the sampling interval is 10.

Stratified sampling

It is a probability sampling method based on dividing a population into homogeneous subpopulations called **strata**. Each stratum splits based on distinctly different properties, such as gender, age, color, and so on. These subpopulations must be distinct so that every member in each stratum has an equal chance of being selected by using simple random sampling. *Figure 1.5* illustrates how stratified sampling is performed to select samples from two subpopulations (a set of numbers and a set of letters):

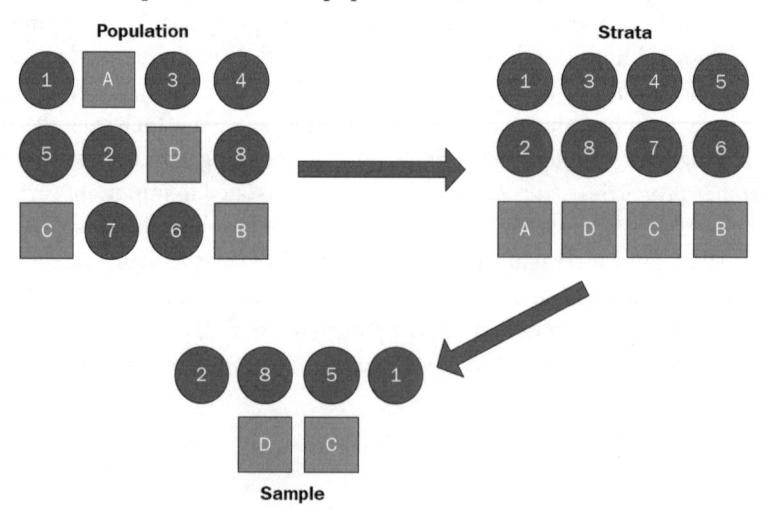

Figure 1.5 – A stratified sample example

The following code sample shows how to implement stratified sampling with numpy using the example shown in *Figure 1.5*. First, the instances are split into the respective strata: numbers and letters. Then, we use numpy to take random samples from each stratum. Like in the previous code example, we utilize the choice method to take the random sample, but the sample size for each stratum is based on the total number of instances in each stratum rather than the total number of instances in the entire population; for example, sampling 50% of the numbers and 50% of the letters:

```python
import numpy as np

# setup generator for reproducibility
random_generator = np.random.default_rng(2020)


population = [
    1, "A", 3, 4,
    5, 2, "D", 8,
    "C", 7, 6, "B"
]
# group strata
strata = {
    'number' : [],
    'string' : [],
}

for item in population:
    if isinstance(item, int):
        strata['number'].append(item)
    else:
        strata['string'].append(item)

# fraction of population to sample
sample_fraction = 0.5
# random sample from stata
sampled_strata = {}

for group in strata:
    sample_size = int(
        sample_fraction * len(strata[group])
    )
    sampled_strata[group] = random_generator.choice(
            strata[group],
            size=sample_size,
            replace=False
    )
```

```
print(sampled_strata)

#{'number': array([2, 8, 5, 1]), 'string': array(['D', 'C'],
dtype='<U1')}
```

The main advantage of this method is that key population characteristics in a sample better represent the population that is studied and are also proportional to the overall population. This method helps to reduce sample selection bias. On the other hand, when classifying each member of a population into distinct subpopulations is not obvious, this method becomes unusable.

Cluster sampling

Here, a population is divided into different subgroups called clusters. Each cluster has homogeneous characteristics. Instead of randomly selecting individual members in each cluster, entire clusters are randomly chosen and each of these clusters has an equal chance of being selected as part of a sample. If clusters are large, then we can conduct a **multistage sampling** by using one of the previous sampling methods to select individual members within each cluster. A cluster sampling example is discussed now. A local pizzeria plans to expand its business in the neighborhood. The owner wants to know how many people order pizzas from his pizzeria and what the preferred pizzas are. He then splits the neighborhood into different areas and selects clients randomly to form cluster samples. A survey is sent to the selected clients for his business study. Another example is related to multistage cluster sampling. A retail chain store conducts a study to see the performance of each store in the chain. The stores are divided into subgroups based on location, then samples are randomly selected to form clusters, and the sample cluster is used as a performance study of his stores. This method is easy and convenient. However, the sample clusters are not guaranteed to be representative of the whole population.

Non-probability sampling

The other type of sampling method is non-probability sampling, where some or all members of a population do not have an equal chance of being selected as a sample to participate in the study. This method is used when random probability sampling is impossible to conduct and it is faster and easier to obtain data compared to the probability sampling method. One of the reasons to use this method is due to cost and time considerations. It allows us to collect data easily by using a non-random selection based on convenience or certain criteria. This method can lead to a higher-biased risk than the probability sampling method. The method is often used in exploratory and qualitative research. For example, if a group of researchers wants to understand clients' opinions of a company related to one of its products, they send a survey to the clients who bought and used the product. It is a convenient way to get opinions, but these opinions are only from clients who already used the product. Therefore, the sample data is only representative of one group of clients and cannot be generalized as the opinions of all the clients of the company.

CUSTOMER SATISFACTION

☑ Excellent

☐ Very Good

☐ Good

☐ Average

Figure 1.6 – A survey study example

The previous example is one of two types of non-probability sampling methods that we want to discuss here. This method is **convenience sampling**. In convenience sampling, researchers choose members the most accessible to the researchers from a population to form a sample. This method is easy and inexpensive but generalizing the results obtained to the whole population is questionable.

Quota sampling is another type of non-probability sampling where a sample group is selected to be representative of a larger population in a non-random way. For example, recruiters with limited time can use the quota sampling method to search for potential candidates from professional social networks (LinkedIn, Indeed.com, etc.) and interview them. This method is cost-effective and saves time but presents bias during the selection process.

In this section, we provided an overview of probability and non-probability sampling. Each strategy has advantages and disadvantages, but they help us to minimize risks, such as bias. A well-planned sampling strategy will also help reduce errors in predictive modeling.

Summary

In this chapter, we discussed installing and setting up the Python environment to run the Statsmodels API and other requisite open-source packages. We also discussed populations versus samples and the requirements to gain inference from samples. Finally, we explained several different common sampling methods used in statistical and machine learning models.

In the next chapter, we will begin a discussion on statistical distributions and their implications for building statistical models. In *Chapter 3, Hypothesis Testing*, we will begin discussing hypothesis testing in depth, expanding on the concepts discussed in the *Observational study* section of this chapter. We will also discuss power analysis, which is a useful tool for determining the sample size based on existing sample data parameters and the desired levels of statistical significance.

2
Distributions of Data

In this chapter, we will cover the essential aspects of data and distributions. We will start by covering the types of data and distributions of data. Having covered the essential measurements of distributions, we will describe the normal distribution and its important properties, including the central limit theorem. Finally, we will cover resampling methods such as permutations and transformation methods such as log transformations. This chapter covers the foundational knowledge necessary to begin statistical modeling.

In this chapter, we're going to cover the following main topics:

- Understanding data types
- Measuring and describing distributions
- The normal distribution and the central limit theorem
- Bootstrapping
- Permutations
- Transformations

Technical requirements

This chapter will make use of Python 3.8.

The code for this chapter can be found here – https://github.com/PacktPublishing/Building-Statistical-Models-in-Python – in the ch2 folder.

Please set up a virtual environment or Anaconda environment with the following packages installed:

- `numpy==1.23.0`
- `scipy==1.8.1`
- `matplotlib==3.5.2`

- `pandas==1.4.2`
- `statsmodels==0.13.2`

Understanding data types

Before discussing data distributions, it would be useful to understand the types of data. Understanding data types is critical because the type of data determines what kind of analysis can be used since the type of data determines what operations can be used with the data (this will become clearer through the examples in this chapter). There are four distinct types of data:

- Nominal data
- Ordinal data
- Interval data
- Ratio data

These types of data can also be grouped into two sets. The first two types of data (nominal and ordinal) are **qualitative data**, generally non-numeric categories. The last two types of data (interval and ratio) are **quantitative data**, generally numeric values.

Let's start with nominal data.

Nominal data

Nominal data is data labeled with distinct groupings. As an example, take machines in a sign factory. It is common for factories to source machines from different suppliers, which would also have different model numbers. For example, the example factory may have 3 of **Model A** and 5 of **Model B** (see *Figure 2.1*). The machines would make up a set of nominal data where **Model A** and **Model B** are the distinct group labels. With nominal data, there is only one operation that can be performed: equality. Each member of a group is equal while members from different groups are unequal. In our factory example, a **Model A** machine would be equal to another **Model A** machine while a **Model B** machine would be unequal to a **Model A** machine.

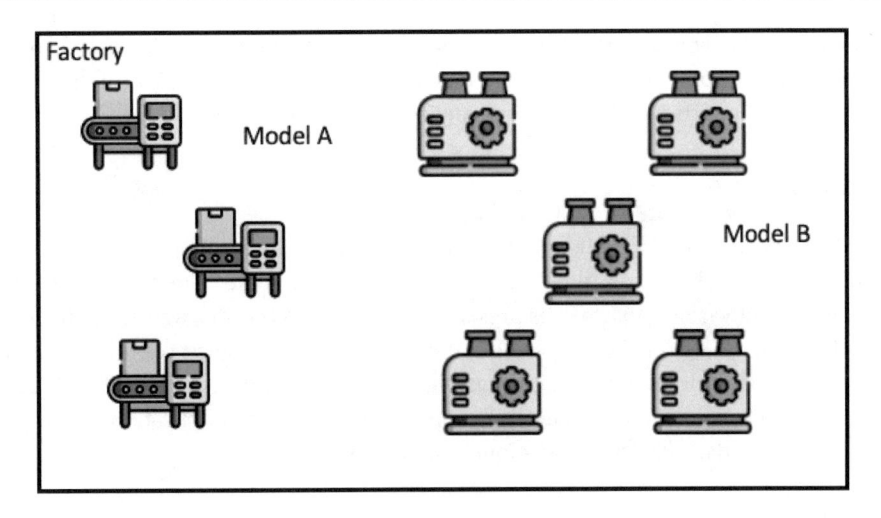

Figure 2.1 – Two groups of machines in a factory

As we can see, with this type of data, we can only group items together under labels. With the next type of data, we will introduce a new feature: order.

Ordinal data

The next type of data is like nominal data but exhibits an order. The data can be labeled into distinct groups and the groups can be ordered. We call this type of data ordinal data. Continuing with the factory example, let's suppose that there is a **Model C** machine, and **Model C** is supplied by the same vendor as **Model B**. However, **Model C** is the high-performance version, which generates higher output. In this case, **Model B** and **Model C** are ordinal data because **Model B** is a lower-output machine, and **Model C** is a higher-output machine, which creates a natural order. For instance, we can put the model labels in ascending order of performance: **Model B**, **Model C**. University education levels are another example of ordinal data with the levels BS, MS, and PhD. As mentioned, the new operation for this type of data is ordering, meaning the data can be sorted. Thus, ordinal data supports order and equality. While this type of data can be ordered in ascending or descending order, we cannot add or subtract the data, meaning **Model B** + **Model C** is not a meaningful statement. The next type of data we will discuss will support addition and subtraction.

Interval data

The next type of data, interval data, is used to describe data that exists on an interval scale but does not have a clear definition of zero. This means the difference between two data points is meaningful. Take the Celsius temperature scale, for example. The data points are numeric, and the data points are evenly spaced at an interval (for example, 20 and 40 are both 10 degrees away from 30).

In this example of the temperature scale, the definition of 0 is arbitrary. For Celsius, 0 happens to be set at water's freezing point, but this is an arbitrary choice made by the designers of the scale. So, the interval data type supports equality, ordering, and addition/subtraction.

Ratio data

The final data type is ratio data. Like interval data, ratio data is ordered numeric data, but unlike interval data, ratio data has an absolute 0. Absolute 0 means that if the value of a ratio-type variable is zero, none of that variable exists or is present. For example, consider wait times for rides at an amusement park. If no one is in line for the ride, the wait time is 0; new guests can ride the amusement ride immediately. There is no meaningful negative measurement for wait times. A wait time of 0 is the absolute minimum value. Ratio data also supports meaningful multiplication/division, making ratio data the type of data with the most supported operations.

Visualizing data types

Data visualization is a critical step for understanding distributions and identifying properties of data. In this chapter (and throughout this book), we will utilize `matplotlib` for visualizing data. While other Python libraries can be used for visualizing data, `matplotlib` is the de facto standard plotting library for Python. In this section, we will begin using `matplotlib` to visualize the four types of data discussed previously.

Plotting qualitative data types

Since the first two types of data are categorical, we will use a bar chart to visualize these distributions of data. Example bar charts are shown in *Figure 2.2*.

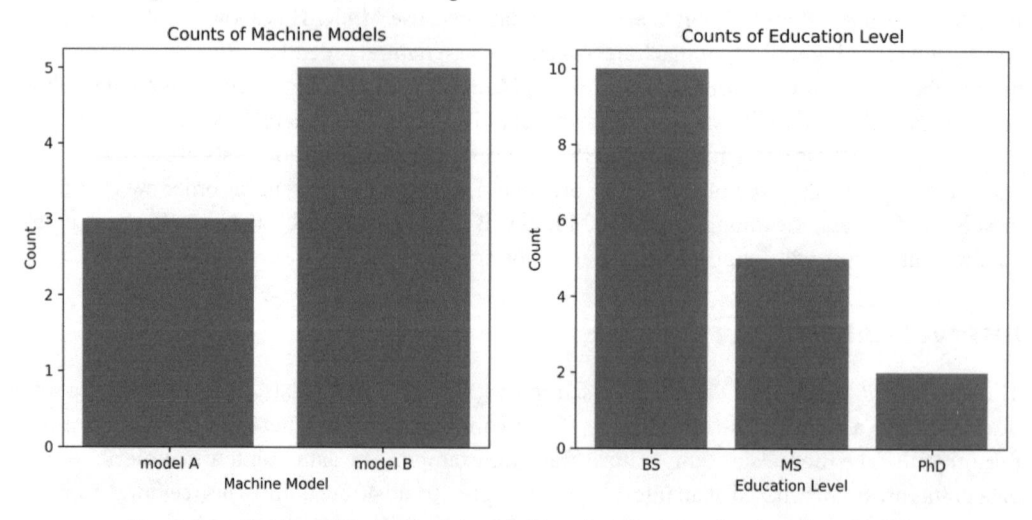

Figure 2.2 – Nominal data in a bar chart (left) and ordinal data in a bar chart (right)

The left bar chart in *Figure 2.2* shows the distribution of the **Model A** machines and **Model B** machines given in the factory example. The right bar chart shows an example distribution of the education levels of a team of engineers. Note that in the education level bar chart, the *x*-axis labels are ordered from the lowest level of education to the highest level of education.

The code used to generate *Figure 2.2* is shown next.

The code has three main parts.

- **The library imports**:

    In this case, we are only importing `pyplot` from `matplotlib`, which is canonically imported as `plt`.

- **The code for data creation**:

    After the `import` statement, there are a few statements to create the data we will plot. The data for the first plot is stored in two Python lists: `label` and `counts`, which contain the machine labels and the number of machines, respectively. It's worth noting that each of these two lists contains the same number of elements (two elements). The education data is stored similarly. While in this example, we are using simple example data, in later chapters, we will have additional steps for retrieving and formatting data.

- **The code for plotting the data**:

    The final step is plotting the data. Since we are plotting two sets of data in this example, we use the `subplots` method, which will create a grid of plots. The first two arguments to `subplots` are the number of rows and the number of columns for the grid of figures. In our case, the number of rows is 1 and the number of columns is 2. The `subplots` method returns two objects; the figure, `fig`, and the axes, `ax`. The first returned object, `fig`, has high-level controls over the figure, such as saving the figure, showing the figure in a new window, and many others. The second object, `ax`, will either be an individual axis object or an array of axis objects. In our case, `ax` is an array of axes objects – since our grid has two plots, indexing into `ax` gives us the axes object. We use the `bar` method of an axes object to create a bar chart. The `bar` method has two required arguments. The first required argument is the list of labels. The second argument is the bar heights that correspond to each label, which is why the two lists must have the same length. The other three methods, `set_title`, `set_ylabel`, and `set_xlabel`, set the values for the corresponding plot attributes: `title`, `ylabel`, and `x-label`.

Finally, the figure is created using `fig.show()`:

```
import matplotlib.pyplot as plt
label = ['model A', 'model B']
counts = [3, 5]
edu_label = ['BS', 'MS', 'PhD']
edu_counts = [10, 5, 2]
fig, ax = plt.subplots(1, 2, figsize=(12, 5))
```

```
ax[0].bar(label, counts)
ax[0].set_title('Counts of Machine Models')
ax[0].set_ylabel('Count')
ax[0].set_xlabel('Machine Model')
ax[1].bar(edu_label, edu_counts)
ax[1].set_title('Counts of Education Level')
ax[1].set_ylabel('Count')
ax[1].set_xlabel('Education Level')
fig.show()
```

Now let's look at how to plot data from the other two data types.

Plotting quantitative data types

Since the last two data types are numeric, we will use a histogram to visualize the distributions. Two example histograms are shown in *Figure 2.3*.

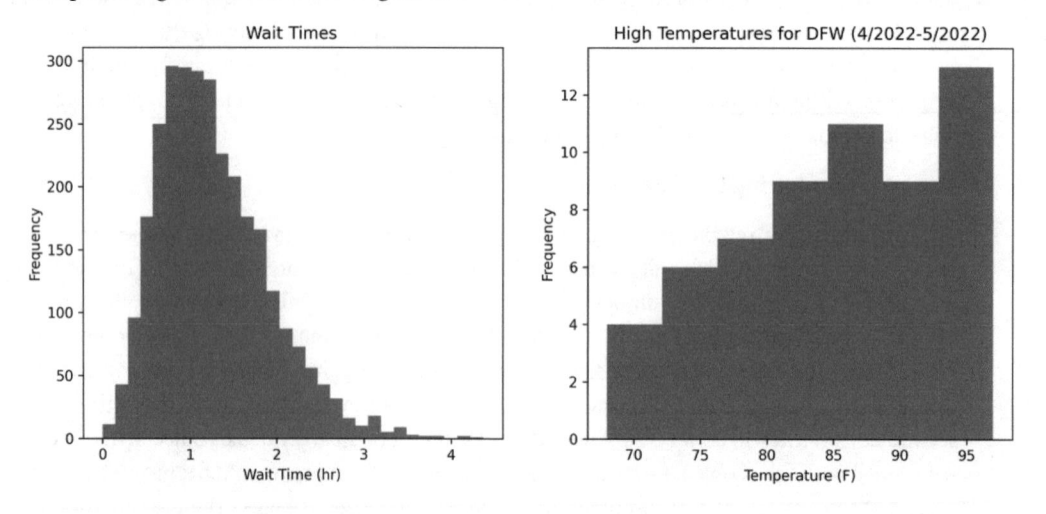

Figure 2.3 – Nominal data in a bar chart (left) and ordinal data in a bar chart (right)

The left histogram is synthetic wait time data (ratio data) that might represent wait times at an amusement park. The right histogram is temperature data (interval data) for the Dallas-Fort Worth area during April and May of 2022 (pulled from https://www.iweathernet.com/texas-dfw-weather-records).

The code used to generate *Figure 2.3* is shown next. Again, the code has three main parts, the library imports, the code for data creation, and the code for plotting the data.

Like in the previous example, `matplotlib` is imported as `plt`. In this example, we also import a function from `scipy`; however, this function is only used for generating sample data to work with and we will not discuss it at length here. For our purposes, just think of `skewnorm` as producing an array of numbers. This code block is very similar to the previous code block.

The main difference is the method used for plotting the data, `hist`, which creates a histogram. The `hist` method has one required argument, which is the sequence of numbers to plot in the histogram. The second argument used in this example is `bins`, which effectively controls the granularity of the histogram – granularity increases with more bins. The bin count of a histogram can be adjusted for the desired visual effect and is generally set experimentally for the data plotted:

```
from scipy.stats import skewnorm
import matplotlib.pyplot as plt
a = 4
x = skewnorm.rvs(a, size=3000) + 0.5
x = x[x > 0]
dfw_highs = [
    85, 87, 75, 88, 80, 86, 90, 94, 93, 92, 90, 92, 94,
    93, 97, 90, 95, 96, 96, 95, 92, 70, 79, 73, 88, 92,
    94, 93, 95, 76, 78, 86, 81, 95, 77, 71, 69, 88, 86,
    89, 84, 82, 77, 84, 81, 79, 75, 75, 91, 86, 86, 84,
    82, 68, 75, 78, 82, 83, 85]
fig, ax = plt.subplots(1,2, figsize=(12, 5))
ax[0].hist(x, bins=30)
ax[0].set_xlabel('Wait Time (hr)')
ax[0].set_ylabel('Frequency')
ax[0].set_title('Wait Times');
ax[1].hist(dfw_highs, bins=7)
ax[1].set_title('High Temperatures for DFW (4/2022-5/2022)')
ax[1].set_ylabel('Frequency')
ax[1].set_xlabel('Temperature (F)')
fig.show()
```

In this section, we had a glimpse into how varied data and distributions can appear. Since distributions of data appear in many shapes and sizes in the wild, it is useful to have methods for describing distributions. In the next section, we will discuss the measurements available for distributions, how those measurements are performed, and the types of data that can be measured.

Measuring and describing distributions

The distributions of data found in the wild come in many shapes and sizes. This section will discuss how distributions are measured and which measurements apply to the four types of data. These measurements will provide methods to compare and contrast different distributions. The measurements discussed in this section can be broken into the following categories:

- Central tendency
- Variability
- Shape

These measurements are called **descriptive statistics**. The descriptive statistics discussed in this section are commonly used in statistical summaries of data.

Measuring central tendency

There are three types of measurement of central tendency:

- Mode
- Median
- Mean

Let's discuss each one of them.

Mode

The first measurement of central tendency we will discuss is the mode. The mode of a dataset is simply the most commonly occurring instance. Using the machines in the factory as an example (see *Figure 2.1*), the mode of the dataset would be model B. In the example, there are 3 of model A and 5 of model B, therefore, making model B the most common – the mode.

A dataset can be one of the following:

- Unimodal – having one mode
- Multimodal – having more than one mode

In the preceding example, the data is unimodal.

Using the factory example again, let's imagine that there are 3 of **Model A**, 5 of **Model B**, and 5 of **Model D** (a new model). Then, the dataset will have two modes: **Model B** and **Model D**, as shown in *Figure 2.4*.

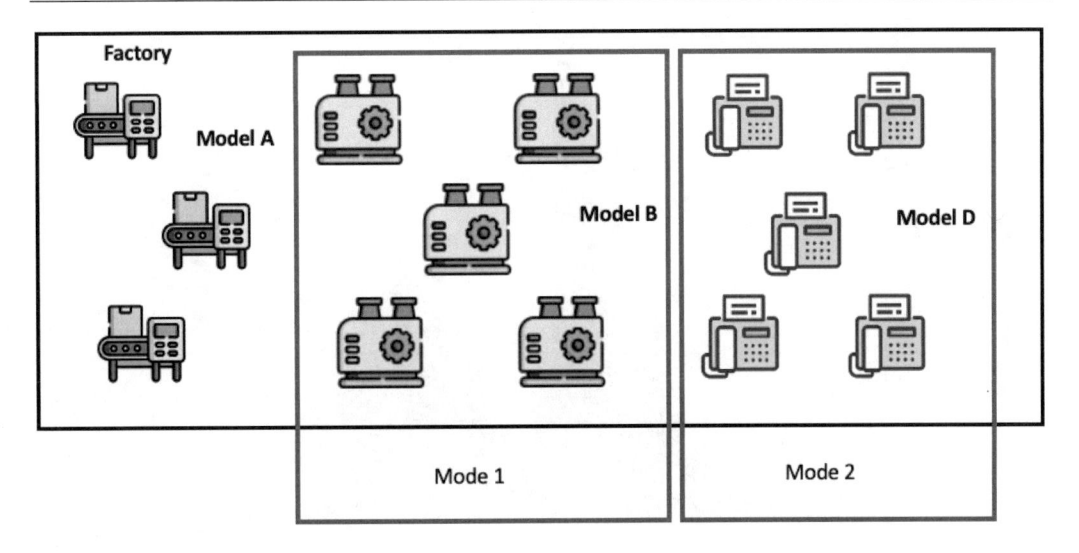

Figure 2.4 – Multimodel distribution of machines in a factory

Therefore, this dataset is multimodal.

> **Mode and Data Types**
>
> These examples of modes have used nominal data, but all four types of data support the mode because all four data types support the equality operation.

While the mode refers to the most common instance, in multimodal cases of continuous data, the term mode is often used in a less strict sense. For example, the distribution in *Figure 2.5* would commonly be referred to as multimodal even though the peaks of the distribution are not the same magnitude. However, with nominal and ordinal data, it is more common to use the stricter definition of *most common* when referring to the modality of a distribution.

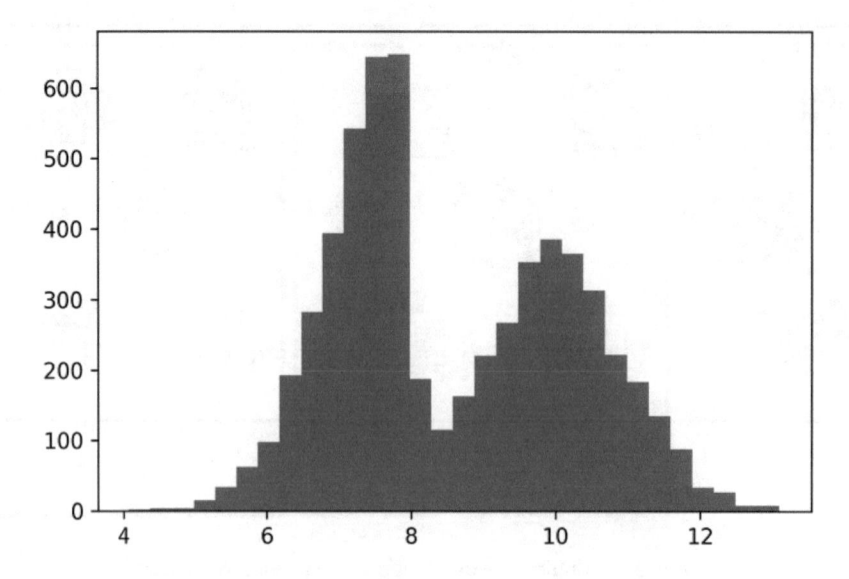

Figure 2.5 – A multimodal distribution of data

Now, we will look at how to calculate the mode with code using `scipy`. The `scipy` library contains functions for calculating descriptive statistics in the `stats` module. In this example, we import mode from `scipy.stats` and calculate the mode of the following numbers, 1, 2, 3, 4, 4, 4, 5, 5:

```
from scipy.stats import mode
m = mode([1,2,3,4,4,4,5,5])
print(
    f"The mode is {m.mode[0]} with a count of"
    f" {m.count[0]} instances"
)
# The mode is 4 with a count of 3 instances
```

The mode function returns a mode object containing mode and count members. Unsurprisingly, the mode and count members contain the modes of the dataset and the number of times the modes appear, respectively. Note that mode and count members are indexable (like lists) because a dataset can contain multiple modes.

Median

The next measure of the center is the median. The median is the middle value occurring when the values are arranged in an order.

> **Median and Data Types**
> This measure can be performed on ordinal data, interval data, and ratio data, but not on nominal data.

We will discuss two cases here.

Finding the median when the number of instances is odd

Finding the median of some numeric data is shown in *Figure 2.6*. The data is sorted, then the median is identified.

$$\text{Raw Data} \quad 85, 99, 70, 71, 86, 88, 94$$

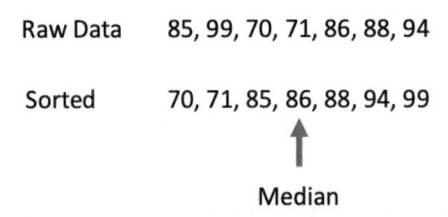

$$\text{Sorted} \quad 70, 71, 85, \underset{\uparrow}{86}, 88, 94, 99$$

$$\text{Median}$$

Figure 2.6 – Identifying the median with an odd number of instances

In the preceding example, the instances are odd in number (7 instances), which have a center value. However, if the number of instances had been even, it would not have been possible to just take the middle number after sorting the values.

Finding the median when the number of instances is even

When there are an even number of instances, the average of the two middle-most values is taken. Unlike the mode, there is no concept of multiple medians for the same series of data. An example with an even number of instances (8 instances) is shown in *Figure 2.7*.

Raw Data	85, 99, 70, 71, 86, 88, 94, 105
Sorted	70, 71, 85, 86, 88, 94, 99, 105
Middle-Most Two	86 88
Average	87 ↑ Median

Figure 2.7 – Identifying the median with an even number of instances

Now, let's see how to calculate the median of a dataset with numpy. Like scipy, numpy contains functions for calculating descriptive statistics. We will calculate the median for the eight numbers listed in the preceding example:

```
import numpy as np
values = [85, 99, 70, 71, 86, 88, 94, 105]
median = np.median(values)
print(f"The median value is {median:.2f}")
# The median value is 87.00
```

The result of the median calculation is 87, as expected. Note that the median function returns a single value, in contrast to the mode function in the previous code example.

Mean

The next center measure is the mean, which is commonly referred to as the average. The mean is defined by the following equation:

$$\bar{x} = \frac{\sum_{i=0}^{n} x_i}{N}$$

Let me explain the equation in words. To calculate the mean, we must add all the values together, then divide the sum by the number of values. Please refer to the following example. The 7 numbers are first added together, which brings the total sum to 593. This sum is then divided by the number of instances, resulting in a value of 84.7.

```
        Raw Data    85, 99, 70, 71, 86, 88, 94
        Total Sum   85 + 99 + 70 + 71 + 86 + 88 + 94 = 593
  Divide By Count   593 / 7 = 84.7
                         ↑
                        Mean
```

Figure 2.8 – Finding the mean

Note that the mean and the median of these values (84.7 and 86, respectively) are not the same value. In general, the mean and median will not be the same value, but there are special cases where the mean and median will converge.

> **Mean and Data Types**
>
> As for the supported data types, the mean is valid for interval and ratio data since the values are added together.

Now, we will look at how to calculate the mean with numpy. The following code example shows the calculation of the mean for the values in the previous example:

```
import numpy as np
values = [85, 99, 70, 71, 86, 88, 94]
mean = np.mean(values)
print(f"The mean value is {mean:.1f}")
# The mean value is 84.7
```

Like the median function, the mean function returns a single number.

Before concluding this section on center measures, it is worth discussing the use of the mean and median in various situations. As mentioned previously, the median and mean will, in general, be different values. This is an effect driven by the shape of the distribution.

> **Shape impacts on Mean and Median**
>
> If the distribution is symmetric, the mean and median will tend to converge. However, if the distribution is not symmetric, the mean and median will diverge.

The degree to which the measures diverge is driven by how asymmetric the distribution is. Four example distributions are given in *Figure 2.6* to show this effect. Distributions 1 and 2 show the mean pulled toward a higher value than the median. The mean is pulled toward values with a larger absolute value. This is an important effect of the mean to be aware of when a dataset contains (or may contain) **outlier values** (often called outliers or influential points), which will tend to pull the mean in their direction. Unlike the mean, the median is not affected by outliers if outliers account for a smaller percentage of the data. Outliers will be discussed further in the *Measuring variability* section.

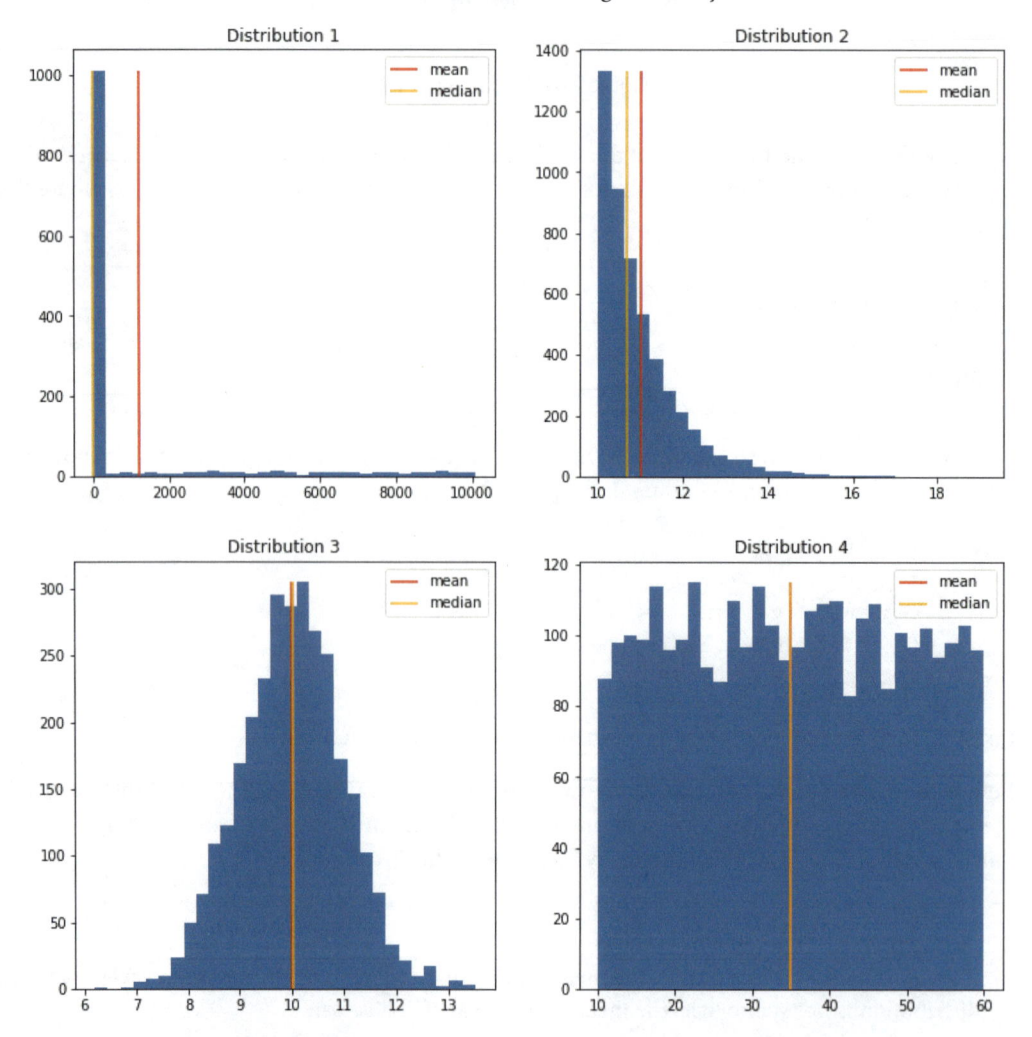

Figure 2.9 – Two asymmetric distributions and two symmetric distributions

The next category of measurements for distributions is measures of variability.

Measuring variability

By variability, we essentially mean how wide a distribution is. The measurements in this category are as follows:

- Range
- Quartile ranges
- Tukey fences
- Variance

Let's discuss each of them.

Range

The range is simply the difference between the maximum value and the minimum value in the distribution. Like the mean, the range will be affected by outliers since it depends on the max and min values. However, there is another variability method that, like the median, is robust to the presence of outliers.

Let's take a look at calculating a range with code with numpy:

```
import numpy as np
values = [85, 99, 70, 71, 86, 88, 94, 105]
max_value = np.max(values)
min_value = np.min(values)
range_ = max_value - min_value
print(f"The data have a range of {range_}"
      f" with max of {max_value}"
      f" and min of {min_value}")
# The data have a range of 35 with max of 105 and min of 70
```

While numpy does not have a range function, the range can be calculated using the min and max functions provided by numpy.

Quartile ranges

The next measures of variability are determined by sorting the data and then dividing the data into four equal sections. The boundaries of the four sections are the quartiles, which are called the following:

- The lower quartile (Q1)
- The middle quartile (Q2)
- The upper quartile (Q3)

An example of quartiles is shown as follows. Like the median, quartiles are robust to outliers so long as the outliers are a small percentage of the dataset. Note that the middle quartile is, in fact, the median. An adjusted range measurement that is less sensitive to outliers than the normal range discussed in the *Range* section is the middle quartile, the **interquartile range** (**IQR**). The IQR is the difference between the upper and lower quartiles (Q3 - Q1). While this range is less sensitive to outliers, it only contains 50% of the data. Thus, making the interquartile range likely to be *less representative of the total variation* of the data.

Raw Data   85, 99, 70, 71, 86, 88, 94

Sorted   70, 71, 85, 86, 88, 84, 99

Lower Quartile   Middle Quartile (Median)   Upper Quartile

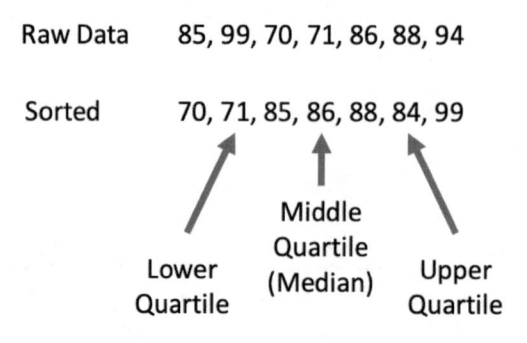

Figure 2.10 – Q1, Q2, and Q3

We can calculate the quartiles and IQR range using numpy and scipy. In the following code example, we use the quantiles function to calculate the quartiles. We will not discuss quantiles here, other than to mention that quantiles are a generalization where the data can be split into any number of equal parts. Since we are splitting the data into four equal parts for quartiles, the quantiles values used for the calculation are 0.25, 0.5, and 0.75. Quartiles Q1 and Q3 could then be used to calculate the IQR. However, we could also use the iqr function from scipy to make the calculation:

```
import numpy as np
from scipy import stats
values = [85, 99, 70, 71, 86, 88, 94]
quartiles = np.quantile(values, [0.25, 0.5, 0.75],
    method="closest_observation")
print(f"The quartiles are Q1: {quartiles[0]},
    Q2: {quartiles[1]}, Q3: {quartiles[2]}")
iqr = stats.iqr(values,interpolation='closest_observation')
print(f"The interquartile range is {iqr}")
# The quartiles are Q1: 71, Q2: 85, Q3: 88
# The interquartile range is 17
```

Note the use of the `method` and `interpolation` keyword arguments in the `quantiles` function and the `iqr` function, respectively. Several options can be used for these keyword arguments, which will lead to different results.

Quartiles are often visualized with a boxplot. The following *Figure 2.11* shows the main parts of a boxplot. A boxplot is made up of two main parts:

- The box
- The whiskers

The box part represents 50% of the data that is contained by the IQR. The whiskers are drawn starting from the edge of the boxes to a length of k * IQR, where k is commonly chosen to be 1.5. Any values beyond the whiskers are considered outliers.

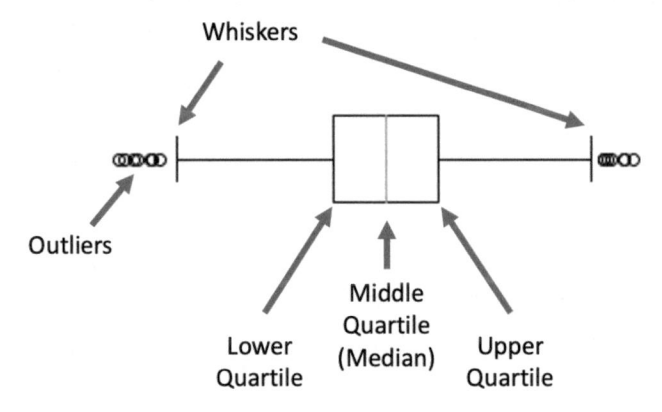

Figure 2.11 – Parts of a box and whisker plot

Figure 2.12 shows how histograms and boxplots visualize the variability of a symmetric and asymmetric distribution. Notice how the boxplot of the asymmetric data is compressed on the left and expanded on the right, while the other boxplot is clearly symmetric. While a boxplot is useful for visualizing the symmetry of the data and the presence of outliers, the modality of the distribution would not be evident.

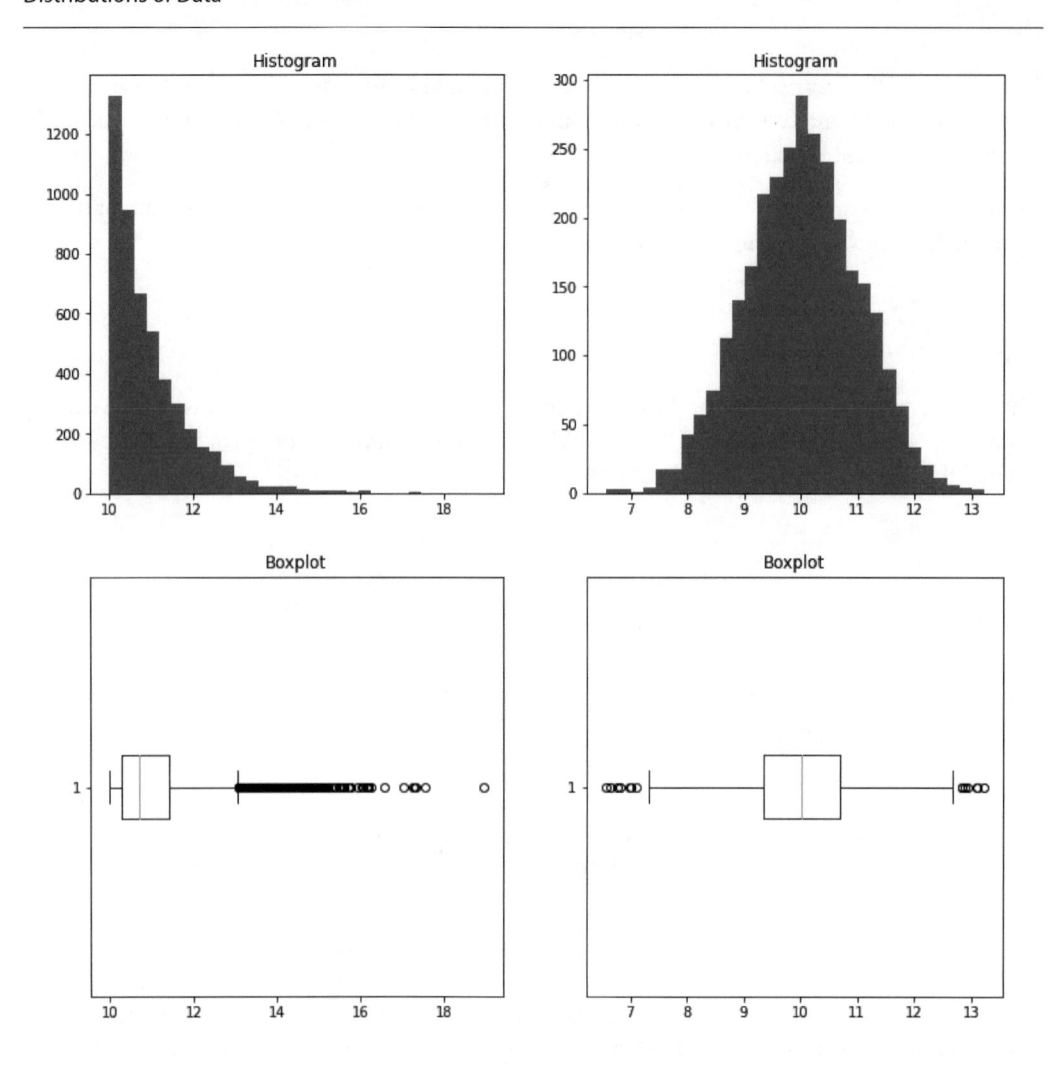

Figure 2.12 – Comparison of boxplots and histograms for asymmetric and symmetric distributions

When exploring, it is common to use multiple visualizations since each type of visualization has its own advantages and disadvantages. It is common to use multiple visualizations since each type of visualization has its own advantages and disadvantages.

Tukey fences

In the last few sections on measurements, the concept of outliers has appeared a few times. Outliers are values that are atypical compared to the main distribution, or anomalous values. While there are methods for classifying data points as outliers, there is no generally robust method for classifying data points as outliers. Defining outliers typically should be informed by the use case of the data analysis, as there will be different factors to consider based on the application domain. However, it is worth mentioning the common technique shown in the boxplot example called Tukey fences. The lower and upper Tukey fences are based on the IQR and defined as follows:

- *Lower fence*: $Q1 - k(IQR)$
- *Upper fence*: $Q3 + k(IQR)$

As mentioned earlier, k is often chosen to be 1.5 as a default value, but there may be a more appropriate value for a given application domain.

Now let's take a look at how to calculate Tukey fences with `numpy` and `scipy`. This code example will build upon the previous example since there is no function to calculate the fences directly. We will again calculate the quartiles and the IQR with `numpy` and `scipy`. Then, we apply these operations to the values listed in the preceding equations:

```
import numpy as np
from scipy import stats
values = stats.norm.rvs(10, size=3000)
q1, q3 = np.quantile(values, [.25, .75],
    method='closest_observation')
iqr = stats.iqr(values,interpolation='closest_observation')
lower_fence = q1 - iqr * 1.5
upper_fence = q3 + iqr * 1.5
# may vary due to randomness in data generation
print(f"The lower fence is {lower_fence:.2f} and the upper
    fence is {upper_fence:.2f}")
# The lower fence is 7.36 and the upper fence is 12.67
```

In this case, we used both `numpy` and `scipy`; however, the `scipy` calculation could be replaced with Q3-Q1 as mentioned previously.

Variance

The last measure of variability that will be covered in this section is variance. Variance is a measure of dispersion that can be understood as how *spread out* the numbers are from the average value. The formula for variance, denoted S^2, is as follows:

$$S^2 = \frac{\Sigma(x_i - \bar{x})^2}{N - 1}$$

In this equation, the term $(x_i - \bar{x})$ is considered the deviation from the mean, which leads to another measure that is closely related to variance – the standard deviation, which is the square root of variance. The formula for standard deviation, denoted σ, is given here:

$$\sigma = \sqrt{S^2} = \sqrt{\frac{\Sigma(x_i - \bar{x})^2}{N - 1}}$$

In general, a wider distribution will have a larger variance and a larger standard deviation, but these values are not as easy to interpret as a range or IQR. These concepts will be covered more in detail in the next section, in the context of the normal distribution, which will provide clearer intuition for what these values measure.

Again, these values will be calculated with code using numpy. The functions for variance and standard deviation are var and std, respectively:

```
import numpy as np
values = [85, 99, 70, 71, 86, 88, 94]
variance = np.var(values)
standard_dev = np.std(values)
print(f"The variance is {variance:.2f} and the standard
    deviation is {standard_dev:.2f}")
# The variance is 101.06 and the standard deviation is 10.05
```

Measuring shape

The next type of measure has to do with the shapes of distributions. They are as follows:

- Skewness
- Kurtosis

Let's discuss each of them.

Skewness

The first measurement is skewness. Put simply, skewness is measurement asymmetry [1]. An example of skewed distributions is shown in *Figure 2.13*.

There are two types of skewed distributions:

- Left-skewed
- Right-skewed

A distribution is skewed in the direction of the dominant tail, meaning that a distribution with a dominant tail to the right is right-skewed and a distribution with a dominant tail to the left is left-skewed (as shown in *Figure 2.13*).

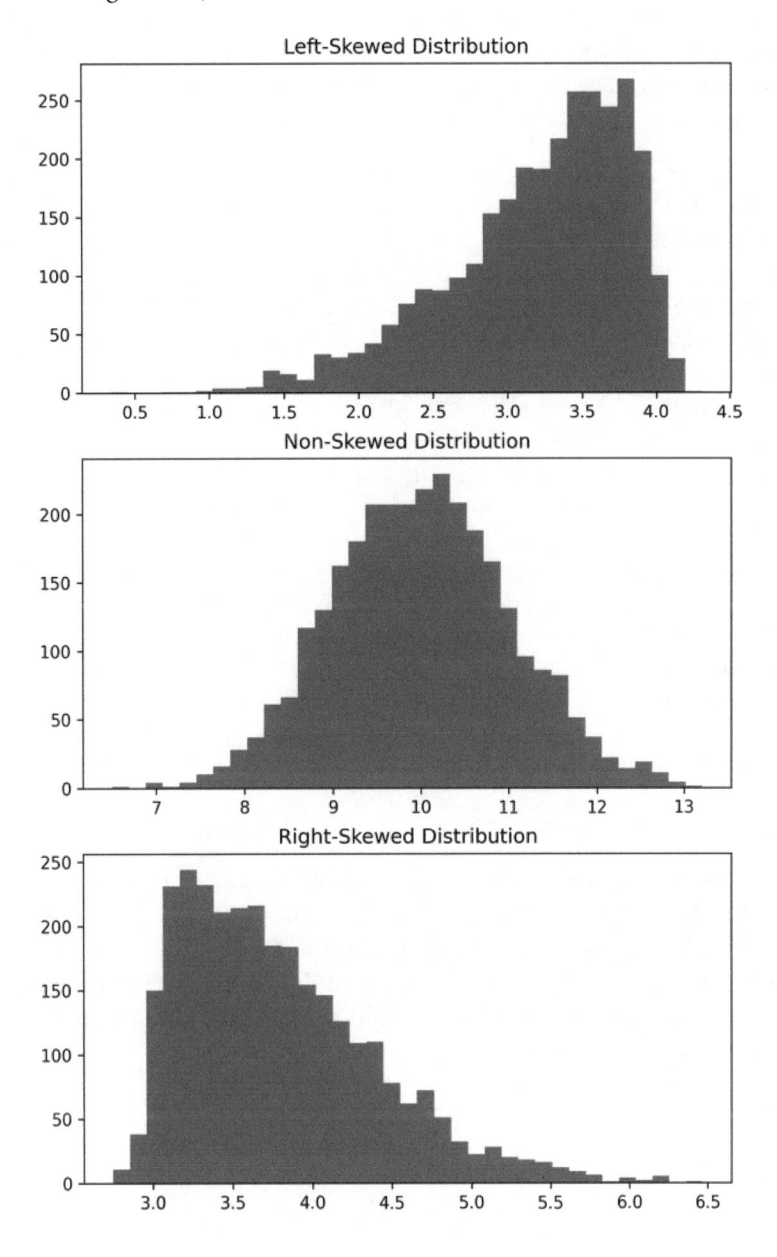

Figure 2.13 – Distributions demonstrating skewness

The formula for skewness will not be shown here since it can be calculated trivially with modern software packages. The output of the skewness calculation can be used to determine the skewness and the direction of the skew. If the skewness value is 0 or near 0, the distribution does not exhibit strong skewness. If the skewness is positive, the distribution is right-skewed, and if the skewness value is negative, the distribution is left-skewed. The larger the absolute value of the skewness value, the more the distribution exhibits skewness. An example of how to calculate skewness with `scipy` is shown in the following code example:

```
from scipy.stats import skewnorm, norm
from scipy.stats import skew as skew_calc
# generate data
skew_left = -skewnorm.rvs(10, size=3000) + 4
skew_right = skewnorm.rvs(10, size=3000) + 3
symmetric = norm.rvs(10, size=3000)
# calculate skewness
skew_left_value = skew_calc(skew_left)
skew_right_value = skew_calc(skew_right)
symmetric_value = skew_calc(symmetric)
# Output may vary some due to randomness of generated data
print(f"The skewness value of this left skewed
    distribution is {skew_left_value:.3f}")
print(f"The skewness value of this right skewed
    distribution is {skew_right_value:.3f}")
print(f"The skewness value of this symmetric distribution
    is {symmetric_value:.3f}")
```

The other shape measurement covered in this section is kurtosis.

Kurtosis

Kurtosis is a measurement of how heavy or light the tail of a distribution is relative to the normal distribution [2]. While the normal distribution has not been covered in depth yet, the idea of kurtosis can still be discussed. A light-tailed distribution means that more of the data is near or around the mode of the distribution. In contrast, a heavy-tailed distribution means that more of the data is at the edges of the distribution than near the mode. A light-tailed distribution, a normal distribution, and a heavy-tailed distribution are shown in *Figure 2.14*.

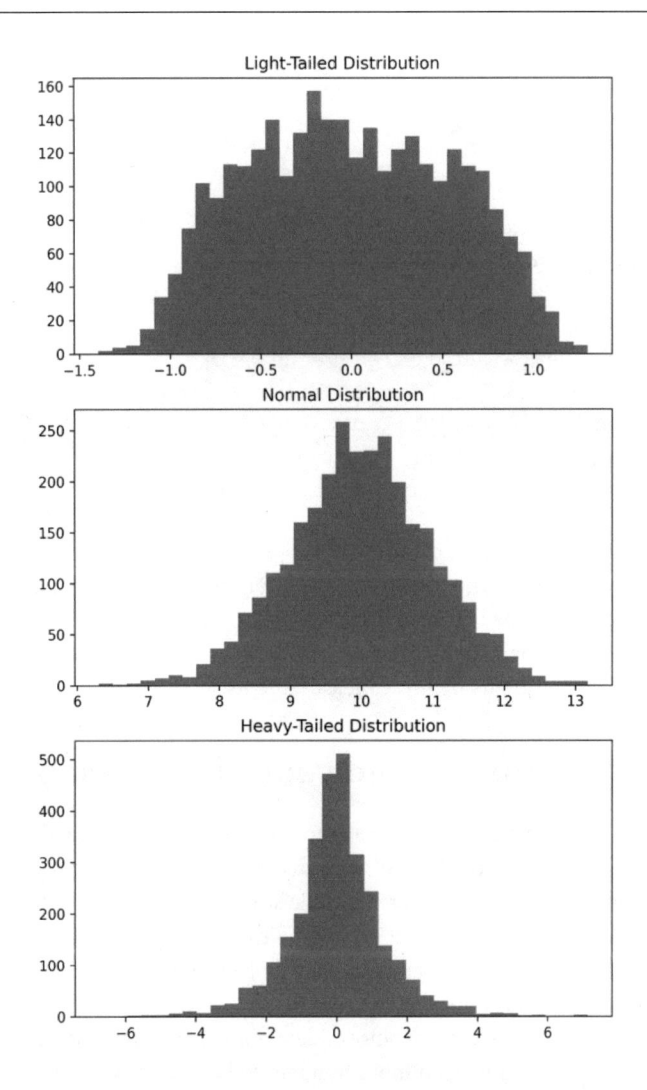

Figure 2.14 – Distributions demonstrating tailedness with reference to a normal distribution

The formula for kurtosis will not be shown here since it can be calculated trivially with modern software packages. If the kurtosis value is 0 or near 0, the distribution does not exhibit kurtosis. If the kurtosis value is negative, the distribution exhibits light-tailedness, and if the kurtosis value is positive, the distribution exhibits heavy-tailedness. An example of how to calculate kurtosis with `scipy` is shown in the following code example:

```
from scipy.stats import norm
from scipy.stats import gennorm
from scipy.stats import kurtosis
# generate data
```

```
light_tailed = gennorm.rvs(5, size=3000)
symmetric = norm.rvs(10, size=3000)
heavy_tailed = gennorm.rvs(1, size=3000)
# calculate skewness
light_tailed_value = kurtosis(light_tailed)
heavy_tailed_value = kurtosis(heavy_tailed)
symmetric_value = kurtosis(symmetric)
# Output may vary some due to randomness of generated data
print(f"The kurtosis value of this light-tailed
    distribution is {light_tailed_value:.3f}")
print(f"The kurtosis value of this heavy_tailed
    distribution is {heavy_tailed_value:.3f}")
print(f"The kurtosis value of this normal
    distribution is {symmetric_value:.3f}")
```

In this section, we walked through the common descriptive statistics that are used for measuring and describing distributions of data. These measurements provide a common language for describing and comparing distributions. The concepts discussed in this chapter are fundamental to many of the concepts discussed in future chapters. In the next section, we will discuss the normal distribution and describe the normal distribution using these measurements.

The normal distribution and central limit theorem

When discussing the normal distribution, we refer to the bell-shaped, **standard normal distribution**, which is formally synonymous with the **Gaussian distribution**, named after Carl Friedrich Gauss, an 18th- and 19th-century mathematician and physicist who – among other things – contributed to the concepts of approximation, and, in 1795, invented the method of least squares and the normal distribution, which is commonly used in statistical modeling techniques, such as least squares regression [3]. The standard normal distribution, also referred to as a **parametric** distribution, is characterized by a symmetrical distribution with a probability of data point dispersion consistent around the mean – that is, the data appears near the mean more frequently than data farther away. Since the location data dispersed within this distribution follows the laws of probability, we can call this a **standard normal probability distribution**. As an aside, a distribution in statistics that is not a probability distribution is generated through non-probability sampling based on non-random selection, whereas a probability distribution is based on random sampling. Both probability-based and non-probability-based distributions can have a standard normal distribution. The standard normal distribution exhibits neither skew nor kurtosis. It has equal variance throughout and frequently occurs in nature. The **Empirical Rule** is used to describe this distribution as having three pertinent standard deviations centered around the mean, μ. There are two distinct assumptions about this distribution:

- The first, second, and third standard deviations contain 68%, 95%, and 99.7% of the measurements dispersed, respectively

- The mean, median, and mode are all equal to each other

Standard Normal Probability Distribution

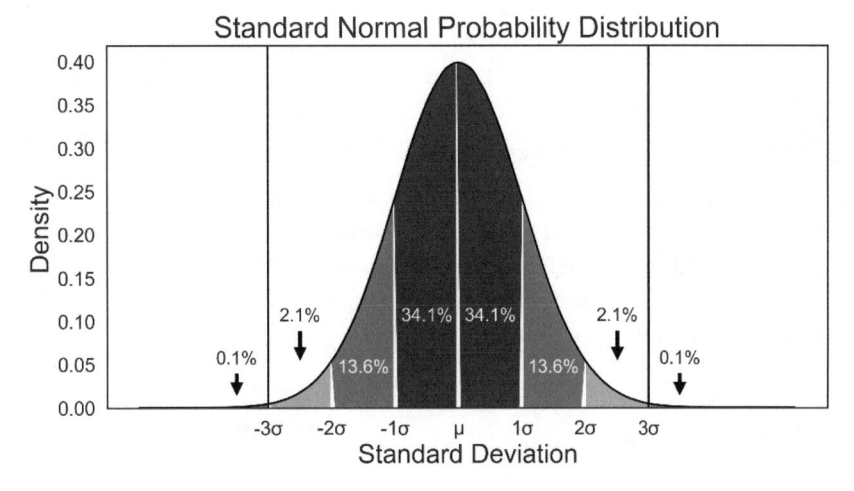

Figure 2.15 – The standard normal distribution

Two common forms of a normal distribution are as follows:

- The probability density distribution
- The cumulative density distribution

As mentioned before, the probability density distribution is based on random sampling, whereas the cumulative density distribution is based on accumulated data, which is not necessarily random.

The two-tailed probability density function of the standard normal distribution is this:

$$f(x) = \frac{e^{-\frac{(x-\mu)^2}{2\sigma^2}}}{\sigma\sqrt{2\pi}}$$

The left-tailed cumulative function of the standard normal distribution is this:

$$f(x) = \int_{-\infty}^{x} \frac{e^{-x^2}}{\sqrt{2\pi}} \frac{}{2}$$

With respect to statistical modeling, the normal distribution represents balance and symmetry. This is important when building statistical models as many models assume normal distribution and are not robust to many deviations from that assumption, as they are built around a mean. Consequently, if variables in such a model are not normally distributed, the model's errors will be increased and inconsistent, thus diminishing the model's stability. When considering multiple variables in a statistical model, their interaction is more easily approximated when both are normally distributed.

In the following *Figure 2.16*, in the left plot, variables X and Y interact with each other and create a centralized dispersion around a mean. In this case, modeling Y using X with a mean line or linear distance can be done reasonably well. However, if the two variables' distributions were skewed, as in the plot on the right, this would result in non-constant variance between the two, resulting in an unequal distribution of errors and unreliable output.

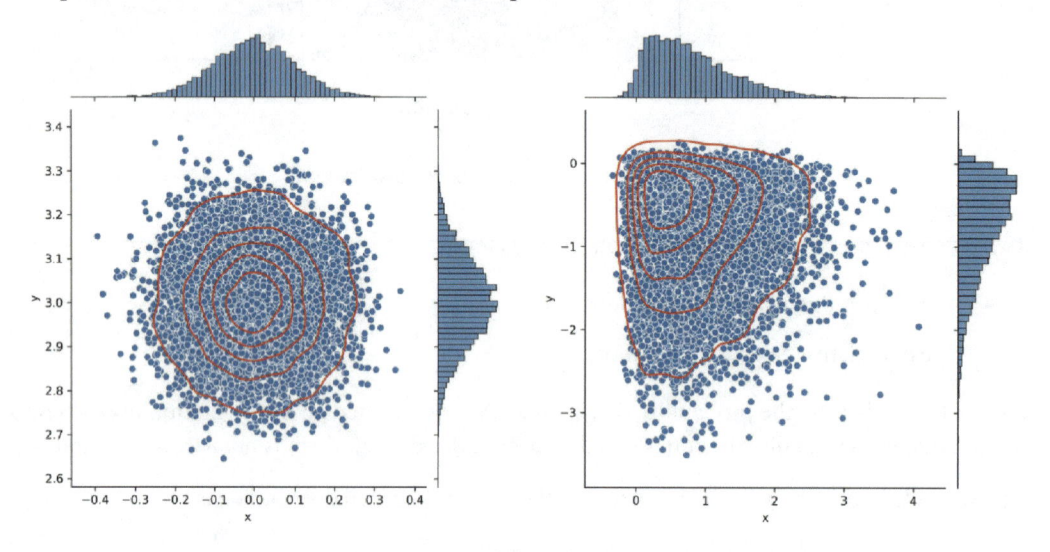

Figure 2.16 – Bivariate normal (left) and skewed (right) distributions

In the case of linear classification and regression models, this will mean some results are better than others while some will likely be very bad. This can be difficult to assess at times using basic model metrics and requires deeper model analysis to prevent trusting what could end up being misleading results. Furthermore, deployment into a production environment would be very risky. More on this will be discussed in *Chapter 6*.

The Central Limit Theorem

When sampling data, it is common to encounter the issue of non-normal data. This may be for multiple reasons, such as the population not having a normal distribution or the sample being misrepresentative of the population. The Central Limit Theorem, which is important in statistical inference, postulates that if random samples of *n* observations are taken from a population that has a specific mean, μ, and **standard deviation**, σ, the sampling distribution constructed from the means of the randomly selected sub-sample distributions will approximate a normal distribution having roughly the same mean, μ, and standard deviation, calculated as $\sqrt{\frac{\Sigma(x_i - \mu)^2}{N}}$, as the population. The next section will use bootstrapping to demonstrate the Central Limit Theorem in action. A later section discussing transformations will provide techniques for reshaping data distributions that do not conform to normal distributions so that tools requiring normal distributions can still be effectively applied.

Bootstrapping

Bootstrapping is a method of resampling that uses random sampling – typically with replacement – to generate statistical estimates about a population by resampling from subsets of the sampled distribution, such as the following:

- Confidence intervals
- Standard error
- Correlation coefficients (Pearson's correlation)

The idea is that repeatedly sampling different random subsets of a sample distribution and taking the average each time, given enough repeats, will begin to approximate the true population using each subsample's average. This follows directly the concept of the Central Limit Theorem, which to be restated, asserts that sampling means begins to approximate normal sampling distributions, centered around the original distribution's mean, as sample sizes and counts increase. Bootstrapping is useful when a limited quantity of samples exists in a distribution relative to the amount needed for a specific test, but inference is needed.

As discussed in *Chapter 1, Sampling and Generalization*, constraints such as time and expense are common reasons for obtaining samples rather than populations. Because the underlying concept of bootstrapping is to make assumptions about the population using samples, it is not beneficial to apply this technique to populations as the true statistical parameters – such as the percentiles and variance – of a population are known. Regarding sample preparation, the balance of attributes in the sample should represent the true approximation of the population. Otherwise, the results will likely be misleading. For example, if the population of species within a zoo is a split of 40% reptiles and 60% mammals and we want to bootstrap their longevity to identify the confidence intervals for their lifespans, it would be necessary to ensure the dataset to which bootstrapping was applied contained a split of 40% reptiles and 60% mammals; a split of 15% reptiles and 85% mammals, for example, would lead to misleading results. In other words, the sample stratification should be balanced in proportion to the population.

Confidence intervals

As mentioned before, one useful application of bootstrapping is to create confidence intervals around sparsely defined or limited datasets – that is to say, datasets with a wide range of values without many samples. Consider an example of bootstrapping to perform a hypothesis test using a 95% confidence interval using the `"Duncan"` dataset in `statsmodels`, which contains incomes by profession, type, education, and prestige. While this is the full dataset, consider this dataset a sample since the sampling method is not mentioned and it is not likely to consider all incomes for all workers of every profession and type. To obtain the dataset, we first load the `matplotlib`, `statsmodels`, `pandas`, and `numpy` libraries. We then download the dataset and store it as a `pandas` DataFrame in the `df_duncan` variable. Following this, we recode the "`prof`", "`wc`", and "`bc`" types as `"professional"`, `"white-collar"`, and `"blue collar"`, respectively. Finally, we create two separate `pandas` DataFrames; one for professional job types and another for blue-collar job types, as these are the two subsets we will analyze using bootstrapping:

```
import matplotlib.pyplot as plt, statsmodels.api as sm, pandas as pd, 
numpy as np, scipy.stats
df_duncan = sm.datasets.get_rdataset("Duncan", 
    "carData").data
df_duncan.loc[df_duncan['type'] == 'prof', 
    'type'] = 'professional'
df_duncan.loc[df_duncan['type'] == 'wc', 
    'type'] = 'white-collar'
df_duncan.loc[df_duncan['type'] == 'bc', 
    'type'] = 'blue-collar'
df_professional = df_duncan.loc[(
    df_duncan['type'] == 'professional')]
df_blue_collar = df_duncan.loc[(
    df_duncan['type'] == 'blue-collar')]
```

	Type	Income	Education	Prestige
accountant	professional	62	86	82
pilot	professional	72	76	83
architect	professional	75	92	90
author	professional	55	90	76
chemist	professional	64	86	90

Figure 2.17 – Table displaying the first five rows of the statsmodels Duncan data

We then build a set of plotting functions, as seen next. In `plot_distributions()`, we denote p=5, meaning the p-value will be significant at a significance level of 0.05 (1.00 - 0.05 = 0.95, hence, 95% confidence). We then divide this value by 2 since this will be a two-sided test, meaning we want to know the full interval rather than just one bound (discussed in *Chapter 1* as a representative test statistic). Regarding the plots, we visualize the data using histograms (the `hist()` function) in `matplotlib` and then plot the 95% sampling confidence intervals using the `axvline()` functions, which we build using the numpy function `percentile()`.

> **Percentile in Bootstrapping**
>
> When applied to the original data, the percentile is only that, but when applied to the bootstrapped sampling distribution, it is the confidence interval.

To state the confidence interval simply, a 95% confidence interval means that for every 100 sample means taken, 95 of them will fall within this interval. In the numpy `percentile()` function, we use p=5 to support that 1-p is the confidence level, where *p* is the level of significance (think *p-value*, where any value at or lower than *p* is significant). Since the test is two-tailed, we divide *p* by 2 and split 2.5 in the left tail and 2.5 in the right since we have a symmetrical, standard normal distribution. The `subplot(2,1,...)` code creates two rows and one column. Axis 0 of the figure is used for professional incomes and axis 1 is used for blue-collar incomes:

```
def plot_distributions(n_replicas, professional_sample, blue_collar_
sample, professional_label, blue_collar_label, p=5):
    fig, ax = plt.subplots(2, 1, figsize=(10,8))
    ax[0].hist(professional_sample, alpha=.3, bins=20)
    ax[0].axvline(professional_sample.mean(),
        color='black', linewidth=5)
# sampling distribution mean
    ax[0].axvline(np.percentile(professional_sample, p/2.),
        color='red', linewidth=3, alpha=0.99)
# 95% CI Lower limit (if bootstrapping)
    ax[0].axvline(np.percentile(professional_sample,
        100-p/2.), color='red', linewidth=3, alpha=0.99)
# 95% CI Upper Limit   (if bootstrapping)
    ax[0].title.set_text(str(professional_label) +
        "\nn = {} Resamples".format(n_replicas))
    ax[1].hist(blue_collar_sample, alpha=.3, bins=20)
    ax[1].axvline(blue_collar_sample.mean(), color='black',
        linewidth=5) # sampling distribution mean
    ax[1].axvline(np.percentile(blue_collar_sample, p/2.),
        color='red', linewidth=3, alpha=0.99)
# 95% CI Lower limit (if bootstrapping)
    ax[1].axvline(np.percentile(blue_collar_sample,
```

```
            100-p/2.), color='red', linewidth=3, alpha=0.99)
# 95% CI Upper Limit (if bootstrapping)
    ax[1].title.set_text(str(blue_collar_label) +
        "\nn = {} Resamples".format(n_replicas))
    if n_replicas > 1:
        print("Lower confidence interval limit: ",
            np.percentile(round(professional_sample,4),
            p/2.))
        print("Upper confidence interval limit: ",
            np.percentile(round(professional_sample,4),
            100-p/2.))
        print("Mean: ", round(professional_sample,
            4).mean())
        print("Standard Error: ",
            round(professional_sample.std() /
            np.sqrt(n_replicas), 4) )
        print("Lower confidence interval limit: ",
            np.percentile(round(blue_collar_sample,4),
            p/2.))
        print("Upper confidence interval limit: ",
            np.percentile(round(blue_collar_sample,4),
            100-p/2.))
        print("Mean: ", round(blue_collar_sample,4).mean())
        print("Standard Error: ",
            round(blue_collar_sample.std() /
            np.sqrt(n_replicas), 4) )
    else:
        print("At least two samples required to create the following
 statistics:\nConfidence Intervals\nMean\nStandard Error")
```

In the original dataset, there are 18 income data points for `professional` job types and 21 data points for `blue-collar` job types. The 95% confidence interval for the professional job type ranges from 29.50 to 79.15 with an average of 60.06. That interval ranges from 7.00 to 64.00 for blue-collar job types with a mean of 23.76. Based on *Figure 2.18*, there is a reasonable overlap between the income differences, which causes the overlapping confidence intervals. Consequently, it would be reasonable to assume there is no statistically significant difference in incomes between blue-collar and professional job types. However, this dataset has a very limited volume of samples:

```
n_replicas=0
plot_distributions(n_replicas=n_replicas,
professional_sample=df_professional['income'],
    blue_collar_sample=df_blue_collar['income'],
    professional_label="Professional",
    blue_collar_label="Blue Collar")
```

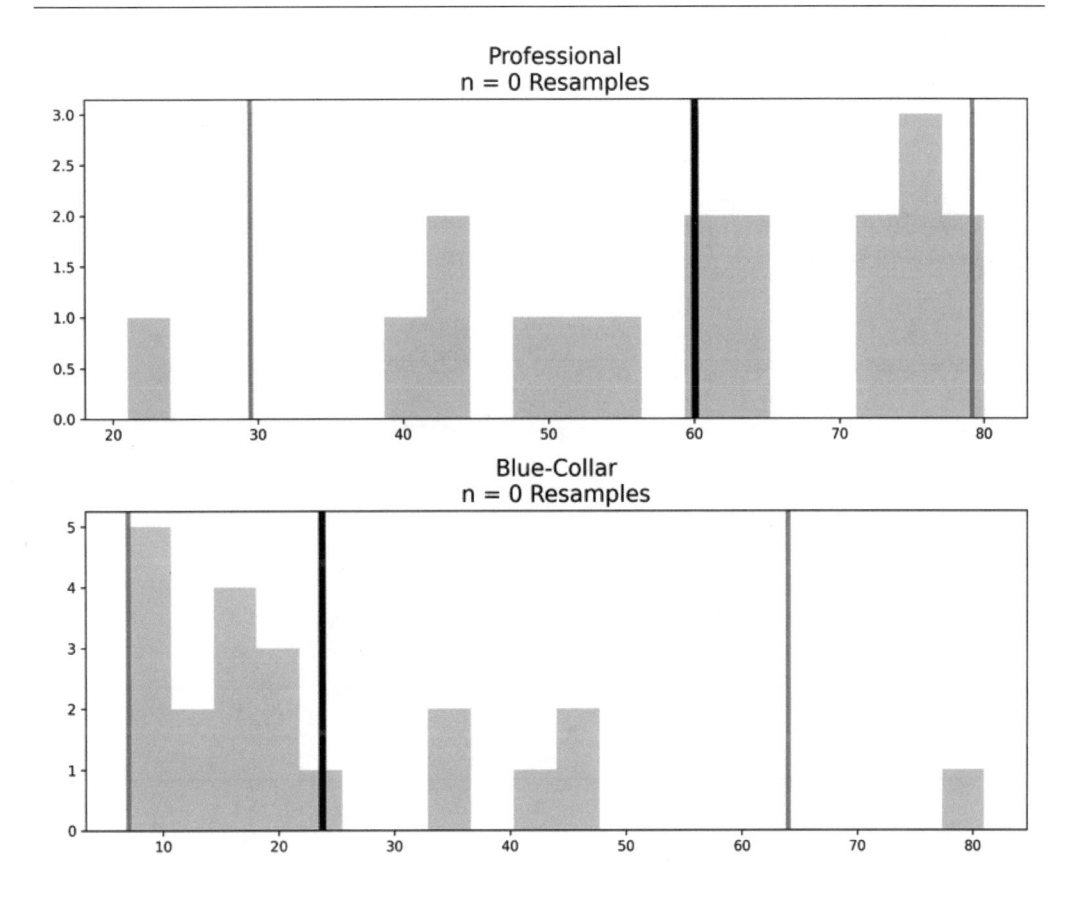

Figure 2.18 – Original data distributions with 95th percentile lines

In the following code, using pandas' .sample() function, we randomly resample 50% (frac=0.5) of the income values from each distribution 1,000 times and calculate a new mean each time, appending it to the Python lists ending with _bootstrap_means. Using those lists, we derive new 95% confidence intervals. *Figure 2.19* shows, with respect to the standard deviations and income values in the dataset, the new sample distributions using the average of each resampled subset. The replace=True argument allows for resampling the same record multiple times (in the event that should randomly occur), which is a requirement of bootstrapping.

After performing the bootstrapping procedure, we can see income has started to distribute in a roughly standard normal, Gaussian form. Notably, from this experiment, the confidence intervals no longer overlap. The implication of the separation of the confidence intervals between the professional and blue-collar groups is that with a 95% level of confidence, it can be shown there is a statistically significant difference between the incomes of the two job types. The confidence interval for the professional income levels is now 48.66 to 69.89 with a mean of 60.04, and for blue-collar, 14.60 to 35.90 with a mean of 23.69:

```
n_replicas = 1000
professional_bootstrap_means = pd.Series(
    [df_professional.sample(frac=0.5, replace=True)
    ['income'].mean() for i in range(n_replicas)])
blue_collar_bootstrap_means = pd.Series(
    [df_blue_collar.sample(frac=0.5, replace=True)
    ['income'].mean() for i in range(n_replicas)])
```

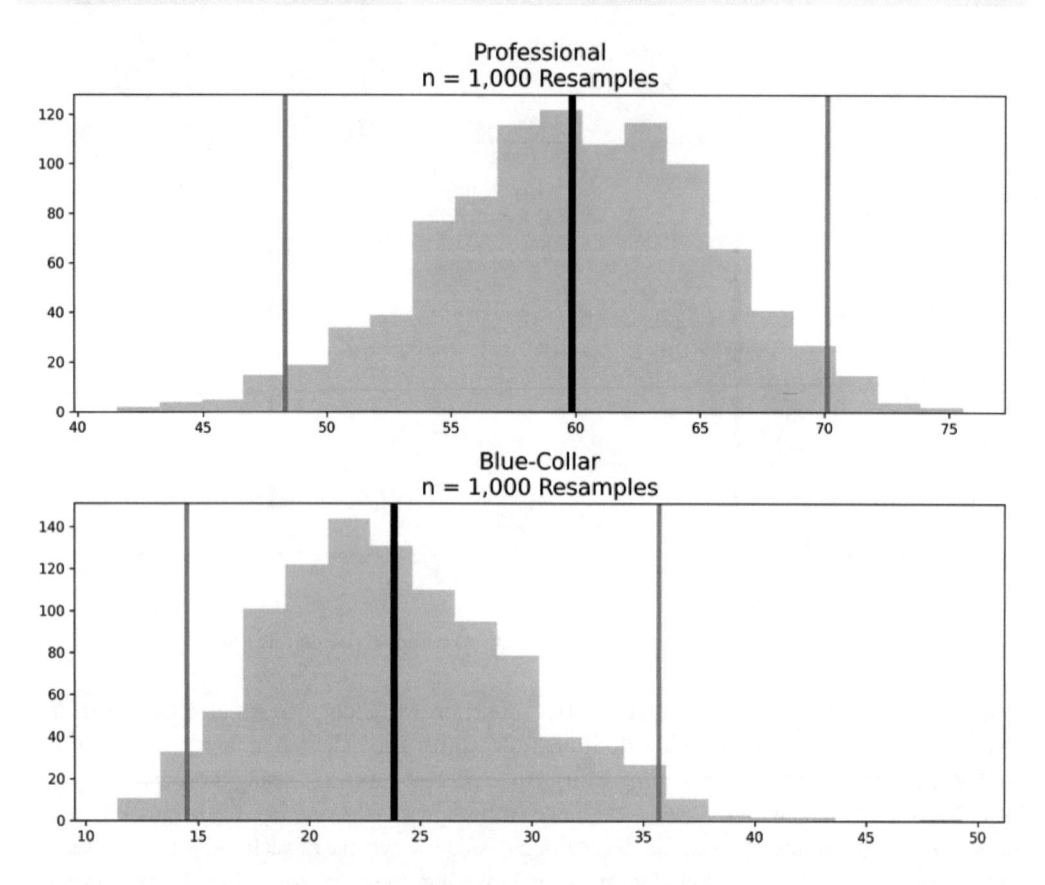

Figure 2.19 – Distributions of the 95% confidence interval for 1,000 bootstrapped sampling means

Here, you can notice the distribution more closely clusters around the mean with tighter confidence intervals.

As mentioned before, bootstrapping can be used to obtain different statistical parameters of the distribution beyond the confidence intervals.

Standard error

Another commonly used metric is the standard error, $\frac{\sigma}{\sqrt{n}}$. We can calculate this using the last variables, `professional_bootstrap_means`, and `blue_collar_bootstrap_means`, as these contain the new distributions of means obtained through the bootstrapping process. We can also see that standard error – calculated by dividing the standard deviation by the square root of the number of samples (or in our case, `n_replicas`, representing the count of averages obtained from each random re-subsample) – decreases as the volume resamples increases. We use the following code to calculate the standard error of the professional and blue-collar type income bootstrapped means. The following table, *Figure 2.20*, shows that the standard error reduces as *n* increases:

```
scipy.stats.sem(professional_bootstrap_means)
scipy.stats.sem(blue_collar_bootstrap_means)
```

n	Professional Standard Error	Blue-Collar Standard Error
10 replicas	0.93	2.09
10,000 replicas	0.03	0.04

Figure 2.20 – Table of standard errors for n = 10 and n = 10,000 bootstrap replicas

Another use case for bootstrapping is Pearson's correlation, which we will discuss in the following section.

Correlation coefficients (Pearson's correlation)

Typically, this is difficult to find using a small sample size since correlation depends on the covariance of two variables. As the variables overlap more significantly, their correlation is higher. However, if the overlap is the result of a small sample size or sampling error, this correlation may be representative. *Figure 2.21* shows a table of correlation at different counts of bootstrap subsamples. As the distributions form more native distinctions, the correlation diminishes from a small positive correlation to an amount approximating zero.

To test correlation on a sample of 10 records from the original dataset, see the following:

```
df_prof_corr = df_professional.sample(n=10)
df_blue_corr = df_blue_collar.sample(n=10)
corr, _ = scipy.stats.pearsonr(df_prof_corr['income'],
    df_blue_corr['income'])
```

To test correlation on samples of bootstrapped means:

```
n_replicas = n_replicas
professional_bootstrap_means = pd.Series([df_prof_corr.
sample(frac=0.5,replace=False).income.mean() for i in range(n_
replicas)])
blue_collar_bootstrap_means = pd.Series([df_blue_corr.sample(frac=0.5,
```

```
replace=False).income.mean() for i in range(n_replicas)])
corr, _ = scipy.stats.pearsonr(
    professional_bootstrap_means,
    blue_collar_bootstrap_means)
print(corr)
```

n	Pearson's Correlation Coefficient
10 samples from original data	0.32
10 replicas	0.22
10,000 replicas	-0.003

Figure 2.21 – Table of Pearson's correlation coefficients alongside the original samples

It is common to run around 1,000 to 10,000 bootstrap replicas. However, this depends on the type of data being bootstrapped. For example, if bootstrapping data from a human genome sequence dataset, it may be useful to bootstrap a sample 10 million times, but if bootstrapping a simple dataset, it may be useful to bootstrap 1,000 times or less. Ultimately, the researcher should perform a visual inspection of the distributions of the means to determine whether the results appear logical compared to what is expected. As common with statistics, it is best to have some domain knowledge or subject-matter expertise to help validate findings, as this will likely be the best for deciding bootstrap replication counts.

Bootstrapping is also used in machine learning, where it underlies the concept of **bootstrap aggregation**, also called **bagging**, a process that combines outputs of predictive models built upon bootstrap subsample distributions. **Random Forest** is one popular algorithm that performs this operation. The purpose of bootstrapping in bagging algorithms is to preserve the low-bias behavior of non-parametric (more to be discussed on this in later chapters) classification, but also reduce variance, thus using bootstrapping as a way to minimize the significance of the bias-variance trade-off in modeling errors.

In the following section, we will consider another non-parametric test called permutation testing using resampling data.

Permutations

Before jumping into this testing analysis, we will review some basic knowledge of permutations and combinations.

Permutations and combinations

Permutations and combinations are two mathematical techniques for taking a set of objects to create subsets from a population but in two different ways. The order of objects matters in permutations but does not matter in combinations.

In order to understand these concepts easily, we will consider two examples. There are 10 people at an evening party. The organizer of the party wants to give 3 prizes of $1,000, $500, and $200 randomly to 3 people. The question is *how many ways are there to distribute the prizes?* Another example is that the organizer will give 3 equal prizes of $500 to 3 people out of 10 at the party. The organizer really does not care which prize is given to whom among the 3 selected people. Huy, Paul, and Stuart are our winners in these two examples but, in the first example, different situations may play out, for instance, if Paul wins the $200 prize, $500 prize, or $1,000 prize.

$1,000	$500	$200
Huy	Paul	Stuart
Paul	Huy	Stuart
Paul	Stuart	Huy
Huy	Stuart	Paul
Stuart	Huy	Paul
Stuart	Paul	Huy

Figure 2.22 – Table of distributed prizes given to Huy, Paul, and Stuart

However, in the second example, because the 3 prizes have the same value of $500, the order of prize arrangements does not matter.

Let us take a closer look at these two permutations and combinations examples. The first example is a permutation example. Since the pool has 10 people, we have 10 possibilities in choosing one person from the pool to give the $1,000 prize. If this person is chosen to win the $1,000 prize, then there are only 9 possibilities in choosing another person to give the $500 prize, and finally, we have 8 possibilities in choosing a person from the pool to give the $200 prize. Then, we have 10*9*8 = 720 ways to distribute the prizes. The mathematical formula for the permutations is this:

$$P(n, r) = \frac{n!}{(n - r)!}$$

Here, $P(n, r)$ is the number of permutations, n is the total number of objects in a set, and r is the number of objects that can be chosen from the set. In this example, $n = 10$ and $r = 3$ so then we see this:

$$P(10, 3) = \frac{10!}{(10 - 3)!} = \frac{10*9*8*7*6*5*4*3*2*1}{7*6*5*4*3*2*1} = 10*9*8 = 720$$

There are 720 ways to select 3 people from the 10 people at the party to whom to distribute the 3 prizes of $1,000, $500, and $200.

In Python, there is a package called `itertools` to help us to find permutations directly. Readers can check out the following link – https://docs.python.org/3/library/itertools.html – for more information related to this package. We need to import this package into the Python environment for permutations:

```
from itertools import permutations
# list of 10 people in the party
people = ['P1','P2','P3','P4','P5','P6','P7','P8','P9','P10']
# all the ways that the 3 prizes are distributed
perm = permutations(people, 3)
list_perm = list(perm)
print(f"There are {len(list_perm)} ways to distribute the prizes!")
```

In the preceding Python code, we created a list, `people`, containing 10 people, `P1` to `P10`, and then use the `permutations` function from `itertools` to get all the ways to distribute the prizes. This method takes a list of 10 people as input and returns an object list of tuples containing all the possibilities in choosing 3 people from this pool of 10 people to whom to distribute the prizes of $1,000, $500, and $200. Because there are 720 ways to distribute the prizes, here we will just print the 10 first ways that the Python code produced:

```
print(f"The 10 first ways to distribute the prizes: \n
    {list_perm[:10]} ")
```

The output of the preceding code is the 10 first ways to distribute the prizes:

```
[('P1', 'P2', 'P3'), ('P1', 'P2', 'P4'), ('P1', 'P2', 'P5'), ('P1', 'P2', 'P6'), ('P1', 'P2', 'P7')]
```

If we have 10 different gifts, each person who participates in the party can take one gift home. How many ways are there to distribute these gifts? There are 3,628,800 ways. That is a really big number! The reader can check with the following code:

```
#list of 10 people in the party
people = ['P1','P2','P3','P4','P5','P6','P7','P8','P9','P10']
# all the ways that the 10 different gifts are distributed
perm = permutations(people)
list_perm = list(perm)
print(f"There are {len(list_perm)}
    ways to distributed the gifts!")
```

Going back to the second example, because the 3 prizes have the same value of $500, the order of the 3 selected people does not matter. Then, if the 3 selected people are Huy, Paul, and Stuart, as in *Figure 2.22*, there are 6 ways to distribute the prizes in the first example. Then, there is only 1 way to distribute the same amount of $500 to Huy, Paul, and Stuart. The mathematical formula of combinations is this:

$$C(n,r) = \frac{n!}{r!(n-r)!}$$

Here, $C(n,r)$ is the number of combinations, n is the total number of objects in a set, and r is the number of objects that can be chosen from the set. Similarly, we can calculate that there are

$$\frac{10!}{3!(10-3)!} = \frac{10.9.8}{1.2.3} = \frac{720}{6} = 120$$

ways to distribute 3 prizes of $500.

In Python, we also use the `itertools` package but, instead of the `permutations` function, we import the `combinations` function:

```
from itertools import combinations
# list of 10 people in the party
people = ['P1','P2','P3','P4','P5','P6','P7','P8','P9','P10']
# all the ways that the 3 prizes are distributed
comb = combinations(people, 3)
list_comb = list(comb)
print(f"There are {len(list_comb)} ways to distribute the prizes!")
```

Permutation testing

Permutation testing is a non-parametric test that does not make the required assumption of normally distributed data. Both bootstrapping and permutations are useful for resampling techniques but best for different uses, one for estimating statistical parameters (bootstrapping) and another for hypothesis testing. Permutation testing is used to test the null hypothesis between two samples generated from the same population. It has different names such as **exact testing**, **randomization testing**, and **re-randomization testing**.

First, we go to see a simple example for better understanding before implementing the code in Python. We suppose that there are 2 groups of people, one group representing children (A) and another group representing people over 40 years old (B) as follows:

$$A = [3,5,4] \text{ and } B = [43,41,56,78,54]$$

The mean difference in age between the two samples A and B is

$$\frac{43+41+56+78+54}{5} - \frac{3+5+4}{3} = 50.4$$

We merge A and B into a single set, denoted as P as follows:

$$P = [3,5,4,43,41,56,78,54].$$

Then, we take a permutation of P, for example, the following:

$$P\_new = [3, 54, 78, 41, 4, 43, 5, 56]$$

Next, we redivide P_new into 2 subsets called A_new and B_new, which have the same size as A and B, respectively:

$$A\_new = [3,54,78] \text{ and } B\_new = [41,4,43,5,56]$$

Then, the mean difference in age between A_new and B_new is 15.2, which is lower than the original mean difference in age between A and B (50.4). In other words, the permuted P_new does not contribute to the p-value. We can observe that only one permutation drawn from all possible permutations of P is greater than or equal to the original mean difference itself, P. Now we will implement the code in Python:

```
import numpy as np
# create permutation testing function
def permutation_testing(A,B,n_iter=1000):
#A, B are 2 lists of samples to test the hypothesis,
#n_iter is number of iterations with the default is 1000
    differences = []
    P = np.array(A+B)
    original_mean = np.array(A).mean()- np.array(B).mean()
    for i in range(n_iter):
      np.random.shuffle(P) #create a random permutation of P
      A_new = P[:len(A)] # having the same size of A
      B_new = P[-len(B):] # having the same size of B
      differences.append(A_new.mean()-B_new.mean())
    #Calculate p_value
    p_value = round(1-(float(len(np.where(
        differences<=original_mean)[0]))/float(n_iter)),2)
    return p_value
```

In the preceding Python code, A and B are two samples and we want to know whether they are from the same larger population; n_ter is the number of iterations that we want to perform; here, 1,000 is the default number of iterations.

Let's perform permutation testing for the two groups of people in the example with 10,000 iterations:

```
A = [3,5,4]
B = [43,41,56,78,54]
permutation_testing(A,B,n_iter=10000)
```

The p-value obtained is 0.98. That means that we fail to reject the null hypothesis, or there is not enough evidence to confirm that samples A and B are from the same larger population.

Next, we will explore an important and necessary step in many statistical tests requiring the normal distribution assumption.

Transformations

In this section, we will consider three transformations:

- Log transformation
- Square root transformation
- Cube root transformation

First, we will import the `numpy` package to create a random sample drawn from a Beta distribution. The documentation on Beta distributions can be found here:

https://numpy.org/doc/stable/reference/random/generated/numpy.random.beta.html

The sample, `df`, has 10,000 values. We also use `matplotlib.pyplot` to create different histogram plots. Second, we transform the original data by using a log transformation, square root transformation, and cube root transformation, and we draw four histograms:

```
import numpy as np
import matplotlib.pyplot as plt
np.random.seed(42) # for reproducible purpose
# create a random data
df = np.random.beta(a=1, b=10, size = 10000)
df_log = np.log(df) #log transformation
df_sqrt = np.sqrt(df) # Square Root transformation
df_cbrt = np.cbrt(df) # Cube Root transformation

plt.figure(figsize = (10,10))
plt.subplot(2,2,1)
plt.hist(df)
plt.title("Original Data")
plt.subplot(2,2,2)
plt.hist(df_log)
plt.title("Log Transformation")
plt.subplot(2,2,3)
plt.hist(df_sqrt)
plt.title("Square Root Transformation")
plt.subplot(2,2,4)
```

```
plt.hist(df_cbrt)
plt.title("Cube Root Transformation")
plt.show()
```

The following is the output of the code:

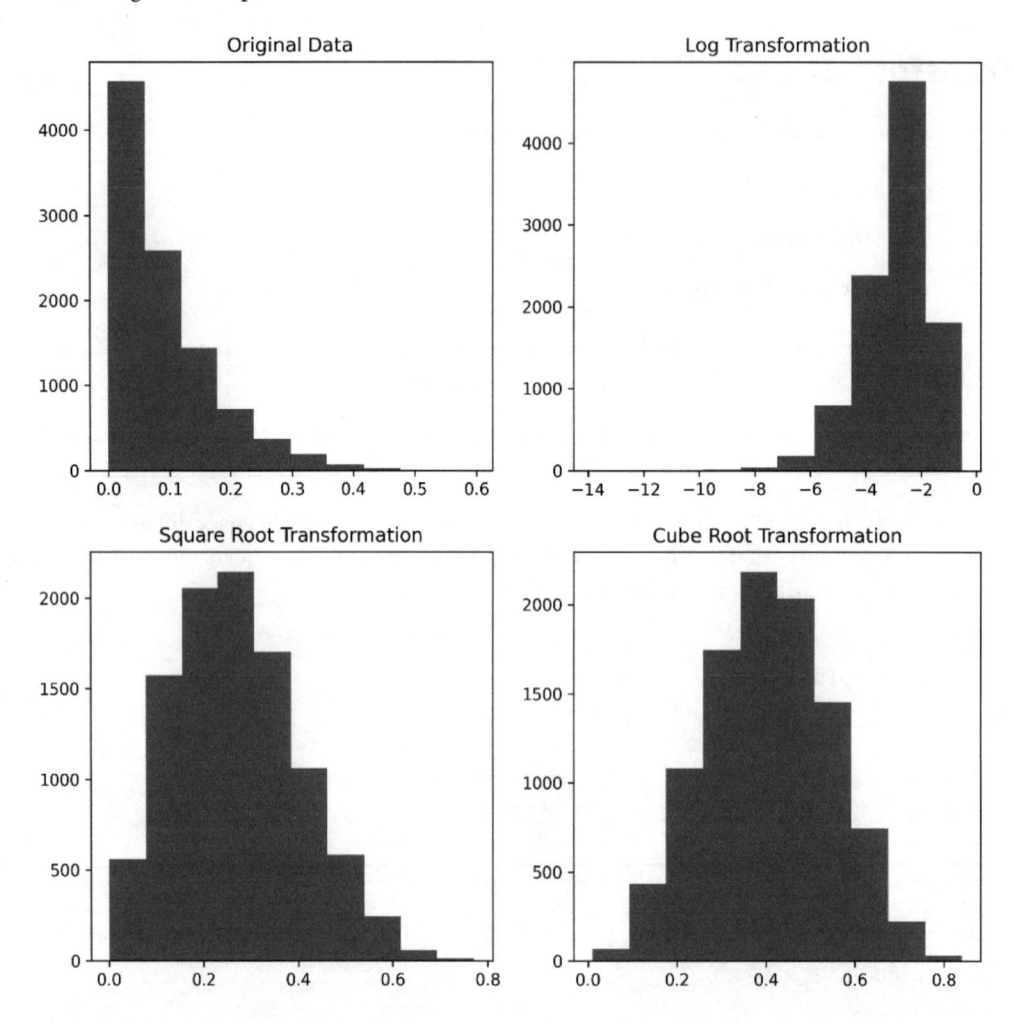

Figure 2.23 – Histograms of the original and transformed data

Using transformation, we can see the transformed histograms are more normally distributed than the original one. It seems that the best transformation in this example is cube root transformation. With real-world data, it is important to determine whether a transformation is needed, and, if so, which transformation should be used.

Other data transformation methods, for example, finding duplicate data, dealing with missing values, and feature scaling will be discussed in hands-on, real-world use cases in Python in the following chapters.

Summary

In the first section of this chapter, we learned about types of data and how to visualize these types of data. Then, we covered how to describe and measure attributes of data distribution. We learned about the standard normal distribution, why it's important, and how the central limit theorem is applied in practice by demonstrating bootstrapping. We also learned how bootstrapping can make use of non-normally distributed data to test hypotheses using confidence intervals. Next, we covered mathematical knowledge as permutations and combinations and introduced permutation testing as another non-parametric test in addition to bootstrapping. We finished the chapter with different data transformation methods that are useful in many situations when performing statistical tests requiring normally distributed data.

In the next chapter, we will take a detailed look at hypothesis testing and discuss how to draw statistical conclusions from the results of the tests. We will also look at errors that can occur in statistical tests and how to select statistical power.

References

- [1] Skewness – https://www.itl.nist.gov/div898/handbook/eda/section3/eda35b.htm
- [2] Kurtosis – https://www.itl.nist.gov/div898/handbook/eda/section3/eda35b.htm#:~:text=Kurtosis%20is%20a%20measure%20of,would%20be%20the%20extreme%20case.
- [3] Normal Distribution – *C.F. GAUSS AND THE METHOD OF LEAST SQUARES, ŚLĄSKI PRZEGLĄD STATYSTYCZNY Silesian Statistical Review*, Nr 12(18), O. Sheynin, Sep. 1999

3
Hypothesis Testing

In this chapter, we will begin discussing drawing statistical conclusions from data, putting together sampling and experiment design from *Chapter 1, Sampling and Generalization* and distributions from *Chapter 2, Distributions of Data*. Our primary use of statistical modeling is to answer questions of interest from data. Hypothesis testing provides a formal framework for answering questions of interest with measures of uncertainty. First, we will cover the goals and structure of hypothesis testing. Then, we will talk about the errors that can occur from hypothesis tests and define the expected error rate. Then, we will walk through the hypothesis test process utilizing the z-test. Finally, we will discuss statistical power analysis.

In this chapter, we're going to cover the following main topics:

- The goal of hypothesis testing
- Type I and type II errors
- Basics of the z-test – the z-score, z-statistic, critical values, and p-values
- One-sample and two-sample z-tests for means and proportions
- Selecting error rate and power analysis
- Applying the power analysis to the z-test

The goal of hypothesis testing

Put simply, the goal of hypothesis testing is to decide whether the data we have is sufficient to support a particular hypothesis. The hypothesis test provides a *formal* framework for testing a hypothesis based on our data rather than attempting to decide based on visual inspection. In this section, we will discuss the process of hypothesis testing. In the next section, *Basics of the z-test – the z-score, z-statistic, critical values, and p-values*, we will put the process to work by walking through an example in detail with the z-test.

Overview of a hypothesis test for the mean

To understand the hypothesis testing process, let's start with a simple example. Suppose we have a factory with machines that produce widgets, and we expect our machines to produce widgets at a certain rate (30 widgets per hour). We start by constructing two hypotheses, the **null hypothesis** and the **alternative hypothesis**. The null hypothesis and alternative hypothesis are given the following symbols, respectively: H_0 and H_a. To create the null hypothesis, we will start by assuming what we want to test is true. In our example, the null hypothesis would be that *the machines output a mean of 30 widgets per hour*. Once we have determined the null hypothesis, we then create the alternative hypothesis, which is just the contradiction of the null hypothesis. In our example, the alternative hypothesis is that *the machines do not output a mean of 30 widgets per hour*. Notice that our hypotheses do not indicate any directionality, that is, the alternative hypothesis contains values both less than and greater than the expected value. This is called a **two-sided test**, meaning there are two alternatives to the null hypothesis. We also have **one-sided tests**. For example, if we had said the null hypothesis was that *the machines output more than a mean of 30 widgets per hour*, the alternative would be that *the machines output less than a mean of 30 widgets per hour*. This set of hypotheses would be a one-sided test.

The two-sided null hypothesis and alternative hypothesis from our example can be stated mathematically as follows:

$$H_0: \bar{x} = 30$$
$$H_a: \bar{x} \neq 30$$

Once we have the null and alternative hypotheses and the data, we will run the test with software. Let's set the implementation details of the test aside for now (these will be covered in detail in the next section). There are two possible outcomes of a statistical test: reject the null hypothesis or fail to reject the null hypothesis. If the mean output of our machines is statistically different from the stated value in the null hypothesis, then we will *reject the null hypothesis*. This means that, given the data, the value stated in the null hypothesis is not a *plausible* value for the mean. However, if the mean output of our machines is not statistically different from the value listed in the null hypothesis, we *fail to reject the null hypothesis*. This means that, given the data, the value stated in the null hypothesis is a *plausible* value for the mean. After running the test for the null hypothesis and determining the conclusion, we will provide a confidence interval (discussed in the next section) and determine the scope of inference.

Scope of inference

The scope of inference is determined by the sampling design discussed in *Chapter 1*. There are two questions to consider – *what is the population, and how was the sample selected from the population?* In this example, let's assume that we are testing the mean output of machines from a large factory (possibly hundreds of machines). The population is then the machines in the factory. If we take a random sample of machines, then our conclusion can be extrapolated to the entire population.

While our current example is realistic, it is rather simple. In other scenarios, there may be additional considerations. For example, the machines in the factory may have different models and different ages, which may impact output. In that case, we could use stratified random sampling and make inferences for each stratum.

Hypothesis test steps

This section provides an overview of hypothesis testing from creating a hypothesis to drawing a conclusion. As we continue through this chapter, keep the following key hypothesis test steps in mind:

- State the null hypothesis and alternative hypothesis
- Perform the statistical test
- Determine the conclusion: reject or fail to reject the null hypothesis
- Provide a statistical conclusion, a confidence interval, and a scope of inference

These steps are applicable to any statistical test, and we will continue to follow this series of steps for hypothesis tests. In the next section, we will discuss the types of errors that can result from hypothesis tests.

Type I and Type II errors

While data can give us a good idea of the characteristics of a distribution, it is possible for a hypothesis test to result in an error. Errors can occur because we are taking a random sample from a population. While randomization makes it less likely that a sample contains sampling bias, *there is no guarantee that a random sample will be representative of the population*. There are two possible errors that could occur as a result of a hypothesis test:

- **Type I error**: Rejecting the null hypothesis when it is actually true
- **Type II error**: Failure to reject the null hypothesis when it is actually false

Type I errors

A type I error occurs when a hypothesis test results in *rejecting the null hypothesis, but the null hypothesis is actually true*. For example, say we have a distribution of data with a population mean of 30. We state our null hypothesis as $H_0: \bar{x} = 30$. We take a random sample for our test, but the random values in the sample happen to be on the higher side of the distribution. Thus, the test result suggests that we should reject the null hypothesis. In this case, we have made a type I error. This type of error is also called a **false positive**.

When we make statistical tests, it is always possible that we will come to an incorrect conclusion due to the data sampled from the target population. The *probability of making a type one error is specified by* α. Said another way, α represents how often we expect to make an error (the expected error rate). This is a free parameter we can select for our test (α is also called the *level of significance*). It is common to use 0.05 for α, but there is no evidential basis for using 0.05; different values may be appropriate in other contexts. Later in the chapter, we will discuss selecting the type I error rate.

Type II errors

The other type of error we can make is called a type II error. In this case, *we fail to reject the null hypothesis when it is actually false*. Let's consider another example. Say we have a distribution of data and we want to test whether the mean of the distribution is 30 or not. We take a random sample for the test and the test suggests that we should not reject the null hypothesis. However, the true population mean is 35. In this case, we have made a type II error. This type of error is also called a **false negative**.

As mentioned previously, it is always possible that a statistical test could lead to an erroneous conclusion. Thus, we want to control the probability of making an error. However, unlike α, the Type II error rate β is not simply a free parameter that we can select. To understand the likeliness of making a type II error, we generally will conduct a power analysis, which will show how various factors, such as sample size, will impact the type II error rate. In the next section, we will discuss selecting the error rate and power analysis.

We can summarize the possible results of a hypothesis test with the help of the table in *Figure 3.1*.

		Null Hypothesis is: True	False
Decision about Null Hypothesis	Don't Reject	Correct Inference (1-α)	Type II Error (β)
	Reject	Type I Error (α)	Correct Inference (1-β)

Figure 3.1 – Results of the hypothesis test

In this section, we have discussed the types of errors that can occur when drawing conclusions from statistical tests. In the next section, we will walk through an example of hypothesis testing with the z-test and later in the chapter, we will discuss how to select an error rate and how to analyze statistical power and related factors.

Basics of the z-test – the z-score, z-statistic, critical values, and p-values

In this section, we will discuss a type of hypothesis test called the z-test. It is a statistical procedure using sample data assumed to be normally distributed to determine whether a statistical statement related to the value of a population parameter should be rejected or not. The test can be performed on the following:

- One sample (a left-tailed z-test, right-tailed z-test, or two-tailed z-test)
- Two samples (a two-sample z-test)
- Proportions (a one-proportion z-test or two-proportion z-test)

The test assumes that the standard deviation is known and the sample size is large enough. In practice, a sample size that is larger than 30 should be considered.

Before going into different types of z-tests, we will discuss the z-score and z-statistic.

The z-score and z-statistic

To measure how far a particular value from a mean is, we could use the z-score or the z-statistic as statistical techniques together with the mean and the standard deviation to determine the relative location.

A z-score is computed with the following formula:

$$z_i = \frac{x_i - \bar{x}}{\sigma}$$

Here, z_i is the z-score for x_i, $\bar{x}$ is the sample mean, and σ is the sample standard deviation. The z-score is also known as the z-value, standardized value, or standard score. Let's consider a few examples. The standard deviation tells us how far a sample is from the mean of the distribution. If $z_i = 1.8$, that point is 1.8 standard deviations away from the mean. Similarly if $z_i = -1.5$, then that point is 1.5 standard deviations away from the mean. The sign of the determines whether it is greater or less than the sample mean. A z_i of -1.5 is less than the mean and a z_i of 1.8 is greater than the mean. Now let us go through an example. In a high school in Dallas (in the US), we ask students to take anonymous IQ tests for some statistical research. The data collected from that school is normally distributed with an IQ score population mean $\mu = 98$ and a population standard deviation $\sigma = 12$. A particular student took an IQ test and his score is 110. He has an IQ score greater than the score mean but he wants to know whether he is in the top 5%. First, we will use the z-score formula to calculate it:

$$z_{student} = \frac{110 - 98}{12} = \frac{12}{12} = 1.$$

The student can check a z-table (*Figure 3.2*), for example, from the website http://www.z-table.com, and get the value of 0.8413. He is in the top 1-0.8413 = 0.1587 or 15.87% of his school IQ scores.

z	.00	.01	.02	.03	.04	.05	.06	.07	.08	.09
0.0	.5000	.5040	.5080	.5120	.5160	.5199	.5239	.5279	.5319	.5359
0.1	.5398	.5438	.5478	.5517	.5557	.5596	.5636	.5675	.5714	.5753
0.2	.5793	.5832	.5871	.5910	.5948	.5987	.6026	.6064	.6103	.6141
0.3	.6179	.6217	.6255	.6293	.6331	.6368	.6406	.6443	.6480	.6517
0.4	.6554	.6591	.6628	.6664	.6700	.6736	.6772	.6808	.6844	.6879
0.5	.6915	.6950	.6985	.7019	.7054	.7088	.7123	.7157	.7190	.7224
0.6	.7257	.7291	.7324	.7357	.7389	.7422	.7454	.7486	.7517	.7549
0.7	.7580	.7611	.7642	.7673	.7704	.7734	.7764	.7794	.7823	.7852
0.8	.7881	.7910	.7939	.7967	.7995	.8023	.8051	.8078	.8106	.8133
0.9	.8159	.8186	.8212	.8238	.8264	.8289	.8315	.8340	.8365	.8389
1.0	.8413	.8438	.8461	.8485	.8508	.8531	.8554	.8577	.8599	.8621
1.1	.8643	.8665	.8686	.8708	.8729	.8749	.8770	.8790	.8810	.8830
1.2	.8849	.8869	.8888	.8907	.8925	.8944	.8962	.8980	.8997	.9015
1.3	.9032	.9049	.9066	.9082	.9099	.9115	.9131	.9147	.9162	.9177
1.4	.9192	.9207	.9222	.9236	.9251	.9265	.9279	.9292	.9306	.9319
1.5	.9332	.9345	.9357	.9370	.9382	.9394	.9406	.9418	.9429	.9441
1.6	.9452	.9463	.9474	.9484	.9495	.9505	.9515	.9525	.9535	.9545
1.7	.9554	.9564	.9573	.9582	.9591	.9599	.9608	.9616	.9625	.9633
1.8	.9641	.9649	.9656	.9664	.9671	.9678	.9686	.9693	.9699	.9706
1.9	.9713	.9719	.9726	.9732	.9738	.9744	.9750	.9756	.9761	.9767
2.0	.9772	.9778	.9783	.9788	.9793	.9798	.9803	.9808	.9812	.9817
2.1	.9821	.9826	.9830	.9834	.9838	.9842	.9846	.9850	.9854	.9857
2.2	.9861	.9864	.9868	.9871	.9875	.9878	.9881	.9884	.9887	.9890
2.3	.9893	.9896	.9898	.9901	.9904	.9906	.9909	.9911	.9913	.9916
2.4	.9918	.9920	.9922	.9925	.9927	.9929	.9931	.9932	.9934	.9936
2.5	.9938	.9940	.9941	.9943	.9945	.9946	.9948	.9949	.9951	.9952

Figure 3.2 – z-table

In Python, we can use the **cumulative distribution function (CDF)** to calculate it:

```
import scipy
round(scipy.stats.norm.cdf(1),4)
# 0.8413
```

We also get the same value in Python as in the z-table check; it is 0.8413 for the z-score = 1 in this example.

Another example here is a simple random sample of 10 scores taken from the IQ survey:

90, 78, 110, 110, 99, 115, 130, 100, 95, 93

To compute the z-score for each IQ score in the sample, we need to calculate the mean and the standard deviation of this sample and then apply the z-score formula. Fortunately, we can use the `scipy` library again as follows:

```
import pandas as pd
import numpy as np
import scipy.stats as stats

IQ = np.array([90, 78,110, 110, 99, 115,130, 100, 95, 93])
z_score = stats.zscore(IQ)

# Create dataframe

data_zscore = {
  "IQ score": IQ,
  "z-score": z_score
}

IQ_zscore = pd.DataFrame(data_zscore)
IQ_zscore
```

We created an array of IQ scores called `IQ` and used `z-score` from `scipy.stats` to compute `z_score`. Finally, we created the following output DataFrame.

	IQ score	z-score
0	90	-0.860663
1	78	-1.721326
2	110	0.573775
3	110	0.573775
4	99	-0.215166
5	115	0.932385
6	130	2.008214
7	100	-0.143444
8	95	-0.502053
9	93	-0.645497

Figure 3.3 – Output DataFrame

Before discussing the z-statistic, we will introduce the notion of sampling distributions. Going back to the last example, as a rule of thumb, we perform the process of selecting a simple random sample of 35 IQ scores from the pool of IQ scores of the high school repeatedly, as many times as needed for the study. We then compute the mean score of each sample, called $\bar{x}$. Because we have various samples selected, we also have various possible values of $\bar{x}$. The expected value of $\bar{x}$ is as follows:

$$E(\bar{x}) = \mu$$

Here, μ is the population mean. $\sigma_{\bar{x}}$ denotes the standard distribution of $\bar{x}$. Practically, in many sampling situations, a population is relatively large compared to small sample sizes. Then, the standard deviation of $\bar{x}$ can be given as follows:

$$\sigma_{\bar{x}} = \frac{\sigma}{\sqrt{n}}$$

Here, σ is the population standard deviation and n is the sample size for finite or infinite populations such that the population is large and the sample size is small, relatively. Note that $\sigma_{\bar{x}}$ is also called the standard error of the mean to help us determine how far the sample mean is from the population mean. Note that $E(\bar{x}) = \mu$, independent of the sample size. Sample size and standard error are inversely correlated: when the sample size is increased, the standard error decreases. Since the sampling distribution of $\bar{x}$ is assumed to be normally distributed, the sample distribution is given by the following formula:

$$z_{\bar{x}} = \frac{\bar{x} - \mu}{\sigma_{\bar{x}}}$$

In hypothesis tests about a population mean, we use test statistics where its formula is given as follows:

$$z = z_{\bar{x}} = \frac{\bar{x} - \mu}{\sigma/\sqrt{n}}$$

Here, $\bar{x}$ is the sample mean, μ is the population mean, σ is the population standard deviation, and n is the sample size.

Consider the IQ test example again. The IQ score data has a mean $\mu = 98$ and a standard deviation $\sigma = 12$. Suppose the data is normally distributed. Let x be the score taken randomly from the IQ data. What is the probability that x is between 95 and 104? We will compute the z-scores when $x = 95$ and $x = 104$ as follows:

$$z_{95} = \frac{95 - 98}{12} = -0.25,$$

$$z_{100} = \frac{104 - 98}{12} = 0.5.$$

Therefore, the probability that the taken score is between 95 and 104 is:

$$P(95 < x < 104) = P(-0.25 < z < 0.5) = 0.6915 - 0.4013 = 0.2902.$$

Then, the probability is about 29.02% that an IQ score taken at random from the data is between 95 and 104. We also can get the values from a z-table as follows.

z	.00	.01	.02	.03	.04	.05	.06	.07	.08	.09
−3.4	.0003	.0003	.0003	.0003	.0003	.0003	.0003	.0003	.0003	.0002
−3.3	.0005	.0005	.0005	.0004	.0004	.0004	.0004	.0004	.0004	.0003
−3.2	.0007	.0007	.0006	.0006	.0006	.0006	.0006	.0005	.0005	.0005
−3.1	.0010	.0009	.0009	.0009	.0008	.0008	.0008	.0008	.0007	.0007
−3.0	.0013	.0013	.0013	.0012	.0012	.0011	.0011	.0011	.0010	.0010
−2.9	.0019	.0018	.0018	.0017	.0016	.0016	.0015	.0015	.0014	.0014
−2.8	.0026	.0025	.0024	.0023	.0023	.0022	.0021	.0021	.0020	.0019
−2.7	.0035	.0034	.0033	.0032	.0031	.0030	.0029	.0028	.0027	.0026
−2.6	.0047	.0045	.0044	.0043	.0041	.0040	.0039	.0038	.0037	.0036
−2.5	.0062	.0060	.0059	.0057	.0055	.0054	.0052	.0051	.0049	.0048
−2.4	.0082	.0080	.0078	.0075	.0073	.0071	.0069	.0068	.0066	.0064
−2.3	.0107	.0104	.0102	.0099	.0096	.0094	.0091	.0089	.0087	.0084
−2.2	.0139	.0136	.0132	.0129	.0125	.0122	.0119	.0116	.0113	.0110
−2.1	.0179	.0174	.0170	.0166	.0162	.0158	.0154	.0150	.0146	.0143
−2.0	.0228	.0222	.0217	.0212	.0207	.0202	.0197	.0192	.0188	.0183
−1.9	.0287	.0281	.0274	.0268	.0262	.0256	.0250	.0244	.0239	.0233
−1.8	.0359	.0351	.0344	.0336	.0329	.0322	.0314	.0307	.0301	.0294
−1.7	.0446	.0436	.0427	.0418	.0409	.0401	.0392	.0384	.0375	.0367
−1.6	.0548	.0537	.0526	.0516	.0505	.0495	.0485	.0475	.0465	.0455
−1.5	.0668	.0655	.0643	.0630	.0618	.0606	.0594	.0582	.0571	.0559
−1.4	.0808	.0793	.0778	.0764	.0749	.0735	.0721	.0708	.0694	.0681
−1.3	.0968	.0951	.0934	.0918	.0901	.0885	.0869	.0853	.0838	.0823
−1.2	.1151	.1131	.1112	.1093	.1075	.1056	.1038	.1020	.1003	.0985
−1.1	.1357	.1335	.1314	.1292	.1271	.1251	.1230	.1210	.1190	.1170
−1.0	.1587	.1562	.1539	.1515	.1492	.1469	.1446	.1423	.1401	.1379
−0.9	.1841	.1814	.1788	.1762	.1736	.1711	.1685	.1660	.1635	.1611
−0.8	.2119	.2090	.2061	.2033	.2005	.1977	.1949	.1922	.1894	.1867
−0.7	.2420	.2389	.2358	.2327	.2296	.2266	.2236	.2206	.2177	.2148
−0.6	.2743	.2709	.2676	.2643	.2611	.2578	.2546	.2514	.2483	.2451
−0.5	.3085	.3050	.3015	.2981	.2946	.2912	.2877	.2843	.2810	.2776
−0.4	.3446	.3409	.3372	.3336	.3300	.3264	.3228	.3192	.3156	.3121
−0.3	.3821	.3783	.3745	.3707	.3669	.3632	.3594	.3557	.3520	.3483
−0.2	.4207	.4168	.4129	.4090	.4052	.4013	.3974	.3936	.3897	.3859
−0.1	.4602	.4562	.4522	.4483	.4443	.4404	.4364	.4325	.4286	.4247
−0.0	.5000	.4960	.4920	.4880	.4840	.4801	.4761	.4721	.4681	.4641

Figure 3.4 – z-table

z	.00	.01	.02	.03	.04	.05	.06	.07	.08	.09
0.0	.5000	.5040	.5080	.5120	.5160	.5199	.5239	.5279	.5319	.5359
0.1	.5398	.5438	.5478	.5517	.5557	.5596	.5636	.5675	.5714	.5753
0.2	.5793	.5832	.5871	.5910	.5948	.5987	.6026	.6064	.6103	.6141
0.3	.6179	.6217	.6255	.6293	.6331	.6368	.6406	.6443	.6480	.6517
0.4	.6554	.6591	.6628	.6664	.6700	.6736	.6772	.6808	.6844	.6879
0.5	.6915	.6950	.6985	.7019	.7054	.7088	.7123	.7157	.7190	.7224
0.6	.7257	.7291	.7324	.7357	.7389	.7422	.7454	.7486	.7517	.7549
0.7	.7580	.7611	.7642	.7673	.7704	.7734	.7764	.7794	.7823	.7852
0.8	.7881	.7910	.7939	.7967	.7995	.8023	.8051	.8078	.8106	.8133
0.9	.8159	.8186	.8212	.8238	.8264	.8289	.8315	.8340	.8365	.8389
1.0	.8413	.8438	.8461	.8485	.8508	.8531	.8554	.8577	.8599	.8621
1.1	.8643	.8665	.8686	.8708	.8729	.8749	.8770	.8790	.8810	.8830
1.2	.8849	.8869	.8888	.8907	.8925	.8944	.8962	.8980	.8997	.9015
1.3	.9032	.9049	.9066	.9082	.9099	.9115	.9131	.9147	.9162	.9177
1.4	.9192	.9207	.9222	.9236	.9251	.9265	.9279	.9292	.9306	.9319
1.5	.9332	.9345	.9357	.9370	.9382	.9394	.9406	.9418	.9429	.9441
1.6	.9452	.9463	.9474	.9484	.9495	.9505	.9515	.9525	.9535	.9545
1.7	.9554	.9564	.9573	.9582	.9591	.9599	.9608	.9616	.9625	.9633
1.8	.9641	.9649	.9656	.9664	.9671	.9678	.9686	.9693	.9699	.9706
1.9	.9713	.9719	.9726	.9732	.9738	.9744	.9750	.9756	.9761	.9767
2.0	.9772	.9778	.9783	.9788	.9793	.9798	.9803	.9808	.9812	.9817
2.1	.9821	.9826	.9830	.9834	.9838	.9842	.9846	.9850	.9854	.9857
2.2	.9861	.9864	.9868	.9871	.9875	.9878	.9881	.9884	.9887	.9890
2.3	.9893	.9896	.9898	.9901	.9904	.9906	.9909	.9911	.9913	.9916
2.4	.9918	.9920	.9922	.9925	.9927	.9929	.9931	.9932	.9934	.9936
2.5	.9938	.9940	.9941	.9943	.9945	.9946	.9948	.9949	.9951	.9952
2.6	.9953	.9955	.9956	.9957	.9959	.9960	.9961	.9962	.9963	.9964
2.7	.9965	.9966	.9967	.9968	.9969	.9970	.9971	.9972	.9973	.9974
2.8	.9974	.9975	.9976	.9977	.9977	.9978	.9979	.9979	.9980	.9981
2.9	.9981	.9982	.9982	.9983	.9984	.9984	.9985	.9985	.9986	.9986
3.0	.9987	.9987	.9987	.9988	.9988	.9989	.9989	.9989	.9990	.9990
3.1	.9990	.9991	.9991	.9991	.9992	.9992	.9992	.9992	.9993	.9993
3.2	.9993	.9993	.9994	.9994	.9994	.9994	.9994	.9995	.9995	.9995
3.3	.9995	.9995	.9995	.9996	.9996	.9996	.9996	.9996	.9996	.9997
3.4	.9997	.9997	.9997	.9997	.9997	.9997	.9997	.9997	.9997	.9998

Figure 3.5 – z-table

In Python, we can implement the code as follows:

```
#calculate z scores at x=95 and 104
zscore_95 = round((95-98)/12,2)
zscore_104 = round((104-98)/12,2)

#calculate cdf and probability
cdf_95 = stats.norm.cdf(zscore_95)
cdf_104 = stats.norm.cdf(zscore_104)
prob = abs(cdf_95-cdf_104)

#print the probability
```

```
print(f"The probability that the taken score between 95 and 104 is
{round(prob*100,2)}%!")
```

Another question raised is what is the probability that the mean score $\bar{x}$ of four scores taken randomly is between 95 and 104? To solve this question, we use the notion of the z-statistic. We assume the mean of $\bar{x}$ is also 98 and $\bar{x}$ has a normal distribution. Then, the standard error of $\bar{x}$ is:

$$\sigma_{\bar{x}} = \frac{\sigma}{\sqrt{n}} = \frac{12}{\sqrt{4}} = 6,$$

then:

$$z = \frac{\bar{x} - \mu_{\bar{x}}}{\sigma_{\bar{x}}} = \frac{\bar{x} - 98}{6}.$$

It is easy to calculate $z_{95} = -0.5$ and $z_{104} = 1$. By using the z-table, we can get the probability that the mean score $\bar{x}$ of six scores taken randomly is between 95 and 104 as follows:

0.8413 − 0.3085 = 0.5328

Or, about 53.28%. To use this idea, we can implement the code in Python as follows:

```
import pandas as pd
import numpy as np
import scipy.stats as stats
import math

# standard error
n= 4
sigma = 12
se = sigma/math.sqrt(n)

#calculate z scores at x=95 and 104
zscore_95 = round((95-98)/se,2)
zscore_104 = round((104-98)/se,2)

#calculate cdf and probability
cdf_95 = stats.norm.cdf(zscore_95)
cdf_104 = stats.norm.cdf(zscore_104)
prob = abs(cdf_95-cdf_104)

#print the probability
print(f"The probability that the taken score between 95 and 104 is
{round(prob*100,2)}%!")
```

Note that in the preceding code, we also need the library called math for calculating the square root function, math.sqrt().

In the following section, we will discuss a z-test for means.

A z-test for means

In this part, one-sample and two-sample z-tests related to a population mean or means of two populations respectively are considered.

A one-sample z-test

A selected sample for research from a population should be normally distributed. The population standard deviation is supposed to be known, at least for practical purposes.

This test is still applicable in cases where the population cannot be assumed to be normally distributed but the sample size needs to be considered large enough by a rule of thumb, based on the experiences of researchers involved in the study. To perform the hypothesis testing, we need to develop the null and alternative hypotheses. The following figure illustrates three null and alternative hypotheses corresponding to left-tailed, right-tailed, and two-tailed z-tests.

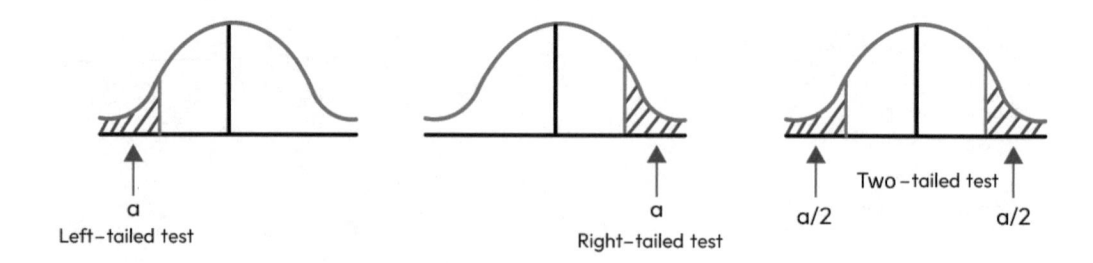

Figure 3.6 – Left-tailed, right-tailed, and two-tailed hypothesis tests

$$H_0: \mu \geq \mu_0 \quad H_0: \mu \leq \mu_0 \quad H_0: \mu = \mu_0$$
$$H_a: \mu < \mu_0 \quad H_a: \mu > \mu_0 \quad H_a: \mu \neq \mu_0$$

Next, we need to specify the level of significance, α, the probability of rejecting the null hypothesis when it is true. In other words, it is the probability of a type I error, as we discussed in the last section. Then, we calculate the value of the test statistic. There are two approaches using a p-value or a critical value for the hypothesis testing.

In the p-value approach, we use the value of the test statistic to calculate a probability, denoted by the p-value, which takes on values as extreme as or more extreme than the test statistic derived from the sample. The smaller the p-value is, the more it indicates evidence against the null hypothesis, or in other words, the probability used to determine whether H_0 should be rejected. The rejection rule (reject H_0) is the p-value being less than or equal to the specified level of significance α in the research. In order to find the p-value based on the value of the test statistic in Python, we use the following syntax:

```
scipy.stats.norm.sf(abs(x))
```

Here, x is the z-score. For example, we want to find the p-value associated with a z-score of -2.67 in a left-tailed test. The Python implementation is as follows:

```
import scipy.stats

#find p-value
round(scipy.stats.norm.sf(abs(-2.67)),4)
```

The output will be 0.0038. Similar Python code is used in the case of a right-tailed test. For a two-tailed test, we need to multiply the value by 2:

```
#find p-value for two-tailed test
scipy.stats.norm.sf(abs(2.67))*2
```

The following figure illustrates the idea of how the p-value is computed in each type of test.

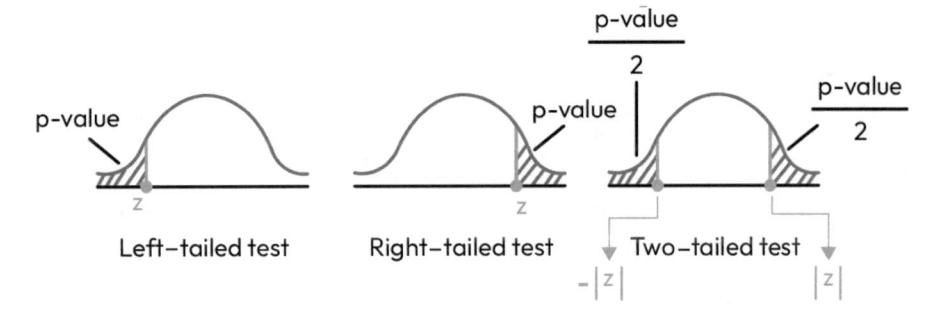

Figure 3.7 – p-values in hypothesis testing

The last step is to interpret the statistical conclusion.

On the other hand, in the critical value approach, we will need to compute a critical value for the test statistic by using the level of significance. Critical values are the boundaries of the critical region where we can reject the null hypothesis.

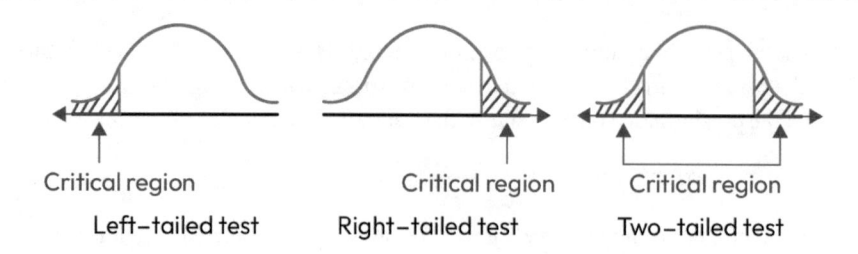

Figure 3.8 – Critical regions in hypothesis testing

To compute the critical value in Python, we use the following syntax:

```
scipy.stats.norm.ppf(alpha)
```

Here, `alpha` is the level of significance to be used. The following is the implementation of the code in Python for left-tailed, right-tailed, and two-tailed tests:

```
import scipy.stats
alpha = 0.05 # level of significance

#find Z critical value for left-tailed test
print(f" The critical value is {scipy.stats.norm.ppf(alpha)}")

#find Z critical value for left-tailed test
print(f" The critical value is {scipy.stats.norm.ppf(1-alpha)}")

##find Z critical value for two-tailed test
print(f" The critical values are {-scipy.stats.norm.ppf(1-alpha/2)} and {scipy.stats.norm.ppf(1-alpha/2)}")
```

Here's the output of the preceding code:

```
The critical value is -1.6448536269514729

The critical value is 1.6448536269514722

The critical values are -1.959963984540054 and 1.959963984540054
```

At the level of significance $\alpha = 0.05$ for the left-tailed test, the critical value is about -1.64485. Since this is a left-tailed test, if the test statistic is less than or equal to this critical value, we reject the null hypothesis. Similarly, for the right-tailed test, if the test statistic is greater than or equal to 1.64485, we reject the null hypothesis. For the two-tailed test, we reject the null hypothesis if the test statistic is greater than or equal to 1.95996 or less than or equal to -1.95996.

After determining whether to reject the null hypothesis, we interpret the statistical conclusion.

Let us discuss the IQ test scores again in the high school in Dallas. The IQ score data has a mean of $\mu = 98$ and a standard deviation of $\sigma = 12$. A researcher wants to know whether IQ scores will be affected by some IQ training. He recruits 30 students and trains them to answer IQ questions 2 hours per day for 30 days and records their IQ levels after finishing the training period. The IQ scores of 30 students after the training section are 95, 110, 105, 120, 125, 110, 98, 90, 99, 100,110, 112, 106, 92, 108, 97, 95, 99, 100, 100,103, 125, 122, 110, 112, 102, 92, 97, 89, and 102. Their average score is 104.17. We can easily implement the calculation in Python as follows:

```
IQscores = [95,110, 105, 120, 125, 110, 98, 90, 99, 100,
            110, 112, 106, 92, 108, 97, 95, 99, 100, 100,
            103, 125, 122, 110, 112, 102, 92, 97, 89, 102]

IQmean = np.array(IQscores).mean()
```

We define the null hypothesis and the alternative hypothesis:

$$H_0 : \mu_{after} = \mu = 98$$

$$H_a : \mu_{after} > \mu = 98$$

We choose the level of significance $\alpha = 0.05$. Previously, the critical value for the right-tailed test was 1.64485. We now calculate the test statistic on the problem:

$$z = \frac{\bar{x} - \mu}{\sigma / \sqrt{n}} = \frac{104.17 - 98}{12 / \sqrt{30}} = 2.8162$$

It is implemented in Python as follows:

```
n=30 #number of students
sigma =12 #population standard deviation
IQmean = 104.17 # IQ mean of 30 students after the training
mu = 98 # population mean

z = (IQmean-mu)/(sigma/math.sqrt(n))
```

Since the test statistic value is 2.8162 > 1.64485, we reject the null hypothesis. This means that the training does affect the IQ levels of these students and helps them improve their IQ scores.

We also can use the `ztest()` function from the `statsmodels` package (*Seabold, Skipper, and Josef Perktold, "statsmodels: Econometric and statistical modeling with python." Proceedings of the 9th Python in Science Conference. 2010*) to perform one- or two-sample z-tests (which we will discuss in the next part). The syntax is as follows:

```
statsmodels.stats.weightstats.ztest(x1, x2=None,
    value=0, alternative='two-sided')
```

Here, we see the following:

- `x1`: The first of the two independent samples
- `x2`: The second of the two independent samples (if performing a two-sample z-test)
- `value`: In the one-sample case, `value` is the mean of `x1` under the null hypothesis. In the two-sample case, `value` is the difference between the mean of `x1` and the mean of `x2` under the null hypothesis. The test statistic is `x1_mean - x2_mean - value`.
- `alternative`: The alternative hypothesis:

    `'two-sided'`: Two-sided test

    `'larger'`: Right-tailed test

    `'smaller'` : Left-tailed test

The following Python code shows how we perform a one-sample z-test:

```
from statsmodels.stats.weightstats import ztest as ztest

#IQ scores after training sections
IQscores = [95,110, 105, 120, 125, 110, 98, 90, 99, 100,
            110, 112, 106, 92, 108, 97, 95, 99, 100, 100,
            103, 125, 122, 110, 112, 102, 92, 97, 89, 102]

#perform one sample z-test
z_statistic, p_value = ztest(IQscores, value=98, alternative = 'larger')

print(f"The test statistic is {z_statistic} and the
    corresponding p-value is {p_value}.")
```

The test statistic is 3.3975 and the p-value = 0.00034 < 0.05, where 0.05 is the level of significance. Therefore, we have enough evidence to reject the null hypothesis. This means that the training does affect the IQ levels of these students.

Two-sample z-test

We consider two normally distributed and independent populations. Let Ω be the hypothesized difference between two population means, μ_1 and μ_2. Similarly, as in the case of the one-sample z-test, we have three forms for the null and alternative hypotheses:

$$H_0: \mu_1 - \mu_2 \geq \Omega \qquad H_0: \mu_1 - \mu_2 \leq \Omega \qquad H_0: \mu_1 - \mu_2 = \Omega$$
$$H_a: \mu_1 - \mu_2 < \Omega \qquad H_a: \mu_1 - \mu_2 > \Omega \qquad H_a: \mu_1 - \mu_2 \neq \Omega$$

In many problems, $\Omega = 0$. That means that for the case of the two-tailed test, the null hypothesis is zero, or in other words, μ_1 and μ_2 are equal. The test statistic for hypothesis tests is computed as follows:

$$z = \frac{(\bar{x}_1 - \bar{x}_2) - \Omega}{\sqrt{\frac{\sigma_1^2}{n_1} + \frac{\sigma_2^2}{n_2}}}$$

Here, $\bar{x}_1$ and $\bar{x}_2$ are the sample means with the sample sizes n_1 and n_2 randomly taken from the two populations with the means μ_1 and μ_2, respectively. σ_1 and σ_2 are the standard deviations for these two populations. With two independent simple random samples, the point estimator $\bar{x}_1 - \bar{x}_2$ has a standard error given as follows:

$$\sigma_{\bar{x}_1 - \bar{x}_2} = \sqrt{\frac{\sigma_1^2}{n_1} + \frac{\sigma_2^2}{n_2}}$$

The distribution of $\bar{x}_1 - \bar{x}_2$ can be considered a normal distribution when the sample sizes are large enough. The step-by-step approach to the two-sample z-test hypothesis test is similar to that of the one-sample z-test.

Let us now consider an example. We study the IQ scores of students from two schools, named A and B, in Dallas, and we want to know whether the mean IQ levels for these two schools are different. A simple random sample of 30 students from each school is recorded.

A= [95,110, 105, 120, 125, 110, 98, 90, 99, 100,110, 112, 106, 92, 108, 97, 95, 99, 100, 100, 103, 125, 122, 110, 112, 102, 92, 97, 89, 102]

B = [98, 90, 100, 93, 91, 79, 90, 100, 121, 89, 101, 98, 75, 90, 95, 99, 100, 120, 121, 95,

96, 89, 115, 99, 95, 121, 122, 98, 97, 97]

The null and alternative hypotheses are:

$$H_0: \mu_1 - \mu_2 = 0$$
$$H_a: \mu_1 - \mu_2 \neq 0$$

We choose the level of significance $\alpha=0.05$. In Python, by using the `ztest()` function of the `statsmodels` package, we perform the following calculation:

```
from statsmodels.stats.weightstats import ztest as ztest

#IQ score
A= [95,110, 105, 120, 125, 110, 98, 90, 99, 100,
    110, 112, 106, 92, 108, 97, 95, 99, 100, 100,
    103, 125, 122, 110, 112, 102, 92, 97, 89,102] #school A
B = [98, 90, 100, 93, 91, 79, 90, 100, 121, 89,
    101, 98, 75, 90, 95, 99, 100, 120, 121, 95,
```

```
              96, 89, 115, 99, 95, 121, 122, 98, 97, 97]   # school B

#perform two- sample z-test
z_statistic, p_value = ztest(A, B, value=0, alternative = 'two-sided')

print(f"The test statistic is {z_statistic} and the corresponding
p-value is {p_value}.")
```

In the preceding code, we chose `alternative = 'two-sided'` related to the null and alternative hypotheses for the study. The test statistic and the p-value produced by the Python code are 1.757 and 0.079, respectively. Using a level of significance of 0.05, since the p-value > 0.05, we fail to reject the null hypothesis. In other words, we do not have enough evidence to show that the IQ mean scores between the students from the two schools are different.

z-test for proportions

We can also test for differences in proportions. Let's take a look at how to perform the z-test for proportions.

A one-proportion z-test

One-proportion z-tests are used to compare the difference between a sample proportion $\bar{p}$ and a hypothesized proportion p_0. Similarly, as in a z-test for means, we have three forms for the null and alternative hypotheses – left-tailed, right-tailed, and two-tailed tests:

$$H_0: \bar{p} \geq p_0 \qquad H_0: \bar{p} \leq p_0 \qquad H_0: \bar{p} = p_0$$
$$H_a: \bar{p} < p_0 \qquad H_a: \bar{p} > p_0 \qquad H_a: \bar{p} \neq p_0$$

The test statistic is computed as follows:

$$z = \frac{\bar{p} - p_0}{\sqrt{\frac{p_0(1-p_0)}{n}}}$$

Here, n is the sample size. Let us consider an example. In a community college in Houston, a researcher wants to know whether students support some changes equal to 80%. He will use a one-proportion z-test at the level of significance $\alpha = 0.05$. To implement the code in Python, he can use `proportions_ztest` from the `statsmodel` library. The syntax is as follows:

```
statsmodels.stats.proportion.proportions_ztest(count,
     nobs, value=None, alternative='two-sided')
```

- `count`: The number of successes
- `nobs`: The number of trials or observations

- `value`: This is the value of the null hypothesis, equal to the proportion in the case of a one-sample test. In the case of a two-sample test, the null hypothesis is `prop[0] - prop[1] = value`, where `prop` is the proportion in the two samples. If not provided, `value = 0`, and the null hypothesis is `prop[0] = prop[1]`.`alternative`: The alternative hypothesis:

  - `'two-sided'`: Two-sided test
  - `'larger'`: Right-tailed test
  - `'smaller'`: Left-tailed test

The researcher gathers a sample of data with an observed sample proportion p = 0.84, the hypothesized population proportion p_0 = 0.8, and the sample size n = 500. The null and alternative hypotheses are as follows:

$$H_0: \bar{p} = p_0,$$
$$H_a: \bar{p} \neq p_0.$$

We will implement a one-proportion two-tailed z-test in Python to calculate the test statistic and p-value:

```
#import proportions_ztest function
from statsmodels.stats.proportion import proportions_ztest

count = 0.8*500
nobs = 500
value = 0.84

#perform one proportion two-tailed z-test
z_statistic, p_value = proportions_ztest(count, nobs,
    value, alternative = 'two-sided')

print(f"The test statistic is {z_statistic} and the
    corresponding p-value is {p_value}.")
```

The test statistic is -2.236067977499786 and the corresponding p-value is 0.0253473186774685. Since the p-value < 0.05 (the level of significance), we reject the null hypothesis. There is enough evidence to suggest the proportion of students who support the changes is different from 0.8.

A two-proportion z-test

This test is used to test the difference between two population proportions. There are also three forms of the null and alternative hypotheses. The test statistic is computed as follows:

$$z = \frac{\bar{P}_1 - \bar{P}_2}{\sqrt{\bar{p}(1-\bar{p})\left(\frac{1}{n_1} + \frac{1}{n_2}\right)}}$$

Here, $\bar{p}_1$ is a sample proportion for a simple random sample from population 1 and $\bar{p}_2$ is a sample proportion for a simple random sample from population 2, n_1 and n_2 are the sample sizes, and $\bar{p}$ is the total pooled proportion, calculated as follows:

$$\bar{p} = \frac{n_1\bar{p}_1 + n_2\bar{p}_2}{n_1 + n_2}.$$

We consider a similar example if there is a difference in the proportion of students from school A who support the changes compared to the proportion of students from school B. Here, $n_1 = 100$, $n_2 = 100$, $\bar{p}_1 = 0.8$, and $\bar{p}_2 = 0.7$. You can use proportions_ztest from the statsmodels library (Seabold, Skipper, and Josef Perktold, "statsmodels: Econometric and statistical modeling with python." Proceedings of the 9th Python in Science Conference. 2010) to perform the hypothesis test. Here, we will compute the p-value directly using the test statistic. The null and alternative hypotheses are as follows:

$$H_0 : \bar{p}_1 = \bar{p}_2 \text{ (The two population proportions are equal)}$$

$$H_a : \bar{p}_1 \neq \bar{p}_2 \text{ (The two population proportions are different)}$$

Next, we specify that the level of significance for the two-tailed test is $\alpha = 0.05$. We then calculate the test statistic and p-value as follows:

```
scipy.stats.norm.sf(abs(z))*2
```

The test statistic is 1.633 and the p-value for a two-tailed test is 0.10. Because the p-value is greater than the specified level of significance, 0.05, we fail to reject the null hypothesis. There is enough evidence to say that the proportion of students who support the changes is different between school A and school B.

Finally, the implemented Python code for the preceding calculation is as follows:

```
import math
import scipy
p_1bar = 0.8
p_2bar = 0.7
n1 = 100.0
n2 = 100.0

p= (p_1bar*n1 + p_2bar*n2)/(n1+n2)    # the total pooled proportion

z = (p_1bar-p_2bar)/math.sqrt(p*(1-p)*(1/n1+1/n2))
pval = scipy.stats.norm.sf(abs(z))*2

print(f"The test statistic is {z} and the p-value for two tailed test is {pval}.")
```

In this section, we went through different important statistical notions such as the z-score, z-statistics, critical values, p-value, and z-test for means and proportions. We will discuss selecting the error rate and power analysis in the next section.

Selecting the error rate and power analysis

Statistics generalizes approximations around which to form acceptable conclusions from behaviors in data. Therefore, errors in statistics are unavoidable. The significance of findings from statistical models is essentially determined by the error rate. One method that can be used to minimize errors in modeling, especially with lower sample volumes, is the power test. Statistical power is the probability of correctly rejecting a null hypothesis, thus minimizing type II errors. Where alpha (α) is a type I error and beta (β) is a type II error, power's formulation is *1 – β*. To refresh, a type I error is the probability of incorrectly rejecting the null hypothesis.

As noted, power is more important with smaller samples because the law of large numbers typically helps minimize errors as the sample size increases if an appropriate sampling method is chosen. Power is also important when the differences being compared are relatively small. One scenario in which power analysis may be particularly useful is when sampling is expensive, for example, with human studies. Power analyses can be used to find an appropriate minimum sample size given the desired power, type I error rate, and effect size. The relationship between these parameters can be explored as needed. Effect size is the difference or similarity of the data being compared in the hypothesis test, such as a standardized difference of means or correlation. As a common level of significance (type I error) in hypothesis testing is anywhere between 0.01 and 0.1, a common level of power is anywhere between 0.8 and 0.9, although these measures are case-dependent (https://www.ncbi.nlm.nih.gov/pmc/articles/PMC7745163/#:~:text=The%20ideal%20power%20of%20a,high%20as%200.8%20or%200.9).

A few notable properties of power are as follows:

- There must be sufficient power to detect a meaningful difference
- Power increases with sample size
- Power increases with the effect size
- Standard deviation (as well as variance and standard error) decreases as power increases

In the following plot, we can see that type I error (α) is where we may falsely assume the null hypothesis and conclude there is no statistically significant difference when, in fact, the data points don't belong to data source 2 (the null hypothesis), but to data source 1. Type II error (β) is where we might make the mistake of assuming the data in that region belongs to data source 1 (the alternate hypothesis) when, in fact, it belongs to data source 2.

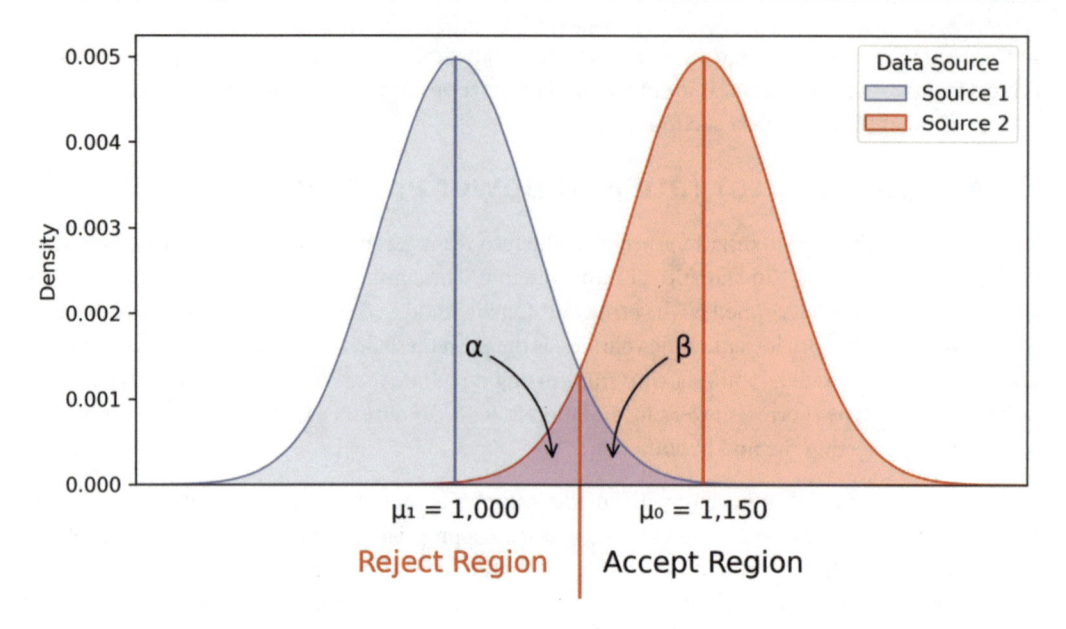

Figure 3.9 – Visualizing error in a left-tailed two-population z-test

As illustrated in the preceding plot, we see a left-tailed t-test comparing the null distribution on the right to the alternate distribution on the left. In this plot, we can see two prominent concepts at hand:

- As α becomes smaller, the critical value slides left, moving more toward distributional outliers, β becomes larger, and vice versa
- Statistical power is the area of data source 2 to the right of β, as *power = (1-β)*

A type I error is determined based on a pre-selected threshold; a researcher may feel most comfortable with a 90, 95, or 99% level of confidence. However, a type II error is based on parameterization (standard deviation, effect size, sample size, and so on). Therefore, we can use a power analysis to identify the sample size, and vice versa. The implementation of power analysis for various hypothesis tests will be implemented in the following two chapters.

Power analysis for a two-population pooled z-test

Let us look at a dataset of salaries for professors of two different disciplines, discipline A and discipline B. We want to know whether there is a statistically significant difference between them based on the data we have. First, we need to perform a power analysis to know whether we have enough samples to trust the results of the z-test we may perform to test this hypothesis. The components of a power analysis we need for this include effect size, the type I error rate, a desired type II error rate, the direction of the alternate hypothesis (group 2 is expected to be larger than, smaller than, or could be either larger or smaller than group 1), and the ratio of observations in the larger sample relative to that of the smaller. Discipline A has 181 salaries and discipline B has 216. Therefore, 216 will be our numerator corresponding to what we will consider as *group 1* (discipline A will be *group 2*).

Let us suppose we are not sure whether one group will be larger or smaller than the other; we will consider this a two-sided hypothesis test. The effect size for this z-test is the difference between the two groups. We will use **Cohen's d**. To calculate that, we need to divide the difference between the means of the two groups by the pooled standard deviation. Calculated here, the pooled standard deviation is the number of samples in group 1 multiplied by the variance of group 1, plus the same for group 2, all divided by the combined sample size for the two groups. The following is the equation for the pooled standard deviation:

$$\sqrt{\frac{n_1 \sigma_1^2 + n_2 \sigma_2^2}{n_1 + n_2}}$$

For effect size, we need to divide the difference of means by the effect size as follows:

$$\frac{|\mu_1 - \mu_2|}{\sqrt{\frac{n_1 \sigma_1^2 + n_2 \sigma_2^2}{n_1 + n_2}}}$$

Note that in the equation for effect size, we take the absolute value of the difference of means. That is because, for Cohen's d, we always need a positive difference of means for this test.

As calculated here, if we want a power of 80% (the type II error rate is 20%) and a type I error rate (a level of significance) of 0.05, we will need 172 samples in group 1 and 145 samples in group 2. However, if we wanted a power of 90% and a level of significance of 0.01 (99% confidence), we would need 325 samples in group 1 and 274 samples in group 2:

```
import statsmodels.api as sm
import math

df_prof = sm.datasets.get_rdataset("Salaries", "carData").data

df_prof_A = df_prof.loc[df_prof['discipline'] == 'A']
df_prof_B = df_prof.loc[df_prof['discipline'] == 'B']

def pooled_standard_deviation(dataset1, dataset2, column) -> float:
    pooledSD = math.sqrt(((len(dataset1) - 1)*(dataset1[column].std()**2)+(len(dataset2) - 1)*(dataset2[column].std()**2))/(len(dataset1) + len(dataset2) - 2))
    return pooledSD;

stdDeviation = pooled_standard_deviation(
    dataset1 = df_prof_A, dataset2=df_prof_B,
    column='salary')
from statsmodels.stats.power import NormalIndPower
```

```python
effect = abs(df_prof_B['salary'].mean() -
    df_prof_A['salary'].mean() ) / stdDeviation
# The difference between two means divided by std if pooled 2-sample
alpha = 0.05
power = 0.8
ratio=1.19 # # of obs in sample 2 relative to sample 1
analysis = NormalIndPower()
result = analysis.solve_power(effect, power=power, nobs1=None,
ratio=ratio, alpha=alpha, alternative='two-sided')
print('Sample Size Required in Sample 1: {:.3f}'.format(
    result*ratio)) # nobs1 is the sample size.
print('Sample Size Required in Sample 2: {:.3f}'.format(
    result)) # nobs2 is the sample size.
```

The output from this code is as follows:

```
Sample Size Required in Sample 1: 171.620
Sample Size Required in Sample 2: 144.218
effect = abs(df_prof_B['salary'].mean() -
    df_prof_A['salary'].mean() ) / stdDeviation
alpha = 0.01
power = 0.9
ratio=1.19 # # of obs in sample 2 relative to sample 1
analysis = NormalIndPower()
result = analysis.solve_power(effect, power=power,
    nobs1=None, ratio=ratio, alpha=alpha,
    alternative='two-sided')
print('Sample Size Required in Sample 1: {:.3f}'.format(
    result*ratio)) # nobs1 is the sample size.
print('Sample Size Required in Sample 2: {:.3f}'.format(
    result)) # nobs2 is the sample size.
```

The output of this code is as follows:

```
Sample Size Required in Sample 1: 325.346
Sample Size Required in Sample 2: 273.400
```

Summary

In this chapter, we introduced the concept of a hypothesis test. We started with a basic outline of a hypothesis test with the four key steps:

- State the hypothesis
- Perform the test
- Determine whether to reject or fail to reject the null hypothesis
- Draw a statistical conclusion with a scope of inference

Then we talked about potential errors that can occur and false positives and false negatives and defined the expected error rate (alpha) of a test and the power (beta) of a test.

We also discussed the statistical procedure called the z-test. This is a type of hypothesis test using sample data assumed to be normally distributed. The z-score and z-statistic were also introduced in the section on different types of z-tests, such as one-sample or two-sample z-tests for means or proportions.

Finally, we discussed the concept and motivation behind the power analysis, which can be used to identify the probability of incorrectly rejecting the null hypothesis and selecting the sample size. We also explored the parameters of the analysis for a two-population pooled z-test. Here, we briefly examined effect size, which is the value of impact (the effect of a treatment) we search for when performing the hypothesis test. We will discuss power analysis in the next two chapters as we iterate over different applications of hypothesis testing.

In the next chapter, we will discuss more parametric hypothesis tests. While these tests will still require distribution assumptions, the assumptions will be less strict than the assumptions of the z-test.

4
Parametric Tests

In the previous chapter, we introduced the concept of a hypothesis test and showed several applications of the z-test. The z-test is a type of hypothesis test in a family of hypothesis tests called parametric tests. Parametric tests are powerful hypothesis tests, but the application of parametric tests requires certain assumptions to be met by the data. While the z-test is a useful test, it is limited by the required assumptions. In this chapter, we will discuss several more parametric tests, which will expand our parametric tool set. More specifically, we will discuss the various applications of the t-test, how to perform tests when more than two subgroups of data are present, and the hypothesis test for Pearson's correlation coefficient. We will complete the chapter with a discussion on power analysis for parametric tests.

In this chapter, we're going to cover the following main topics:

- Assumptions of **parametric tests**
- **T-test**—a parametric hypothesis test
- Tests with more than two groups and **analysis of variance** (**ANOVA**)
- **Pearson's correlation coefficient**
- **Power analysis** examples

Assumptions of parametric tests

Parametric tests make assumptions about population data that require the statistics practitioner to perform analysis of data prior to modeling, especially when using sample data because the sample statistics are leveraged as estimates for the population parameters when the true population parameters are unknown. These are the three primary assumptions of parametric hypothesis tests:

- Normally distributed population data
- Samples are independent
- Equal population variances (when comparing two or more groups)

In this chapter, we discuss the z-test, t-test, ANOVA, and Pearson's correlation. These tests are used on continuous data. In addition to these assumptions, Pearson's correlation requires data to contain paired samples. In other words, there must be an equal number of samples in each group being compared as Pearson's correlation is based on pairwise comparisons.

While these assumptions are ideal, there are many occasions where these cannot be ensured. Consequently, it is useful to understand there is some robustness to these assumptions, depending on the test.

Normally distributed population data

Because in parametric hypothesis tests we are interested in gaining inferences about population parameters, such as the mean or standard deviation, we must assume the parameter of choice is representative of the distribution and that it is safe to assume a central tendency in the data. We must also assume the statistic (parametric value taken from a sample or sampling distribution) is representative of its respective population parameter. Therefore, since we assume in parametric hypothesis tests that the population is normally distributed, the sample should also be normally distributed as well. Otherwise, it is not safe to assume the sample is representative of the population.

Parametric hypothesis tests rely heavily on the mean and assume it is strongly representative of the data's central point (all population data is centrally distributed around the mean). Consider where the means of two distributions are being compared to test if there is a statistically significant difference between them. If the distributions are skewed, the mean will not be the center point of the data and, consequently, cannot represent the distributions very well. Since this would be the case, inference obtained from a test comparing the means would not be reliable.

Robustness to normally distributed data

Many hypothesis tests specify degrees of freedom when using samples to make estimates about populations. Degrees of freedom force models to assume there is extra variance in the distributions used than actually present. While the statistical parameters in the analysis remain the same, the assumed extra variance forces measures of central tendency closer. Stated differently, using degrees of freedom forces measures of central tendency to be more centrally representative of the distributions from which they are calculated. The reason for this is that it is assumed samples—while representative of their overall populations—represent their populations with a margin of error. Consequently, parametric hypothesis tests using degrees of freedom have some robustness to violations of the requirement for normally distributed data.

In the plots shown in *Figure 4.1*, we have a slightly skewed distribution. One applies degrees of freedom while the other does not. We can see the mean and median have the same distance between them whether degrees of freedom are used or not. However, the distribution using degrees of freedom takes on more errors (more variance):

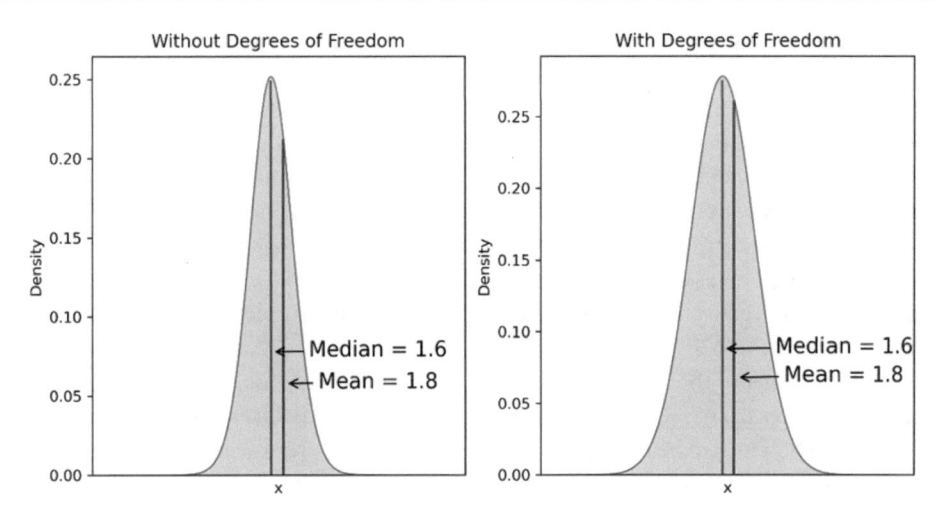

Figure 4.1 – Visualizing the influence of degrees of freedom

When using a hypothesis test that considers the mean, we can see that the mean, while not centered (as is the median), approximates the center of the distribution much more closely, relative to all data points, when degrees of freedom are used. Since parametric hypothesis tests use the mean as the central point, this is important for the usefulness of the model as the mean is more representative of the central point of the data when degrees of freedom are used. This is a primary reason there is some robustness to normality. Some other robustness is in the statistical interpretation, such as in choosing the level of confidence; if a distribution is not perfectly normally distributed, it may be beneficial to use a 90% level of confidence rather than a 99% level of confidence, for example.

Testing for normally distributed data

There are multiple methods for determining whether a distribution is normally distributed and thus can be used in parametric hypothesis testing. Generally, the level of adherence to normality is up to the discretion of the researcher. The methods in this section leave some margin for debate on normality based on visual inspection as well as levels of statistical significance applied.

Visual inspection

The best tests to identify whether a distribution is normally distributed or not are based on visual inspection. We can use **Quantile-Quantile (QQ)** plots and histograms—among other tests—to visually inspect the distributions.

In the following code snippet, we generate plots of the original data as well as the QQ plots using the `scipy.stats` module's `probplot` function:

```
import matplotlib.pyplot as plt
import scipy.stats as stats
import numpy as np
```

```
mu, sigma = 0, 1.1
normally_distributed = np.random.normal(mu, sigma, 1000)
```

In *Figure 4.2*, we can see in the first column a histogram of exponentially distributed data and, beneath it, its QQ plot. As the points are very far from approximating adherence to the 45-degree red line, which represents a pure normal distribution, we can conclude the data is not normally distributed. By visually inspecting the data in the second column, we can see the histogram exhibits an approximately normally distributed dataset. This is backed up by the QQ plot below it, where the points mostly approximate the 45-degree red line. With respect to the tails of the QQ plot, these data points represent the density of skewness. We expect with a normally distributed dataset that the bulk of data points will tend toward the center of the red line. With the exponential distribution, we can see a heavy density toward the left, lower tail of the red line, and a sparse scattering of points toward the upper-right side of the line. The QQ plot can be read left to right, mirroring the spread seen in the histogram, where the smallest values appear on the left-hand side of the x axis and the largest on the right-hand side:

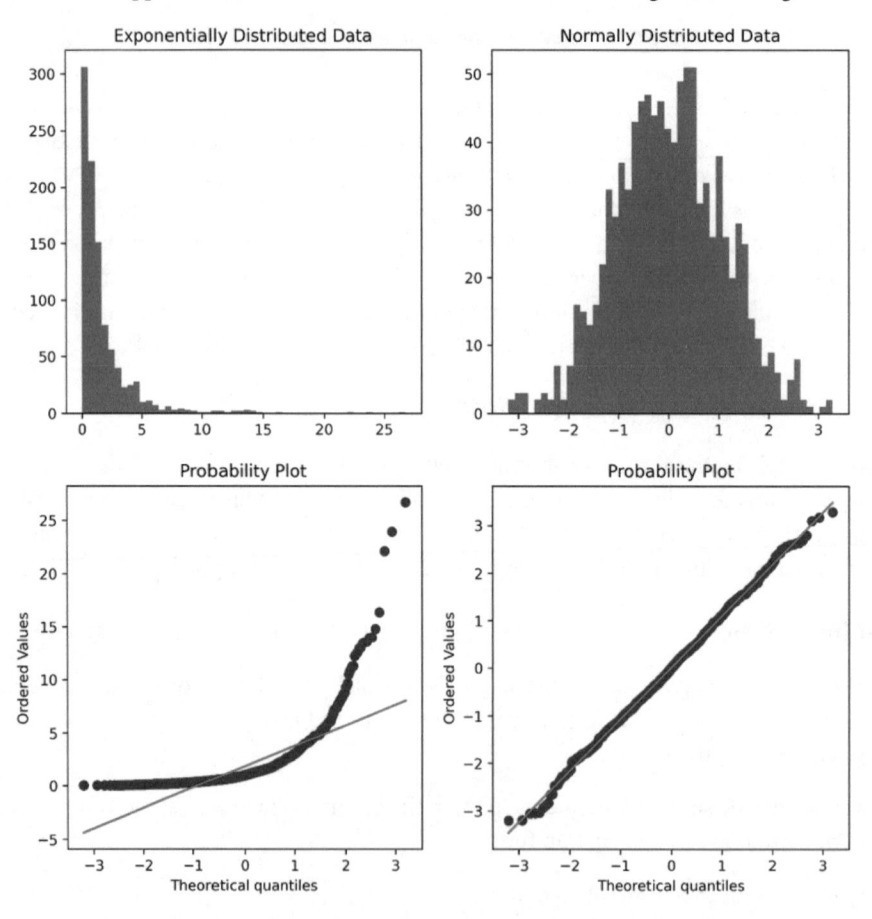

Figure 4.2 – Visually assessing normality with QQ and histogram plots

Visual inspection of the QQ plots and histograms should be enough to help a researcher conclude whether the normality assumption has been violated or not. However, in cases where one might not want to perform visual inspection—such as when constructing a data science pipeline—there are alternative approaches that provide specific measurements of normality. Three of the most commonly used tests are the **Kolmogorov-Smirnov**, **Anderson-Darling**, and **Shapiro-Wilk** tests.

The Kolmogorov-Smirnov test focuses more on the centrality of the data. Consequently, however, the test has less power if there is a wide variance around the center of the data. Anderson-Darling focuses more on the tails of the data than the center and is more likely to identify non-conformity to normality if data is heavy-tailed with extreme outliers. These two tests perform well on large sample sizes but do not have as much power when sample sizes are lower. The third test we consider, Shapiro-Wilk, is more general than the Kolmogorov-Smirnov and Anderson-Darling tests and therefore more robust to small sample sizes. Based on these traits, it may be more useful to use Shapiro-Wilk tests in an automated pipeline. Alternatively, it may be better to lower the level of confidence for the test being applied.

Kolmogorov-Smirnov

The **Kolmogorov-Smirnov** test can be used to test the null hypothesis that a given sample distribution is normally distributed. This version of the Kolmogorov-Smirnov test is the one-sample goodness-of-fit test, which performs analysis against a benchmark cumulative density distribution. When running the `kstest` function in the `scipy.stats` module, using `stats.norm.cdf` (scipy's cumulative density function) performs this one-sample version of the test. The two-sample version tests against a specified distribution to determine whether the two distributions match. In the two-sample case, the distribution to be tested must be provided as a `numpy` array instead of the `stats.norm.cdf` function used in the code snippet shown below *Figure 4.3*. However, this is outside of the scope of testing for normality, so we will not look at this.

Kolmogorov-Smirnov measures a calculated test statistic against a table-based critical value (`kstest` calculates this internally). As with other hypothesis tests, if the test statistic is larger than the critical value, the null hypothesis that the given distribution is normally distributed can be rejected. This can also be assessed if the p-value is low enough to be significant. The test statistic is calculated as the absolute value of the maximum distance between all data points in the given distribution against the cumulative density function.

> **Kolmogorov-Smirnov special requirement**
>
> The Kolmogorov-Smirnov test requires data to be centered around zero and scaled to a standard deviation of one. All data must be transformed for the test, but inference can be applied to the pre-transformed distribution; the centered and scaled distribution does not need to be the distribution used in further statistical testing or analysis.

In the following code snippet, we test to confirm whether a normally distributed dataset, `normally_distributed`, is normally distributed. The dataset has a mean of 0 and a standard deviation of 1. The output confirms the data is normally distributed. The plots in *Figure 4.3* show the normally distributed distribution centered around a mean of 0 with a standard deviation of 1, and on the right of it is the exponentially transformed version of the same distribution:

```
from scipy import stats
import numpy as np
mu, sigma = 0, 1
normally_distributed = np.random.normal(mu, sigma, 1000)
```

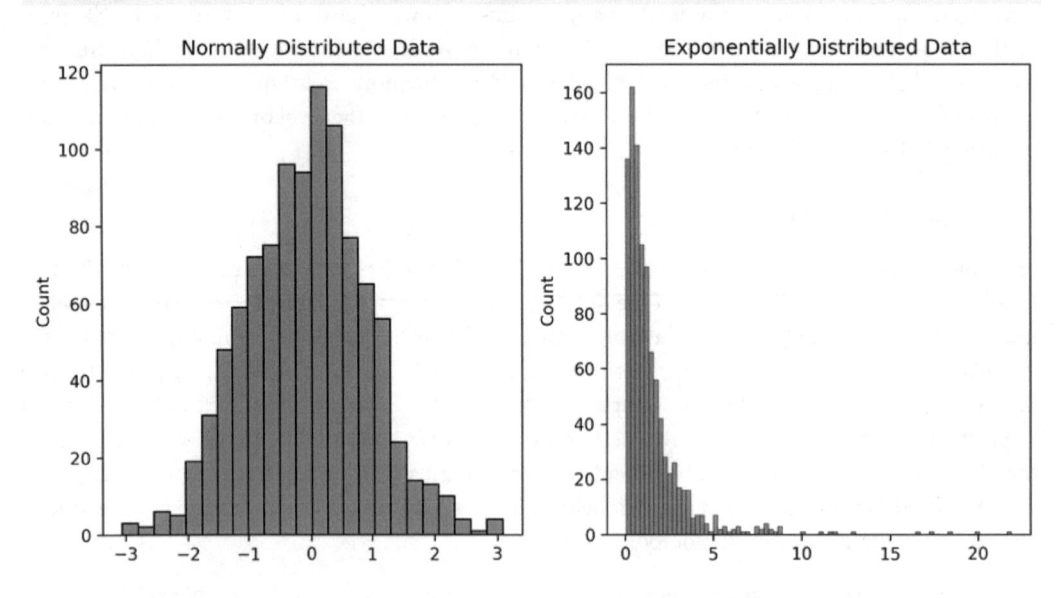

Figure 4.3 – Normally distributed and exponential data

Here, we run the Kolmogorov-Smirnov test:

```
stats.kstest(normally_distributed,
            stats.norm.cdf)
```

The `statsmodels` Kolmorogov-Smirnov test yielded the following results for our data:

`KstestResult(statistic=0.0191570377833315, pvalue=0.849436919292824)`

If we use the same data, but transform it exponentially to be right-skewed, the same test indicates the data is no longer normally distributed:

```
stats.kstest(np.exp(normally_distributed), stats.norm.cdf)
```

The signficant p-value confirms non-normality:

```
KstestResult(statistic=0.5375205782404135, pvalue=9.59979841227121e-271)
```

Next, let us take a distribution of 1,000 samples with a mean of 100 and a standard deviation of 2. We need to center it to a mean of 0 with unit variance (standard deviation of 1). In the following code snippet, we generate the data, then perform the scaling and save it to the `normally_distributed_scaled` variable:

```
mu, sigma = 100, 2
normally_distributed = np.random.normal(mu, sigma, 1000)
normally_distributed_scaled = (
    normally_distributed-normally_distributed.mean()) /
    normally_distributed.std()
```

Now that the data is centered and scaled as required, we check it using the Kolmogorov-Smirnov test. As expected, the data is confirmed normally distributed:

```
stats.kstest(normally_distributed_scaled, stats.norm.cdf)
```

This is the output:

```
KstestResult(statistic=0.02597307287070466, pvalue=0.5016041053535877)
```

Anderson-Darling

Similar to the Kolmogorov-Smirnov test, the **Anderson-Darling** test measures a given distribution against a normally distributed distribution. In `scipy`'s `anderson` test, we can test against other distributions, but the default argument specifying a normal distribution, `dist="norm"`, assumes a null hypothesis that the given distribution is statistically the same as a normally distributed distribution. For each distribution tested against, a different set of critical values must be calculated.

> **Anderson-Darling compared to Kolmogorov-Smirnov**
>
> Note that while both the Anderson-Darling and Kolmogorov-Smirnov tests use the cumulative density frequency distributions to test for normality, the Anderson-Darling test is different from the Kolmogorov-Smirnov test because it weights the variance in the tails of the cumulative density frequency distribution more than the middle. This is because the variance in the tails can be measured in smaller increments than in the middle of the distribution. Consequently, the Anderson-Darling test is more sensitive to tails than the Kolmogorov-Smirnov test. In line with the Kolmogorov-Smirnov test, a test statistic is calculated and measured against a critical value. If the test statistic is larger than the critical value, the null hypothesis that the given distribution is normally distributed can be rejected at the specified level of significance.

Here, we are using the Anderson-Darling test to test a random normal probability distribution generated with a mean of 19 and a standard deviation of 1.7. We also test an exponentially transformed version of this data:

```
import matplotlib.pyplot as plt
import seaborn as sns
import numpy as np
mu, sigma = 19, 1.7
normally_distributed = np.random.normal(mu, sigma, 1000)
not_normally_distributed = np.exp(normally_distributed);
```

Figure 4.4 shows plots of the data:

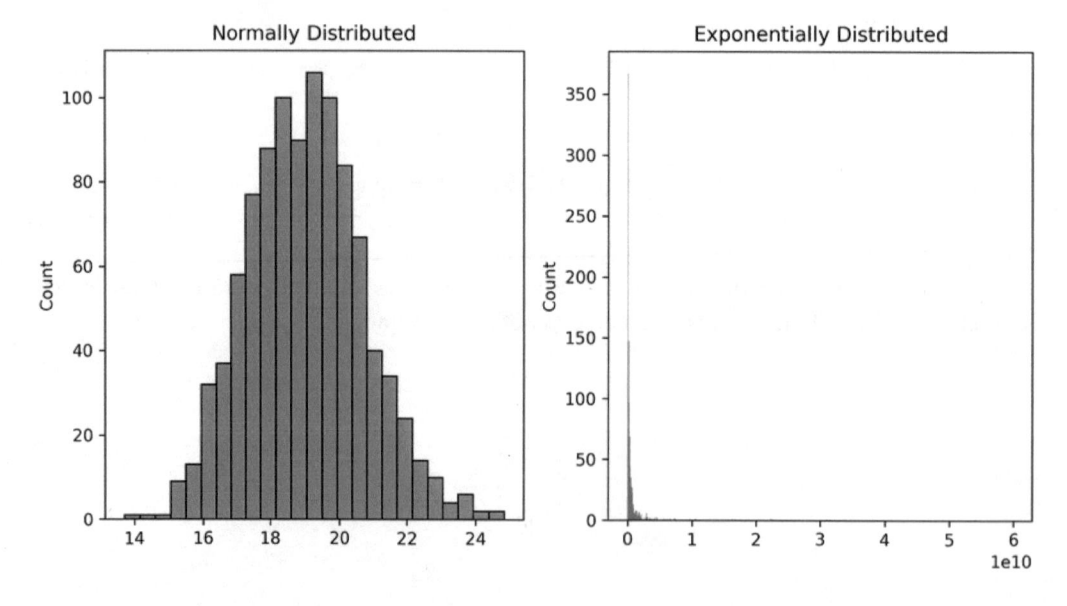

Figure 4.4 – Normal distribution versus heavy-tailed exponential distribution

In the code and output shown next, in *Figure 4.5*, we can see the distribution is normally distributed at all levels of significance. Recall that the level of significance is the p-value (that is, a level of significance = 15.0 means a p-value of 0.15 or smaller is significant):

```
from scipy import stats
import pandas as pd
import numpy as np
def anderson_test(data):
    data = np.array(data)
    test_statistic, critical_values, significance_levels = stats.anderson(normally_distributed, dist='norm')

    df_anderson = pd.DataFrame({'Test Statistic':np.repeat(test_statistic, len(critical_values)), 'Critical Value':critical_values, 'Significance Level': significance_levels})
    df_anderson.loc[df_anderson['Test Statistic'] >= df_anderson['Critical Value'], 'Normally Distributed'] = 'No'
    df_anderson.loc[df_anderson['Test Statistic'] <df_anderson['Critical Value'], 'Normally Distributed'] = 'Yes'
    return df_anderson;
mu, sigma = 19, 1.7
normally_distributed = np.random.normal(mu, sigma, 1000)
anderson_test(normally_distributed)
```

Here, the data generated through the `numpy random.normal` function is tested with the Anderson-Darling method and confirmed to be normally distributed:

Test statistic	Critical value	Significance level	Normally distributed
0.191482344	0.574	15	Yes
0.191482344	0.653	10	Yes
0.191482344	0.784	5	Yes
0.191482344	0.914	2.5	Yes
0.191482344	1.088	1	Yes

Figure 4.5 – Anderson-Darling results for normally distributed data

Here, we test an exponential transformation of the normally distributed data to check for normality. The data is exponentially distributed and should reject at all levels of significance. However, we see in *Figure 4.6* that it has failed to reject at the 0.01 level of significance (99% confidence). Therefore, depending on the use case, it may be prudent to check all levels of significance, use a different test, or make a decision based on multiple tests:

```
not_normally_distributed = np.exp(normally_distributed)
anderson_test(not_normally_distributed)
```

Our Anderson-Darling test of non-normally distributed data outputs are as follows in *Figure 4.6*:

Test statistic	Critical value	Significance level	Normally distributed
0.96277351	0.574	15	No
0.96277351	0.653	10	No
0.96277351	0.784	5	No
0.96277351	0.914	2.5	No
0.96277351	1.088	1	Yes

Figure 4.6 – Anderson-Darling results for non-normally distributed data

Shapiro-Wilk

The **Shapiro-Wilk** test is a goodness-of-fit test that checks whether a given distribution is normally distributed. The test checks how closely a distribution of observed values centered on 0 and scaled to a unit variance of 1 approximates an observed centered and scaled standard normal distribution. This centering and scaling (called **standardizing**) are performed within the function in the `scipy.stats shapiro` module, so input data does not need to be altered prior to testing. The level of significance for this test in `scipy` is 0.05.

> **Shapiro-Wilk compared to Kolmogorov-Smirnov and Anderson-Darling**
>
> Shapiro-Wilk is ideal, compared to Kolmogorov-Smirnov and Anderson-Darling, for testing small sample sizes of roughly less than 50. However, one drawback is that since Shapiro-Wilk uses repeated sampling and testing for the calculated test statistic by applying Monte Carlo simulation, the law of large numbers poses a risk that as the sample size increases, there is an inherent increase in the risk of encountering a *type II* error (a loss of power) and failing to reject the null hypothesis, where the null hypothesis states the given distribution is normally distributed.

Using the same distributions as in the Anderson-Darling test, we test with Shapiro-Wilk. We can see with the random normal distribution with a mean of 19 and a standard deviation of 1.7, the Shapiro-Wilk test has confirmed with a p-value of 0.99 that the null hypothesis that the input distribution is normally distributed should not be rejected:

```
mu, sigma = 19, 1.7
normally_distributed = np.random.normal(mu, sigma, 1000)
stats.shapiro(normally_distributed)
```

This is the output:

`ShapiroResult(statistic=0.9993802905082703, pvalue=0.9900037050247192)`

When testing using the exponentially transformed version of the normally distributed data, we find a significant p-value (p = 0.0), indicating we have enough evidence to reject the null hypothesis and conclude the distribution is not normally distributed:

```
not_normally_distributed = np.exp(normally_distributed)
stats.shapiro(not_normally_distributed)
```

This is the output:

`ShapiroResult(statistic=0.37320804595947266, pvalue=0.0)`

Independent samples

In parametric hypothesis testing, the independence of samples is another important assumption. Two effects can occur from non-independent sampling. One effect occurs when subgroup sampling is performed. The issue here is that responses in one subgroup of the population may be different than responses from another subgroup of the same population or even more similar to those of a different population. However, when sampling representative of the overall population is taken, this type of subgroup difference may not be very representative of the population.

Another effect of non-independent sampling is when samples are taken close enough together in time that the occurrence of one precludes or excludes the occurrence of another. This is called serial (or auto-) correlation.

Parametric tests are not typically robust to violations of this requirement as it has direct, categorical implications on the interpretability of test outcomes. With respect to subgroup sampling, this can be prevented through a well-structured sampling approach such as those outlined in *Chapter 1, Sampling and Generalization*. However, as regards the serial effect, we can test for autoregressive correlation (also called serial correlation) in the data.

Durbin-Watson

One of the most common tests performed to assess a lack of independence in sampling is the first-order (also referred to as lag-one) autoregressive test called the **Durbin-Watson** test. **Autoregressive** means previous data points are used to predict the current data point. First-order means the last sampled data point (lag one) is the point most significantly correlated to the most recently sampled data point (lag zero) in a sequence of sampled data. In first-order autocorrelation, the correlation for each data point is strongest with the previous data point. The Durbin-Watson test does not test whether any value is correlated to the value before it, but instead if, overall, there is a strong enough relationship between each value and the value before it to conclude there is significant autocorrelation. In that sense, there is some robustness to non-independent sampling such that an accident or two may not completely invalidate a hypothesis test, but a consistent recurrence of this type of violation will.

A Durbin-Watson value of 2 indicates no significant autocorrelation, a value between 0 and 2 represents positive (direct) autocorrelation, and a value between 2 and 4 represents negative (inverse) autocorrelation.

In the following example, we have two distributions, each with 1,000 samples. The distribution on the left is a sinusoidal distribution that exhibits strong autoregressive correlation, and the distribution on the right is a set of randomly generated data displaying as white-noise variance (random points centered around a mean of 0). Using the `durbin_watson()` function from the `statsmodels.stats` module, we are able to confirm direct, positive lag-one autocorrelation in the sinusoidal pattern (a very small Durbin-Watson value) and a Durbin-Watson statistic of 2.1 with the random noise, indicating no autocorrelation. Therefore, in *Figure 4.7*, the plot on the left is not composed of independent samples whereas the plot on the right is:

```
from statsmodels.stats.stattools import durbin_watson
import matplotlib.pyplot as plt
import numpy as np
mu, sigma = 0, 1.1
independent_samples = np.random.normal(mu, sigma, 1000)
correlated_samples = np.linspace(-np.pi, np.pi, num=1000)
fig, ax = plt.subplots(1,2, figsize=(10,5))
ax[0].plot(correlated_samples, np.sin(correlated_samples))
ax[0].set_title('Durbin Watson = {}'.format(
    durbin_watson(correlated_samples)))
ax[1].plot(independent_samples)
ax[1].set_title('Durbin Watson = {}'.format(
    durbin_watson(independent_samples)))
```

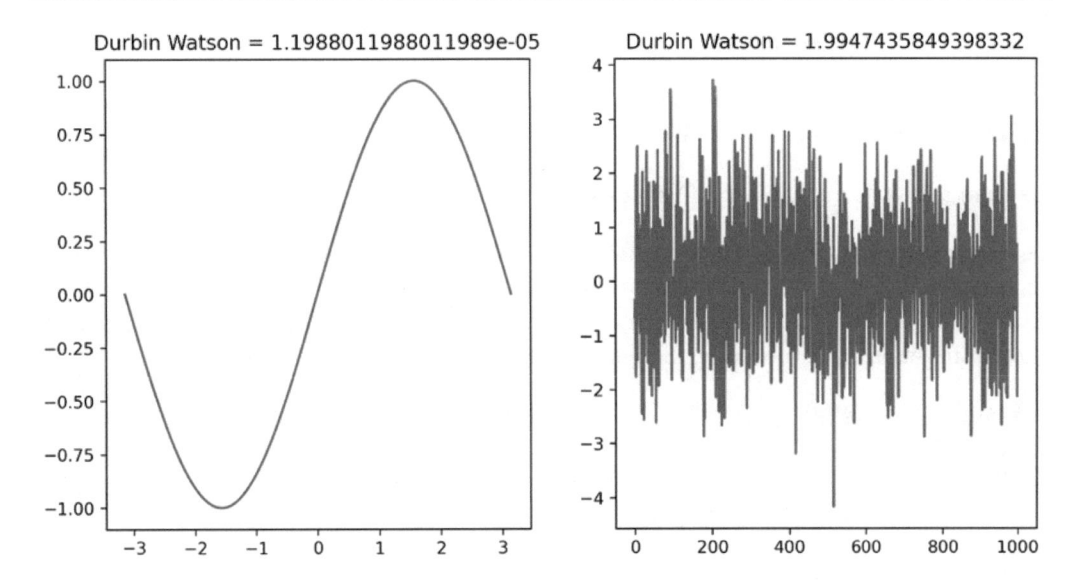

Figure 4.7 – Serially correlated and normally distributed sequence data

Equal population variance

Similar to the assumption of normally distributed data, the assumption of equal population variance—also referred to as homogeneity of variance—is about the shape of the physical properties of the distributions being compared. Assuming equal population variance helps increase the power of a parametric test. This is because there is confidence when means are identified as being different; we also know the degree of potential distribution overlap. When a test has an intuition about the location of the full distributions—in effect, true knowledge of the effect size—power increases. Conversely, as variances diverge, power decreases.

Robustness to equal population variance

While equal population variance is useful in parametric testing, modifications to these tests exist that help results be robust to deviance from equal variance. One prominent modified version of these tests uses the **Welch-Satterthwaite** adjustment to the degrees of freedom used. Because applying the same degree of freedom to each group when each group has a different variance would result in a misrepresentation of the data, the Welch-Satterthwaite adjustment accounts for variance differences when allocating degrees of freedom to parametric tests that assume equal variance. Two common tests that use the Welch-Satterthwaite adjustment are Welch's t-test and Welch's ANOVA test. When used on small samples, these tests may not be reliable, but when used on sample sizes large enough to have sufficient power, the results should be approximately the same as their non-Welch counterparts.

Testing for equal variance

When testing for equal variance among distributions, we have two prominent tests: **Levene's test for equality of variances** and **Fisher's F-test**.

Levene's test for equality of variances

Levene's test for equality of variances is useful when testing for homogeneity of variance of two or more groups. In the code snippet shown below *Figure 4.8*, we test with three distributions, each having a sample size of 100, a mean of 0, and standard deviations of 0.9, 1.1, and 2. *Figure 4.8* is a plot of the three distributions generated using the data output from the code above *Figure 4.8*.

```
from scipy.stats import levene
np.random.seed(26)
mu1, sigma1, mu2, sigma2, mu3, sigma3 = 0,0.9,0,1.1,0,2
distro1, distro2, distro3 = pd.DataFrame(), pd.DataFrame(),
    pd.DataFrame()
distro1['x'] = np.random.normal(mu1, sigma1, 100)
distro2['x'] = np.random.normal(mu2, sigma2, 100)
distro3['x'] = np.random.normal(mu3, sigma3, 100)
```

We can see how their different standard deviations impact their range.

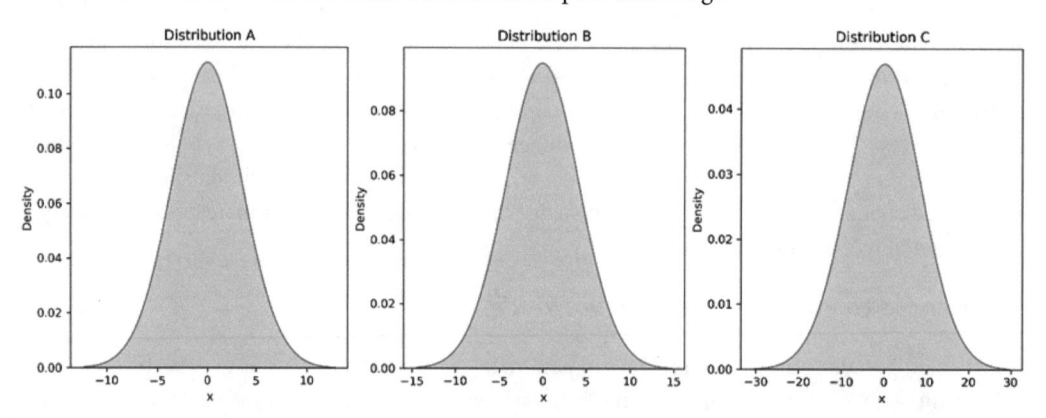

Figure 4.8 – Distributions for multiple equality of variance testing

We can see the test is sensitive to violations of non-homogenous variance because the result of this is a statistically significant p-value indicating non-homogenous variance:

```
f_statistic, p_value = levene(distro1['x'], distro2['x'],
distro3['x'])
if p_value <= 0.05:
    print('The distributions do not have homogenous variance.
P-value = %.4f, F-statistic = %.4f'%(p_value, f_statistic))
```

```
else:
    print('The distributions have homogenous variance.P-value = 
%.4f, F-statistic = %.4f'%(p_value, f_statistic))
```

This is the output:

```
The distributions do not have homogenous variance. P-value = 0.0000
```

Fisher's F-test

Fisher's F-test is useful when testing for homogeneity of variance for two groups at a time. This test compares a test statistic to a critical value to determine whether the variances are statistically the same or not. The calculated F-statistic is the variance of group one divided by the variance of group two. Group one is always the group with the larger variance. Using the preceding data, let us compare distribution 1 with distribution 3. Distribution 3 has a larger variance of 2, so that group's variance will be the numerator when calculating the F-statistic. Since each group has a sample size of 100, their degrees of freedom for the table lookup will each be 99. However, since we will use the `scipy` Python package to compute the test, here, the table lookup is not needed as `scipy` does this for us with the `f.cdf()` function. In line with the results of the Levene test, the F-test indicates distribution 1 and distribution 3 do not have homogenous variance:

```
from scipy.stats import f
def f_test(inputA, inputB):
    group1 = np.array(inputA)
    group2 = np.array(inputB)
    if np.var(group1) > np.var(group2):
        f_statistic = np.var(group1) / np.var(group2)
        numeratorDegreesOfFreedom = group1.shape[0] - 1
        denominatorDegreesOfFreedom = group2.shape[0] - 1
    else:
        f_statistic = np.var(group2)/np.var(group1)
        numeratorDegreesOfFreedom = group2.shape[0] - 1
        denominatorDegreesOfFreedom = group1.shape[0] - 1
    p_value = 1 - f.cdf(f_statistic,numeratorDegreesOfFreedom,
denominatorDegreesOfFreedom)
    if p_value <= 0.05:
        print('The distributions do not have homogenous 
variance. P-value = %.4f, F-statistic = %.4f'%(p_value, f_
statistic))
    else:
        print('The distributions have homogenous variance. 
P-value = %.4f, F-statistic = %.4f'%(p_value, f_statistic))
f_test(distro3['x'], distro1['x'])
```

This F-test output is as follows:

```
The distributions do not have homogenous variance. P-value = 0.0000,
F-statistic = 102622.9745
```

T-test – a parametric hypothesis test

In the last chapter, the z-test for means was applied when population standard deviations were known. However, in the real world, it is not easy (or virtually impossible) to obtain the population standard deviation. In this section, we will discuss another hypothesis test called the **t-test**, which is used when the population standard deviations are unknown. The mean and the standard deviation of a population are estimated by taking the mean and the standard deviation of sample data representative of this population.

Broadly speaking, the method for the t-test for means is very similar to the one for the z-test for means, but the calculations for the test statistic and p-value are not the same as for the z-test. The test statistic is computed by the following formula:

$$t = \frac{\bar{x} - \mu}{s/\sqrt{n}}$$

Here, $\bar{x}$, μ, s, and n are the sample mean, population mean, sample standard deviation, and sample size, respectively, which has a *t*-**distribution** when the sample data, x, is normally distributed. The following code illustrates the standard normal distribution (blue curve) with 1,000 samples and t-distribution (green and red curves) with two sample sizes—3 and 16 samples:

```
# libraries
import numpy as np
import scipy.stats as stats
# creating normal distribution
x =np.linspace(-5, 5, 1000) #create 1000 point from -5 to 5
y = stats.norm.pdf(x) # create probability density for each point x - normal distribution

# creating Student t distributions for 2 sample sizes n =3 and n =15
degree_freedom1 = 2
t_dis1 = stats.t.pdf(x, degree_freedom1)

degree_freedom2 = 15
t_dis2 = stats.t.pdf(x, degree_freedom2)
```

The following visualization is for these 3 distributions considered.

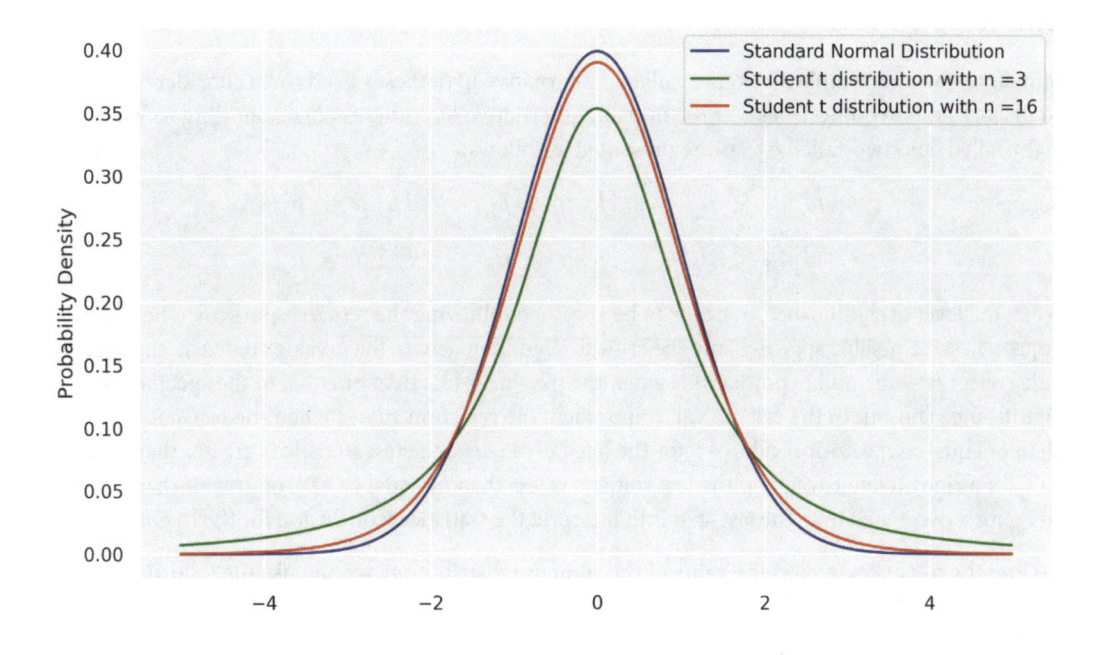

Figure 4.9 – Normal and t-distributions

Observe that the three curves have similar symmetry and shapes but there is more variability (or, in other words, heavier tails) for a sample with a smaller size. Historically, researchers considered a sample standard deviation to represent the population when the sample size was greater than 30, that is, the red curve approximates the blue curve when $n > 30$. It was also common to use the z-test if the sample distribution overlapped the standard normal distribution. This practice has some reasoning behind it because, previously, critical value tables were stored up to a sample size of 50, but nowadays, with the power of computation and the internet, the *t* values can be obtained easily with any sample size.

T-test for means

One-sample and two-sample t-tests related to a population mean or means of two populations where the population variances or population standard deviations are unknown will be considered in this part.

To perform a t-test, the following assumptions need to be satisfied:

Normality: The sample is normally distributed

Independence: Observations are randomly selected from a population to form a sample or, in other words, they are independent

Let us consider the one-sample t-test in the following section.

One-sample t-test

Similar to the one-sample z-test, the null and alternative hypotheses need to be considered in order to perform the hypothesis test. Three null and alternative hypotheses corresponding to left-tailed, right-tailed, and two-tailed t-tests are presented as follows:

$$H_0: \mu \geq \mu_0 \qquad H_0: \mu \leq \mu_0 \qquad H_0: \mu = \mu_0$$
$$H_a: \mu < \mu_0 \qquad H_a: \mu > \mu_0 \qquad H_a: \mu \neq \mu_0$$

Next, the level of significance, α, needs to be specified following the research purpose. There are two approaches: the p-value approach and the critical value approach. In the p-value approach, the rejection rule (reject H_0—the null hypothesis) is when the p-value is less than or equal to the specified level of significance chosen. In the critical value approach, the rejection rule is when the test statistic is less than or equal to the critical value $-t_\alpha$ for the left-tailed t-test, the test statistic is greater than or equal to t_α for a right-tailed t-test, and the test statistic is less than or equal to $-t_{\alpha/2}$ or greater than or equal to $t_{\alpha/2}$ for a two-tailed test. The last step is to interpret the statistical conclusion for the hypothesis test.

To find the p-value based on the value of the student t distribution, we can use the following syntax:

```
scipy.stats.t.sf(abs(x), df)
```

Here, x is the test statistic and df is the degree of freedom (df = n-1 where n is the sample size) in the formula.

For example, to find the p-value associated with a t-score of 1.9 with the degree of freedom 14 in a left-tailed test, this would be the Python implementation:

```
import scipy.stats
round(scipy.stats.t.sf(abs(1.9), df=14),4)
```

The output would be 0.0391. If the level of significance $\alpha = 0.05$, then we reject the null hypothesis because the p-value is less than α. For a right-tailed t-test, similar Python code as in the left-tailed t-test is implemented to find the p-value. For a two-tailed test, we need to multiply the value by 2, as follows:

```
scipy.stats.t.sf(abs(t), df)*2
```

Here, t is the test statistic and df is the degree of freedom (df = n-1 where n is the sample size).

To compute the critical value in Python, we use the following syntax:

```
scipy.stats.t.ppf(q, df)
```

Here, q is the level of significance and df is the degree of freedom to be used in the formula. Here is the implementation of the code in Python for left-tailed, right-tailed, and two-tailed tests:

```
import scipy.stats as stats
alpha = 0.05 # level of significance
df= 15 # degree of freedom
#find t critical value for left-tailed test
print(f" The critical value is {stats.t.ppf(q= alpha, df =df)}")
#find t critical value for right-tailed test
print(f" The critical value is {stats.t.ppf(q= 1-alpha, df =df)}")
##find t critical value for two-tailed test
print(f" The critical values are {-stats.t.ppf(q= 1-alpha/2, df =df)} and {stats.t.ppf(q= 1-alpha/2, df =df)}")
```

This is the output of the preceding code:

- The critical value is -1.7530503556925552
- The critical value is 1.7530503556925547
- The critical values are -2.131449545559323 and 2.131449545559323

At the level of significance $\alpha = 0.05$, for the left-tailed test, the critical value is about -1.753. Since this is a left-tailed test, if the test statistic is less than or equal to this critical value, we reject the null hypothesis. Similarly, for the right-tailed test, if the test statistic is greater than or equal to 1.753, we reject the null hypothesis. For the two-tailed test, we reject the null hypothesis if the test statistic is greater than or equal to 2.1314 or less than or equal to -2.1314. Finally, we interpret the statistical conclusion for the hypothesis testing.

Let us randomly choose 30 students from a high school and score their IQ. We would like to test the claim that the mean IQ score of the distribution of the students from this high school is higher than 100. This means that we will perform a right-tailed t-test. The IQ scores of 30 students are given here:

```
IQscores = [113, 107, 106, 115, 103, 103, 107, 102, 108, 107, 104,
104, 99, 102, 102, 105, 109, 97, 109, 103, 103, 100, 97, 107,116, 117,
105, 107, 104, 107]
```

Before conducting the hypothesis testing, we will check normality and independence assumptions. The assumption of independence is satisfied if the sample is randomly selected from the population of high school students at this school. For normality, we will check the histogram and QQ plots of IQ score data:

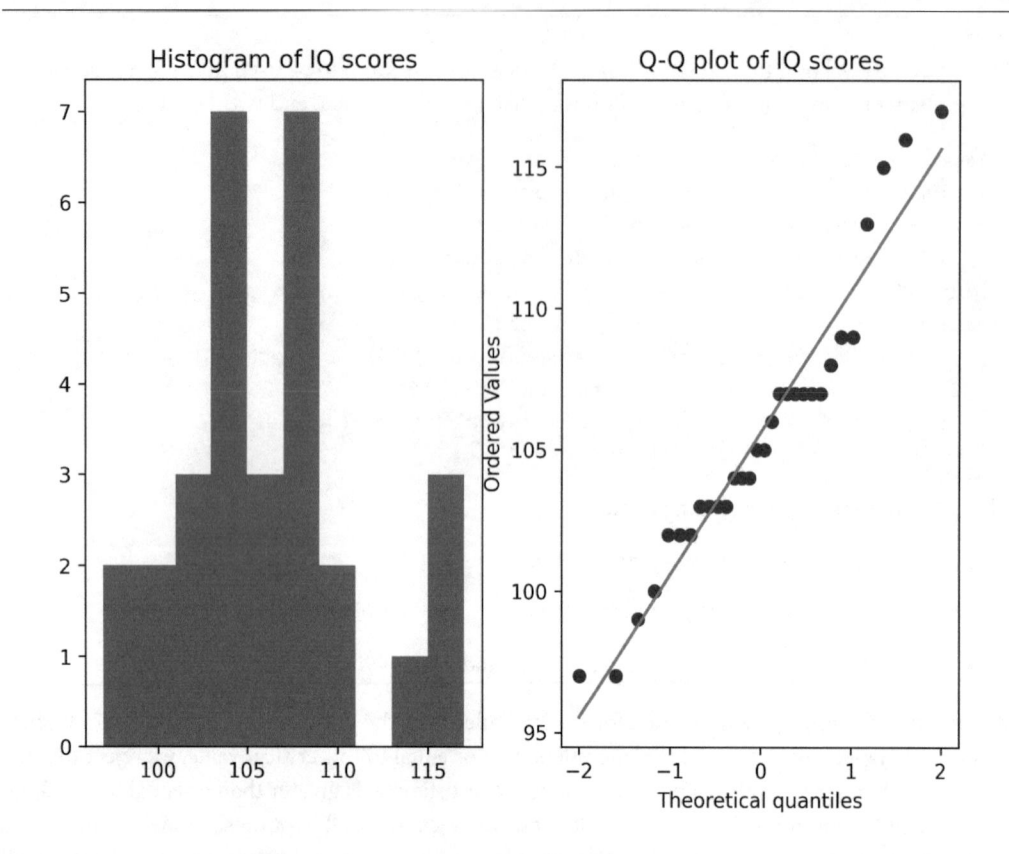

Figure 4.10 – Visually assessing normality of student IQ scores

There is little to no evidence from the histogram and QQ plot that the population IQ score distribution of the students at the high school is not normal. Since the distribution is assumed to be normal, we will proceed with the t-test.

First, we define the null hypothesis and the alternative hypothesis:

$$H_0: \mu \leq 100$$

$$H_a: \mu > 100$$

We choose the level of significance $\alpha=0.05$. You can calculate the test statistic by using its mathematical formula by hand or by implementing Python. For the critical value and p-value, the implemented code was shown in the previous part. Here, we will use another function from the `scipy` library to find the test statistic and p-value:

```
scipy.stats.ttest_1samp(data, popmean, alternative='greater')
```

Here, the following applies:

- `data`: The observations from the sample
- `popmean`: The expected value in the null hypothesis
- `alternative`: `'two-sided'` for a two-tailed t-test, `'less'` for a left-tailed t-test, and `'greater'` for a right-tailed t-test

The Python code is implemented as follows:

```
import scipy.stats as stats
#perform one sample t-test
t_statistic, p_value = stats.ttest_1samp(IQscores, popmean =100, axis=0, alternative='greater')
print(f"The test statistic is {t_statistic} and the corresponding p-value is {p_value}.")
```

This is the output:

```
The test statistic is 6.159178830896832 and the corresponding p-value is 5.15076734562176e-07.
```

Because the p-value < 0.05 where 0.05 is the level of significance, we have enough evidence to reject the null hypothesis and conclude that the true mean IQ scores of the students from this school is higher than 100.

In addition, with 95% confidence, the mean IQ score lies between 104.08 and 107.12. We can perform the calculation for the confidence interval in Python as follows:

```
IQmean = np.array(IQscores).mean() # sample mean
IQsd = np.array(IQscores).std() # sample standard deviation
sample_size = len(np.array(IQscores)) # sample size
df = sample_size-1 # degree of freedom
alpha = 0.05 # level of significance
t_crit = stats.t.ppf(q=1-alpha, df =df) # critical
confidence_interval = (IQmean-IQsd*t_crit/np.sqrt(sample_size), IQmean+IQsd*t_crit/np.sqrt(sample_size))
```

The steps to perform a hypothesis test in Python using the left-tailed t-test are similar to those of the right-tailed and two-tailed t-tests.

Two-sample t-test – pooled t-test

Similar to what was covered in *Chapter 3, Hypothesis Testing*, (two-sample z-test for means), the two-sample t-test for means has three forms for the null and alternative hypotheses. Some assumptions need to be satisfied before conducting the test, as follows:

- **Normality**: Two samples are drawn from their normally distributed populations
- **Independence**: The observations of one sample are independent of one another
- **Homogeneity of variance**: Both populations are assumed to have similar standard deviations

For normality, we use visual histograms and also QQ plots of the two samples and compare them. Let us assume independence is satisfied. In order to check equal standard deviations between the two samples, we could use visualization by observing their histograms and also use an F-test to have additional evidence if the visualization is inconclusive. This is a hypothesis test to check whether two sample variances are equal.

Let us look at the IQ scores between two high schools, A and B. The following are the scores of 30 students from each school, randomly selected:

```
IQscoresA=[113, 107, 106, 115, 103, 103, 107, 102,108, 107,
          104, 104, 99, 102, 102, 105, 109, 97, 109, 103,
          103, 100, 97, 107, 116, 117, 105, 107, 104, 107]

IQscoresB = [102, 108, 110, 101, 98, 98, 97, 102, 102, 103,
            100, 99, 97, 97, 94, 100, 104, 98, 92, 104,
            98, 95, 92, 111, 102, 112, 100, 103, 103, 100]
```

The histograms and QQ plots shown in Figure 4.11 are generated by the IQ data above.

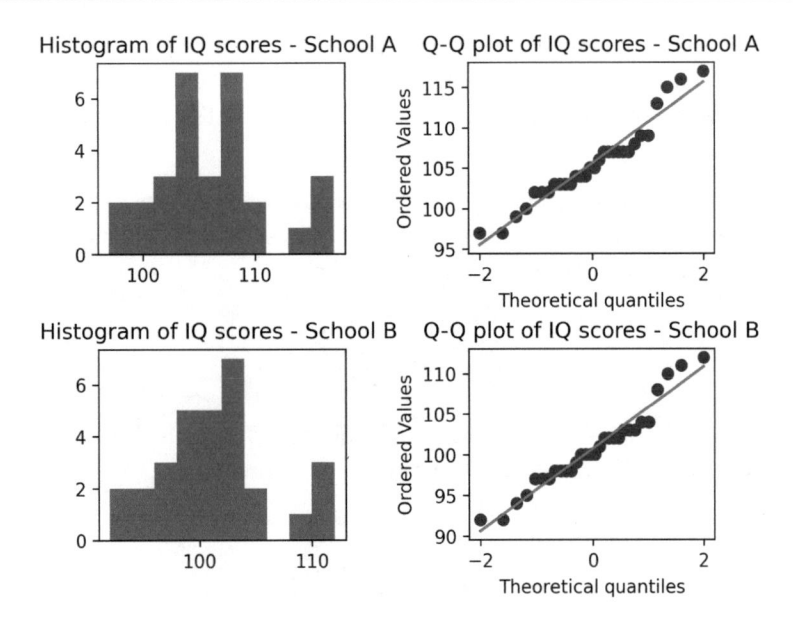

Figure 4.11 – Assessing normality of two schools' IQ scores

We can see that the normality assumption is satisfied. We also can assume by observing the histograms that the equal variance assumption is supported. Another F-test to check the equal variance assumption (if necessary) follows:

```
# F-test
import numpy as np
import scipy.stats as stats
IQscoresA = np.array(IQscoresA)
IQscoresB = np.array(IQscoresB)
f = np.var(IQscoresA, ddof=1)/np.var(IQscoresB, ddof=1) # F statistic
dfA = IQscoresA.size-1 #degrees of freedom A
dfB = IQscoresB.size-1 #degrees of freedom B
p = 1-stats.f.cdf(f, dfA, dfB) #p-value
```

The output of the preceding code tells us the F-test statistic is 0.9963 and the corresponding p-value is 0.50394 > 0.05 (0.05 is the level of significance), then we fail to reject the null hypothesis. This means that there is enough evidence to say that the standard deviations of these two samples are equal.

We now define the null hypothesis and the alternative hypothesis:

$$H_0: \mu_A = \mu_B,$$
$$H_a: \mu_A \neq \mu_B.$$

We choose the level of significance $\alpha=0.05$. We use the `statsmodels.stats.weightstats.ttest_ind` function to conduct the t-test. The documentation can be found here: https://www.statsmodels.org/dev/generated/statsmodels.stats.weightstats.ttest_ind.html.

We can use this function to perform three forms of the alternate hypothesis with `alternative='two-sided'`, `'larger'`, or `'smaller'`. In the pooled-variance t-test, when the assumption for equal variances is satisfied, the test statistic is computed as follows:

$$t = \frac{(\bar{x}_1 - \bar{x}_2) - (\mu_1 - \mu_2)}{S_p \sqrt{\frac{1}{n_1} + \frac{1}{n_2}}}$$

Here, $\bar{x}_1, \bar{x}_2, \mu_1, \mu_2, n_1,$ and n_2 are sample means, population means, and sample sizes of two samples, 1 and 2 respectively, and the pooled standard deviation is given here:

$$S_p = \sqrt{\frac{(n_1 - 1) s_1^2 + (n_2 - 1) s_2^2}{n_1 + n_2 - 2}}$$

The degree of freedom is shown here:

$$df = n_1 + n_2 - 2.$$

Let's go back to the example:

```
from statsmodels.stats.weightstats import ttest_ind as ttest
t_statistic, p_value, degree_freedom = ttest(IQscoresA,
    IQscoresB, alternative='two-sided', usevar='pooled')
```

The output returns the test statistic 3.78 and the p-value 0.00037, and the degrees of freedom used in the t-test are 58 (each sample size has 30 observations, then the degrees of freedom are calculated as 30 + 30 - 2 = 58).

Because the p-value <0.05, we reject the null hypothesis. There is sufficient evidence to suggest that there is a difference in mean IQ score between students at high schools A and B. To perform the confidence level, you can adapt the Python code in the last part of the one-sample t-test.

Recall that in some situations, if the histograms and QQ plots show some evidence of skewness, we can consider testing the medians instead of the means for hypothesis testing. As shown in *Chapter 2, Distributions of Data*, we can perform a data transformation (for example, log transformation) to obtain the normality assumption. After the transformation, the median of `log(data)` is equal to the mean of `log(data)`. This means that the test is performed on means of the transformed data.

Two-sample t-test – Welch's t-test

This is a practical two-sample t-test when the data is normally distributed but the population standard deviations are unknown and unequal. We have the same assumption for normality as with a pooled t-test but we can relax the assumption for equal variance when performing Welch's t-test. Let us consider the following example where we have two sample datasets:

```
sample1 = np.array([2,3,4,2,3,4,2,3,5,8,7,10])
sample2 = np.array([30,26,32,34,28,29,31,35,36,33,32,27])
```

We assume the independence is satisfied, but we will check the normality and equal standard deviation assumptions for these two samples, as follows:

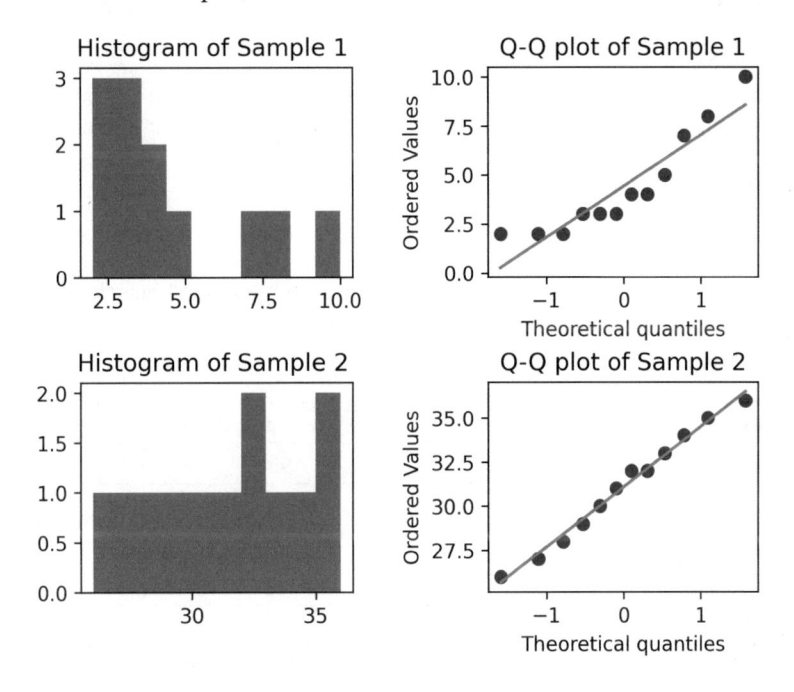

Figure 4.12 – Checking equal variance for Welch's t-test

There is strong visual evidence against equal standard deviations by looking at the *x* axis scale of the histograms in *Figure 4.12*. Let us assume the normality assumption is satisfied. In this case, a two-sample pooled t-test is not a good idea, but Welch's t-test would suffice. The null and alternative hypotheses are given here:

$$H_0: \mu_1 = \mu_2$$
$$H_a: \mu_1 \neq \mu_2$$

We specify the level of significance 0.05. To calculate the test statistic and p-value, we implement the code as follows:

```
import scipy.stats as stats
t_statistic, p_value = stats.ttest_ind(sample1, sample2,
    equal_var = False)
```

The test statistic is -22.47 and the p-value is <0.05 (the level of significance). We reject the null hypothesis. There is strong evidence to suggest the mean of sample data 1 is different from the mean of sample data 2.

Paired t-test

The paired t-test is also known as a matched pairs or dependent t-test and is used in studies when each element in a sample is tested twice (pre-test and post-test or repeated measures) and when the researcher thinks that there are some similarities, such as family. The assumptions are set out here:

- Differences are normally distributed
- Differences are independent between observations but dependent from one test to another test

The paired t-test is used in many studies, especially in medical reasoning tests related to pre- and post-treatments. Let's go back to IQ test scores—a researcher recruits a number of students to see whether there is a score difference before and after a training section, as represented in the following table:

Students	Pre-training score	Post-training score	Differences
A	95	95	0
B	98	110	12
C	90	97	7
D	115	112	-3
E	112	117	5

Figure 4.13 – Pre-training and post-training scores

In this case, we should not use an independent two-sample t-test. The mean of the differences should be tested here. We can check the assumption about normal distribution by using histogram and QQ plots, as follows:

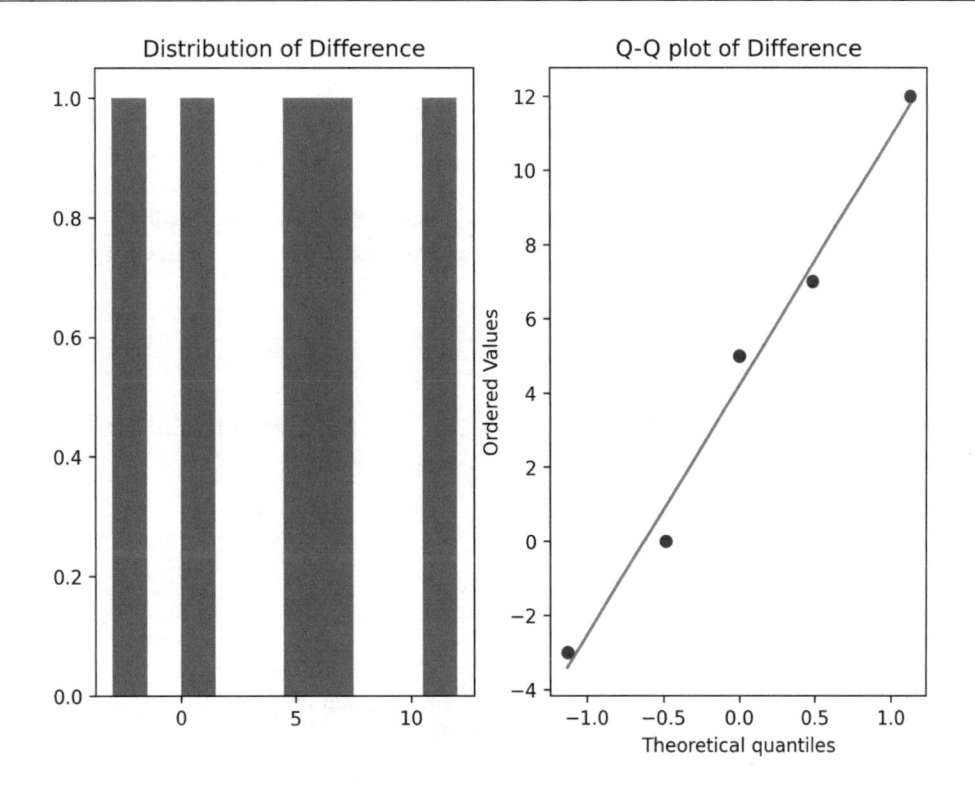

Figure 4.14 – Checking normality for the paired t-test

Evidence for the data being normally distributed is more obvious by looking at the QQ plot than the histogram.

The differences are assumed to be independent. The null and alternative hypotheses are given here:

$$H_0: \mu_{pos} - \mu_{pre} = 0$$

$$H_a: \mu_{pos} - \mu_{pre} > 0$$

d_i denotes the difference between the pre-training score and the post-training score of each student. The null and alternative hypotheses can be rewritten as follows:

$$H_0: \mu_d = 0$$

$$H_a: \mu_d > 0$$

Then, the test statistic is computed like so:

$$t = \frac{\bar{d} - \mu_d}{\frac{s_d}{\sqrt{n}}}$$

Here, $\bar{d}$ is the sample mean of the differences and s_d is the sample standard deviation of differences. In other words, a paired t-test is reduced to a one-sample t-test. However, we can use the following function in `scipy` directly:

```
stats.ttest_rel(data_pos, data_pre, alternative = {'two-sided', 'less', 'greater'})
```

The alternative hypothesis corresponds to a left-tailed, right-tailed, or two-tailed test. Here is the Python implementation for the IQ test score study example:

```
from scipy import stats
IQ_pre = [95, 98, 90, 115, 112]
IQ_pos = [95, 110, 97, 112, 117]
t_statistic, p_value = stats.ttest_rel(IQ_pos, IQ_pre, alternative = 'greater')
```

The test statistic is 1.594 and the p-value is 0.093. Therefore, given the p-value is <0.05 and the level of significance $\alpha = 0.05$, we reject the null hypothesis. There is sufficient evidence to suggest that training has a significant effect on IQ scores.

Tests with more than two groups and ANOVA

In the previous chapter and previous sections, we covered tests between two groups. In this section, we will cover two methods for testing differences between groups, as follows:

- Pairwise tests with the **Bonferroni correction**
- ANOVA

When testing for differences between more than two groups, we will have to use multiple tests, which will affect our *type I* error rate. There are several methods to control the error rate. We will see how to utilize the Bonferroni correction to control the *Type I* error rate. We will also discuss ANOVA in this section, which is used to test for a difference in means of multiple groups.

Multiple tests for significance

In the previous sections, we looked at making a comparison between two groups. In this section, we will consider how to perform tests when there are more than two groups present. Let's again consider the factory example where we have several models (model A, model B, and model C) of machines on a factory floor, and these machines are used to perform the same operation in the factory. A plausible question of interest is: *Does one machine model have a higher mean output than the other two models?* To make this determination, we would need to do three tests comparing the difference in means of each model to the other models, testing that the difference in means is different than zero. These are the null hypotheses we would need to test:

$$\mu_{output,A} - \mu_{output,B} = 0$$
$$\mu_{output,B} - \mu_{output,C} = 0$$
$$\mu_{output,A} - \mu_{output,C} = 0$$

When performing multiple tests, we will need to apply p-value corrections for our expected error rate. Recall that in *Chapter 3, Hypothesis Testing*, we defined the expected error rate for a hypothesis test as α. This is the rate at which we expect a *single* hypothesis test to result in a *Type I* error. In our example with factory machines, we are making three hypothesis tests, which means *we are three times more likely to see a Type I error*. While our example specifically considers multiple tests for differences in means, this applies to any type of hypothesis test. In these situations with multiple tests, we will generally define a **familywise error rate** (**FWER**) and apply p-value corrections to control for the FWER. The FWER is the probability of making a *Type I* error from a group of hypothesis tests. The error rate from tests within the group is the **individual error rate** (**IER**). We will define the IER and FWER as follows:

- **IER**: The expected *Type I* error rate for an individual hypothesis test
- **FWER**: The expected *Type I* error rate for a group of hypothesis tests

We will discuss one method for p-value correction in this section to provide intuition for the rationale.

The Bonferroni correction

One method for adjusting the p-value to control multiple hypothesis tests is the Bonferroni correction. The Bonferroni correction controls the FWER by uniformly reducing the significance level of each individual test in the family of tests. Given that we have m tests in a family each with the p-value p_i, then the p-value correction is given as follows:

$$p_i \leq \frac{\alpha}{m}$$

Taking our previous example of three models of machines, we have a family of three tests, making $m = 3$. If we let FWER be 0.05, then, with the Bonferroni correction, the level of significance for the three individual tests is this:

$$\frac{0.05}{3} = 0.0167$$

Thus, in this example, any of the individual tests would be required to have a p-value of 0.0167 to be considered significant.

> **Effects on Type I and Type II errors**
>
> As discussed, the Bonferroni correction reduces the significance levels of individual tests to control the *Type I* error rate at the family level. We should also consider how this change impacts the *Type II* error rate. In general, reducing the significance level of individual tests will increase the chance of making a *Type II* error for that test (as is done in the Bonferroni correction). While we have only discussed the Bonferroni correction in this section, there are other methods for p-value correction that provide different trade-offs. Check the documentation of `multipletests` in `statsmodels` to see a list of p-value corrections implemented in `statsmodels`.

Let's take a look at an example using the miles per gallon (MPG) data from the *Auto MPG* dataset from the *UCI Machine Learning Repository*, which can be found at this link: https://archive.ics.uci.edu/ml/datasets/Auto+MPG [1]. This dataset contains various attributes, including `origin`, `mpg`, `cylinders`, and `displacement`, for vehicles manufactured between 1970 and 1982. We will show an abbreviated form of the analysis here; the full analysis is included in the notebook in the code repository for this chapter.

For this example, we will use the `mpg` and `origin` variables, and test whether there is a difference in mpg from the different origins with a significance level of 0.01. The group means are shown in the following table (`origin` is an integer-encoded label in this dataset).

Vehicle origin (`origin`)	MPG (`mpg`)
1	20.0
2	27.9
3	30.5

Figure 4.15 – Vehicle MPG means for each origin group

Running a t-test to compare each mean, we get the following p-values:

Null hypothesis	Uncorrected p-value
$\mu_1 - \mu_2 = 0$	7.946116336281346e-12
$\mu_1 - \mu_3 = 0$	4.608511957238898e-19
$\mu_2 - \mu_3 = 0$	0.0420926104552266

Figure 4.16 – Uncorrected p-values for t-tests on the difference between each group

Applying the Bonferroni correction to the p-values, we get the following p-values:

Null hypothesis	Corrected p-value (Bonferroni)
$\mu_1 - \mu_2 = 0$	2.38383490e-11
$\mu_1 - \mu_3 = 0$	1.38255359e-18
$\mu_2 - \mu_3 = 0$	0.126277831

Figure 4.17 – Corrected p-values for t-tests on the difference between each group

With the preceding p-values, reject the null hypothesis for a difference in the means of groups 1 and 2 and a difference in the means of groups 1 and 3, but fail to reject the null hypothesis for a difference in the means of groups 2 and 3 at our significance level of 0.01.

> **Other p-value correction methods**
>
> In this section, we only discussed one method for p-value correction—the Bonferroni correction—to provide intuition for the rationale of p-value corrections. However, there are other correction methods available that might be better suited for your problem. To see a list of p-value correction methods implemented within `statsmodels`, check the documentation of `statsmodels.stats.multitest.multipletests`.

ANOVA

In the previous section on multiple tests for significance, we saw how to perform multiple tests to determine whether means differed between groups. When dealing with means of groups, a useful first task is to conduct an analysis of variance. ANOVA is a statistical test for determining whether there is a difference between means of several groups. The null hypothesis is there is no difference in means, and the alternative hypothesis is the means are not all equal. Since ANOVA tests for a difference in means, it is commonly used before testing for a difference in means with pairwise hypothesis tests. If the ANOVA null hypothesis fails to be rejected, then there is no need to perform the pairwise tests. However, if the ANOVA null hypothesis is rejected, then pairwise tests can be performed to determine which specific means differ.

> **ANOVA versus pairwise tests**
>
> While pairwise testing is a general procedure for testing for differences between groups, ANOVA can only be used to test for differences in means.

In this example, we will again consider the MPG of vehicles from the *Auto MPG* dataset. Since we have already run pairwise tests and found a significant difference in the mean mpg of vehicles based on origin, we expect that ANOVA will provide a positive test result (reject the null hypothesis). Performing the ANOVA calculation, we get the following output. The small p-value suggests that we should reject the null hypothesis:

```
anova = anova_oneway(data.mpg, data.origin, use_var='equal')
print(anova)
# statistic = 98.54179491075868
# pvalue = 1.915486418412936e-35
```

The ANOVA analysis shown here is abbreviated. For the full code, see the associated notebook in the code repository for this chapter.

In this section, we covered methods for performing hypothesis tests for more than two groups of data. The first method was pairwise testing with p-value correction, which is a general method that can be used for any type of hypothesis test. The other method we covered was ANOVA, which is a specific test for differences in the means of groups. This is not a general method such as pairwise testing but can be used as a first step before performing pairwise tests for differences in means. In the next section, we cover another type of parametric test that can be used to determine whether two sets of data are correlated.

Pearson's correlation coefficient

Pearson's correlation coefficient, also called Pearson's *r* (or Pearson's rho (ρ) when applied to population data) or the **Pearson product-moment sample coefficient of correlation (PPMCC)**, is a bivariate test that measures the linear correlation between two variables. The coefficient produces a value ranging from -1 to 1 where -1 is a strong, inverse correlation and 1 is a strong, direct correlation. A zero-valued coefficient indicates no correlation between the two variables. Weak correlation is generally considered to be correlation between +/- 0.1 and +/- 0.3, moderate correlation is between +/- 0.3 and +/- 0.5, and strong correlation is between +/- 0.5 to +/- 1.0.

This test is considered parametric but does not require assumptions of normal distribution or homogeneity of variance. It is, however, required that data be independently sampled (both randomly selected and without serial correlation), have finite variance—such as with a distribution that has a very heavy tail—and be of a continuous data type. The test does not indicate an input variable and a response variable; it is simply a measure of the linear relation between two variables. The test uses standardized covariance to derive correlation. Recall that standardization requires dividing a value by the standard deviation.

The equation for the population Pearson's coefficient, ρ, is shown here:

$$\rho = \frac{\sigma_{xy}}{\sigma_x \sigma_y}$$

Here, σ_{xy} is the population covariance, calculated as follows:

$$\sigma_{xy} = \frac{\sum_{i=1}^{N}(x_i - \mu_x)(y_i - \mu_y)}{N}$$

The equation for the sample Pearson's coefficient, r, is shown here:

$$r = \frac{S_{xy}}{S_x S_y}$$

Here, S_{xy} is the sample covariance, calculated as follows:

$$S_{xy} = \frac{\sum_{i=1}^{n}(x_i - \bar{x})(y_i - \bar{y})}{n-1}$$

In Python, we can perform this test using the scipy `scipy.stats.pearsonr` function. In the following code snippet, we generate two normally distributed datasets of random numbers using numpy. We want to test the hypothesis that there is a correlation between the two groups since there is some significant overlap:

```
from scipy.stats import pearsonr
import matplotlib.pyplot as plt
import scipy.stats as stats
import seaborn as sns
import pandas as pd
import numpy as np
mu1, sigma1 = 0, 1.1
normally_distributed_1 = np.random.normal(mu1, sigma1, 1000)
mu2, sigma2 = 0, 0.7
normally_distributed_2 = np.random.normal(mu2, sigma2,
    1000)
df_norm = pd.DataFrame({'Distribution':['Distribution 1' for
i in range(len(normally_distributed_1))] + ['Distribution
2' for i in range(len(normally_distributed_2))], 'X':np.
concatenate([normally_distributed_1, normally_distributed_2])})
```

In *Figure 4.18*, we can observe the overlapping variance of the two correlated distributions:

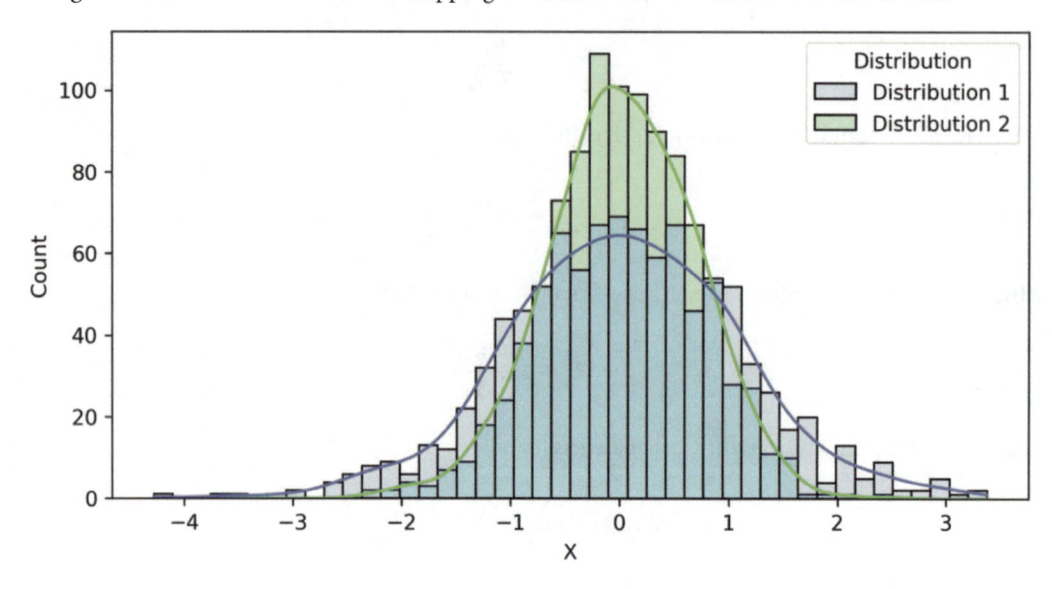

Figure 4.18 – Correlated distributions

In the plot shown in *Figure 4.18*, we can see the overlap of the populations. Now, we want to test the correlation using the `pearsonr()` function, as follows:

```
p, r = pearsonr(df_norm.loc[df_norm['Distribution'] == 
'Distribution 1', 'X'], df_norm.loc[df_norm['Distribution'] == 
'Distribution 2', 'X'])
print("p-value = %.4f"%p)
print("Correlation coefficient = %.4f"%r)
```

The following output indicates that at a 0.05 level of significance, we have a 0.9327 level of correlation (p-value is 0.0027):

p-value = 0.0027

Correlation coefficient = 0.9327

To frame the correlation differently, we could say the level of variance explained (r^2, also called **goodness-of-fit** or the **coefficient of determination**) in distribution 2 by distribution 1 is 0.9327^2 = 87%, assuming we know that distribution 2 is a response to distribution 1. Otherwise, we could simply say there is a correlation of 0.93 or an 87% level of variance explained in the relationship between the two variables.

Now, let us look at the *Motor Trend Car Road Tests* dataset from R, which we import using the `statsmodels datasets.get_rdataset` function. Here, we have the first five rows, which have the variables for miles per gallon (`mpg`), number of cylinders (`cyl`), engine displacement (`disp`), horsepower (`hp`), rear-axle gear ratio (`drat`), weight (`wt`), minimum time to drive a quarter of a mile (`qsec`), engine shape (`vs=0` for v-shaped and `vs=1` for inline), transmission (`am=0` for automatic and `am=1` for manual), number of gears (`gear`), and number of carburetors (`carb`) (if not fuel injected):

```
import statsmodels.api as sm
df_cars = sm.datasets.get_rdataset("mtcars","datasets").data
```

In Figure 4.19, we can see the first five rows of the data set, which contains data suitable for Pearson's correlation analysis.

mpg	cyl	disp	hp	drat	wt	qsec	Vs	am	gear	carb
21	6	160	110	3.9	2.62	16.46	0	1	4	4
21	6	160	110	3.9	2.875	17.02	0	1	4	4
22.8	4	108	93	3.85	2.32	18.61	1	1	4	1
21.4	6	258	110	3.08	3.215	19.44	1	0	3	1
18.7	8	360	175	3.15	3.44	17.02	0	0	3	2

Figure 4.19 – first five rows from the mtcars dataset

Using the dataset, we can plot a correlation matrix with the following code, which shows each pairwise correlation for all features in the dataset to see how they relate to one another:

```
sns.set_theme(style="white")
corr = df_cars.corr()
f, ax = plt.subplots(figsize=(15, 10))
cmap = sns.diverging_palette(250, 20, as_cmap=True)
sns.heatmap(corr, cmap=cmap, vmax=.4, center=0,
            square=True, linewidths=.5, annot=True)
```

Suppose we are curious about the variables most meaningful for quarter-mile time (`qsec`). In *Figure 4.20*, we can see by looking at the line for `qsec` that `vs` (v-shaped) is positively correlated at 0.74. Since this is a binary variable, we can assume, based on this dataset, that inline engines are faster than v-shaped engines. However, there are other covariates involved with significant correlation. For example, almost as strongly correlated with speed as engine shape is horsepower, such that as horsepower goes up, quarter-mile runtime goes down:

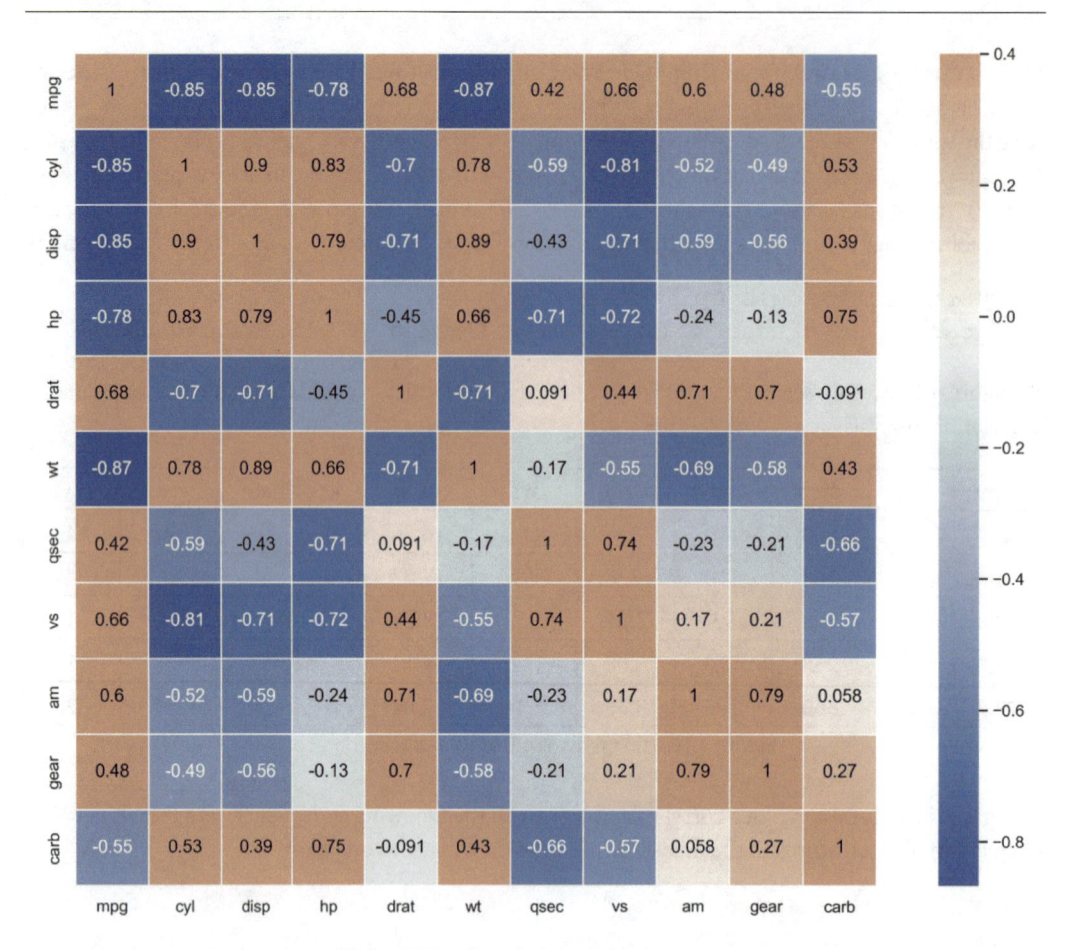

Figure 4.20 – Correlation matrix heatmap

A correlation matrix is useful for exploring the relationships between multiple variables at one time. It is also a useful tool for feature selection when building statistical and **machine learning** (**ML**) models, such as linear regression.

Power analysis examples

Power analysis is a statistical method for identifying an appropriate sample size required for a hypothesis test to have sufficient power in preventing *Type II* errors – or failing to reject the null hypothesis when the null hypothesis should be rejected. Power analysis can also be used for identifying, based on sample size, a detectable effect size (or difference) between samples tested. In other words, based on a specific sample size and distribution, a power analysis can provide the analyst with a specific minimum difference the researcher may be able to reliably identify with a given test. In this section, we will demonstrate a power analysis using a one-sample t-test.

One-sample t-test

Let's assume a manufacturer sells a type of machine capable of producing 100,000 units per month with a standard deviation of 2,800 units. A company has bought a number of these machines and has found them to only be producing 90,000 units. The company wants to know how many of the machines are needed to determine with a high level of confidence the machines are not capable of producing 100,000 units. The following power analysis indicates that for a t-test, a sample of three machines is required to prevent, with an 85% probability, failing to identify a statistically significant difference in actual versus marketed machine performance when there is one:

```
from statsmodels.stats.power import TTestPower
import numpy as np
# Difference of distribution mean and the value to be assessed 
divided by the distribution standard deviation
effect_size = abs(100000-90000) / 2800
powersTT = TTestPower()
result = powersTT.solve_power(effect_size, nobs=3, alpha=0.05, 
alternative='two-sided')
print('Power based on sample size:{}'.format(round(result,2)))
# Power based on sample size: 0.85
```

Additional power analysis example

To see additional examples of power analysis in Python, please refer to this book's GitHub repository. There, we have examples for additional t-tests and F-tests, which focus on analyzing variance between sample groups.

Summary

This chapter covered topics of parametric tests. Starting with the assumptions of parametric tests, we identified and applied methods for testing the violation of these assumptions and discussed scenarios where robustness can be assumed when the required assumptions are not met. We then looked at one of the most popular alternatives to the z-test, the t-test. We iterated through multiple applications of this test, covering one-sample and two-sample versions of this test using pooling, pairing, and Welch's non-pooled version of the two-sample analysis. Next, we explored ANOVA techniques, where we looked at using data from multiple groups to identify statistically significant differences between them. This included one of the most popular adjustments to the p-value for when a high volume of groups is present—the Bonferroni correction, which helps prevent inflating the *Type I* error when performing multiple tests. We then looked at performing correlation analysis on continuous data using Pearson's correlation coefficient and how to visualize correlation using a correlation matrix and accompanying heatmap. Finally, we briefly overviewed power analysis, with an example of performing this with a one-sample t-test. In the next chapter, we will discuss non-parametric hypothesis testing, including new tests in addition to those that pair with the parametric tests in this chapter for when assumptions cannot be safely assumed.

References

[1] *Dua, D.* and *Graff, C. (2019). UCI Machine Learning Repository* [`http://archive.ics.uci.edu/ml`]. *Irvine, CA: University of California, School of Information and Computer Science.*

5
Non-Parametric Tests

In the previous chapter, we discussed parametric tests. Parametric tests are useful when test assumptions are met. However, there are cases where those assumptions are not met. In this chapter, we will discuss several non-parametric alternatives to the parametric tests presented in the previous chapter. We start by introducing the concept of a non-parametric test. Then, we will discuss several non-parametric tests that can be used when t-test or z-test assumptions are not met.

In this chapter, we're going to cover the following main topics:

- When parametric test assumptions are violated
- The rank-sum test
- The signed-rank test
- The Kruskal-Wallis test
- The chi-square test
- Spearman's correlation analysis
- Chi-square power analysis

When parametric test assumptions are violated

In the previous chapter, we discussed parametric tests. Parametric tests have strong statistical power but also require adherence to strong assumptions. When the assumptions are not satisfied, the test results are not valid. Fortunately, we have alternative tests that can be used when the assumptions of a parametric test are not satisfied. These tests are called **non-parametric** tests, meaning that they make *no assumptions about the underlying distribution of the data*. While non-parametric tests do not require distributional assumptions, these *tests will still require the samples to be independent*.

Permutation tests

For the first non-parametric test, let's look more deeply at the definition of a p-value. A p-value is the *probability of obtaining a test statistic at least as extreme as the observed value* under the assumption of the null hypothesis. Then, to calculate a p-value, we need the null distribution and an observed statistic. The p-value is the proportion of samples with a test statistic more extreme than the observed statistic. It turns out that we can construct the null distribution using permutations from data. Let's see how to construct the null distribution using the following dataset. This dataset could represent counts of machine failures at low and high temperatures. We assume that the samples are independent:

```
low_temp = np.array([0, 0, 0, 0, 0, 1, 1])
high_temp = np.array([1, 2, 3, 1])
```

To construct the null hypothesis, we first need to decide on a statistical measure. In this case, we will look for a difference in the mean of the two distributions. So, our statistical measure will be as follows:

$$\bar{x}_{lowtemp} - \bar{x}_{hightemp}$$

Now, to calculate the distribution values, calculate the statistical measure for all permutations of the dataset. Here are a couple of examples of the permutations of the dataset.

Label	Observed	P1	P2	P3	P4	P5	...
low	0	1	0	1	0	1	...
low	0	1	3	0	3	0	...
low	0	0	0	1	1	3	...
low	0	1	1	0	2	0	...
low	0	0	1	2	0	1	...
low	1	1	1	3	0	1	...
low	1	0	1	1	1	0	...
high	1	0	2	0	1	2	...
high	2	3	0	0	0	1	...
high	3	0	0	1	1	0	...
high	1	2	0	0	0	0	...
mean difference	-1.46	-0.68	0.5	0.89	0.5	0.11	...

Figure 5.1 – First five permutations of the observed data and the mean difference

The table shows the observed data with five randomly generated permutations of the values. We calculate the difference in the mean for each permutation. The differences in means are the values of the null distribution. Once we have the distribution, we can calculate the p-value as the proportion of values more extreme than the observed value.

> **Scaling of permutation calculations**
>
> In general, permutation tests can be expensive to calculate because the number of permutations grows quickly with the size of the distribution. For example, dataset sizes of 3, 5, and 7 samples correspond to permutation sizes of 6, 120, and 5,040. The dataset shown here has more than 39 million permutations! The runtime performance of a permutation test on a large dataset will likely be slow due to the number of permutations necessary to compute the null distribution.

We can perform a permutation test in Python using the `permutation_test` function from `scipy`. This function calculates the test statistic, the null distribution, and the p-value:

```
def statistic_function(set_one, set_two):
    return np.mean(set_one) - np.mean(set_two)

random_gen=42

perm_result = sp.stats.permutation_test(
    (low_temp, high_temp) ,
    statistic_function, random_state=random_gen,

)
```

This function produces the following distribution from the dataset.

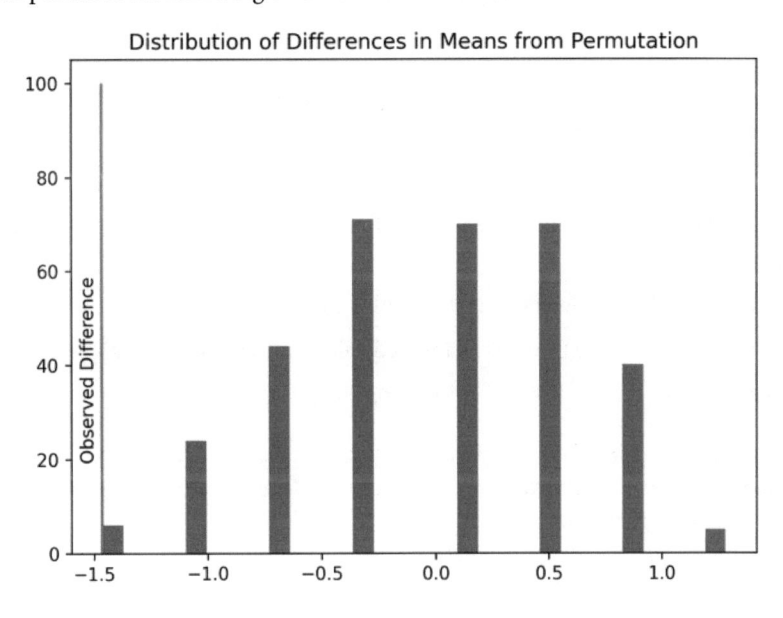

Figure 5.2 – Null distribution from the permutation test

The null distribution from the permutation test and the observed difference in the mean of the two groups is shown in *Figure 5.2*. The p-value from the permutation test is 0.036. The permutation test is the first of several non-parametric tests that will be covered in this section. Again, these types of tests are useful when the assumptions for parametric tests are not met. *However, if a parametric test can be used, it should be used*; non-parametric tests should not be used as default methods.

In this section, we introduced non-parametric tests with permutation tests. The permutation test is a widely applicable non-parametric test, but the computations for permutations grow quickly with the size of the dataset, which may make its use impractical in some situations. In the following sections, we will cover several other non-parametric tests that do not require computing a null distribution.

The Rank-Sum test

When the assumptions of the t-test are not met, the Rank-Sum test is often a good non-parametric alternative test. While the t-test can be used to test for the *difference between the means of two distributions*, the Rank-Sum test is used to test for the *difference between the locations of two distributions*. This difference in the test utility is due to the lack of parametric assumptions in the Rank-Sum test. The null hypothesis of the Rank-Sum test is that the distribution underlying the first sample is the same as the second sample. If the sample distributions appear to be similar, this allows us to use the Rank-Sum test to test for the difference in the locations of the two samples. As stated, the Rank-Sum test cannot specifically be used for testing the difference between means because it does not require assumptions about the sample distributions.

The test statistic procedure

The test procedure is straightforward. The process is outlined here and an example is shown in the following table:

1. Combine all sample values into one set and sort the samples in ascending order, keeping track of their labels.
2. Assign ranks to all samples starting with rank 1 for the lowest sample value.
3. Where ties occur, replace the rank of the tied values with the mean rank of the tied values.
4. Sum the ranks for the smallest sample group, which is the test statistic T.

Once the test statistic is calculated, the p-value can be calculated. The p-value can be done with a normal approximation or with an exact method. Generally, the exact method is only used when the sample size is small (less than 8 samples).

Normal approximation

Once the *T* statistic is calculated for the Rank-Sum test, we can determine a p-value with an exact method or with a normal approximation. We will cover the approximation method here as the exact method requires a permutation test, which will require the use of software. We approximate the p-value with a z-score (recall the z-score from *Chapter 3, Hypothesis Testing*):

$$Z = \frac{T - Mean(T)}{STDEV(T)}$$

where

$$Mean(T) = n_T \bar{R}$$

and

$$STDEV(T) = s_R \sqrt{\frac{n_T n_O}{n_T + n_O}}$$

In the equations here, n_T is the number of samples in the group used to calculate *T*, n_O is the number of samples in the other group, $\bar{R}$ is the mean of the corrected ranks, and s_R is the standard deviation of the corrected ranks. Once *Z* is calculated, the corresponding p-value can be looked up for the z-distribution. Having described the method, let's look at an example. The data shown in the following table can be found in the accompanying Jupyter notebook.

Rank-Sum example

The following table shows this process performed on a set of data for two groups labeled with "L" and "H," where a Rank-Sum test will be used to test for the difference in location of the two sample distributions. The test statistic for this table is 42.5, which is the sum of the corrected ranks of group L. This test statistic corresponds to an approximate p-value of 0.00194 (a two-sided test). The data for this table was downloaded from `https://github.com/OpenIntroStat/openintro/raw/master/data/gpa_iq.rda`.

IQ	Group	Rank	Corrected Rank
77	L	1	1
79	L	2	2
93	L	3	3
96	L	4	4
104	L	5	5
105	H	6	6
106	H	7	7
107	L	8	8
109	L	9	9

IQ	Group	Rank	Corrected Rank
111	H	10	10.5
111	L	11	10.5
112	H	12	12
116	H	13	13
118	H	14	14
124	H	15	15
126	H	16	16
127	H	17	17
128	H	18	18.5
128	H	19	18.5

Figure 5.3 – Corrected ranks for Rank-Sum test

We can also perform this test with software using `mannwhitenyu` from `scipy`. For the following code sample, the values corresponding to L and H from the preceding table are contained in the `lower_score_iqs` and `higher_score_iqs` variables, respectively:

```
mannwhitneyu(higher_score_iqs, lower_score_iqs).pvalue
# 0.00222925216588146
```

In this section, we discussed the Rank-Sum test, which is a non-parametric alternative to the t-test. The Rank-Sum test is used to test for a difference in the locations of two distributions of sample data. In the next section, we will look at a similar rank-based test, which is used to compare paired data like the paired t-test from *Chapter 4, Parametric Tests*.

The Signed-Rank test

The Wilcoxon Signed-Rank test is a non-parametric alternative version of the paired t-test that is used when the assumption of normality is violated. This test is robust to outliers because of the use of ranks and medians instead of means in the null and alternative hypotheses. As indicated by the name of the test, it uses the magnitudes of differences between two stages and their signs.

In research, a null hypothesis considers that the median difference between stage 1 and stage 2 is zero. Similarly, as in a paired t-test, for the alternative hypothesis, for a two-tailed test, the median difference between Stage 1 and Stage 2 is considered not to be zero, or for a one-tailed test, the median difference between Stage 1 and Stage 2 is greater (or less) than zero.

Though the normality requirement is relaxed, the test requires independence between paired observations and these observations to be from the same population. In addition, the dependent variable is required to be continuous.

The Signed-Rank test

To compute the test statistic, there are the following procedures:

1. Calculate the differences in each pair between the two stages.
2. Drop pairs with zero difference, if they exist.
3. Take the absolute difference between each pair and rank them from smallest to largest.
4. Calculate the signed-rank statistic S by summing the ranks with a positive sign.

Let us consider the simple example that follows. We consider generic data for two stages (before and after the treatment) of nine samples.

Pair	Before the treatment	After the treatment	Absolute difference	Sign	Rank
1	37	38	1	-	1
2	14	17	3	-	2.5
3	22	19	3	+	2.5
4	12	7	5	+	4
5	24	15	9	+	5
6	35	25	10	+	6
7	35	24	11	+	7
8	51	38	13	+	8
9	39	19	20	+	9

Figure 5.4 – Demonstration of rank calculation for sign-rank test

By observing the preceding table, we can see the second and the third pairs have the same absolute difference. Therefore, their ranks are computed by averaging their original rank, (2+3)/2 = 2.5. The null and alternative hypotheses for a one-sided test are as follows:

H_0: The median difference between before and after treatment is zero

H_a: The median difference between before treatment and after treatment is positive

The signed-rank statistic is the sum of ranks for positive differences and it is given as follows:

$$S = 2.5 + 4 + 5 + 6 + 7 + 8 + 9 = 41.5.$$

The mean of S is:

$$Mean(S) = \frac{n(n+1)}{4} = \frac{9 \cdot 10}{4} = 22.5$$

And the standard deviation of S is:

$$SD(S) = \sqrt{\frac{n(n+1)(2n+1)}{24}} = \sqrt{\frac{9*10*19}{24}} = 8.44.$$

Then, the test statistic is calculated by the following formula:

$$Z_{statistic} = \frac{S - Mean(S)}{SD(S)} = \frac{41.5 - 22.5}{8.44} = 2.2511.$$

Referring to *Chapter 3, Hypothesis Testing*, we could use `scipy.stats.norm.sf()` to calculate the approximate one-sided `p-value` `0.012` using the $Z_{statistic}$.

At $\alpha = 0.05$ – level of significance – with `p-value` $< \alpha$, we reject the null hypothesis. There is strong evidence that the median before the treatment is greater than the median after the treatment. In Python, it is simple to implement the test as follows:

```
import scipy.stats as stats
import numpy as np

before_treatment = np.array([37, 14, 22, 12, 24, 35, 35, 51, 39])
after_treatment = np.array([38,17, 19, 7, 15, 25, 24, 38, 19])

# Signed Rank Test
stats.wilcoxon(before_treatment, after_treatment, alternative = 'greater')
```

The documentation for this test can be found at the following link:

https://docs.scipy.org/doc/scipy/reference/generated/scipy.stats.wilcoxon.html

The Kruskal-Wallis test

Another non-parametric test we will now discuss is the Kruskal-Wallis test. It is an alternative to the one-way ANOVA test when the normality assumption is not satisfied. It uses the medians instead of the means to test whether there are statistically significant differences between two or more independent groups. Let us consider a generic example of three independent groups:

```
group1 = [8, 13, 13, 15, 12, 10, 6, 15, 13, 9]
group2 = [16, 17, 14, 14, 15, 12, 9, 12, 11, 9]
group3 = [7, 8, 9, 9, 4, 15, 13, 9, 11, 9]
```

The null and alternative hypotheses are stated as follows.

H_0 : The medians are equal among these three groups

H_a : The medians are not equal among these three groups

In Python, it is easy to implement by using the `scipy.stats.kruskal` function. The documentation can be found at the following link:

https://docs.scipy.org/doc/scipy/reference/generated/scipy.stats.kruskal.html

```
from scipy import stats
group1 = [8, 13, 13, 15, 12, 10, 6, 15, 13, 9]
group2 = [16, 17, 14, 14, 15, 12, 9, 12, 11, 9]
group3 = [7, 8, 9, 9, 4, 15, 13, 9, 11, 9]
#Kruskal-Wallis Test
stats.kruskal(group1, group2, group3)
```

The output of the preceding code is as follows:

`KruskalResult(statistic=5.7342701722574905, pvalue=0.056861597028239855)`

As α = 0.05 – the level of significance – with `p-value` > α, we fail to reject the null hypothesis. There is no strong evidence to show that the medians are not equal across these three groups.

Chi-square distribution

Researchers are often faced with the need to test hypotheses on categorical data. The parametric tests covered in *Chapter 4, Parametric Tests*, are often not very helpful for this type of analysis. In the last chapter, we discussed using an F-test to compare sample variances. Extending that concept, we can consider the non-parametric and non-symmetric chi-square probability distribution, which is a distribution useful for comparing the means of sampling distribution variances to their population variances, specifically when the mean of a sampling distribution of sample variances is expected to equal the population variance under the null hypothesis. Because variance cannot be negative, the distribution starts at an origin of 0. Here, we can see the **chi-square distribution**:

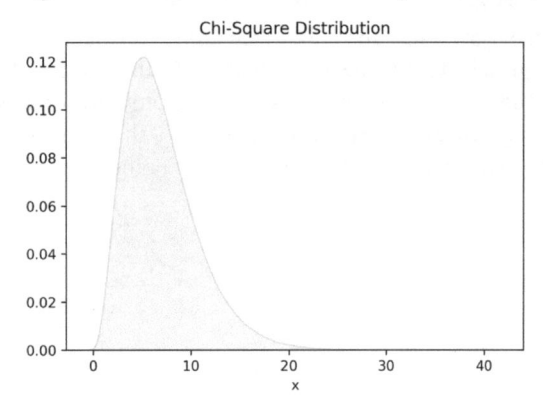

Figure 5.5 – Chi-square distribution with seven degrees of freedom

The shape of the chi-square distribution does not represent an assumption that percentiles are fixed to standard deviations, as with standard normal distribution; it is expected to change with each additional sample variance calculated. When the original population data from which the sampling distribution of variances is calculated can be assumed to be normally distributed, the chi-square standardized test statistic is calculated as:

$$\chi^2 = \frac{(n-1)s^2}{\sigma^2}$$

Where (n-1) is used to calculate the degrees of freedom, which are used to explain errors in sampling when building the test statistic. The critical values are based on table lookups using degrees of freedom and desired levels of significance. The null hypothesis is always $H_0: \sigma_x^2 = \sigma_0^2$, where σ_x^2 is the observed distribution variance and σ_0^2 is the expected variance. The alternative hypothesis for a one-tailed test is $H_a: \sigma_x^2 \geq \sigma_0^2$ when testing the right tail and $H_a: \sigma_x^2 \leq \sigma_0^2$ when testing the left tail. The alternative hypothesis for a two-tailed test is $H_a: \sigma_x^2 \neq \sigma_0^2$. This test, when the population data can be assumed to be normally distributed, is similar to the F-test in that the two variances are being compared and the result is 1 when both are the same. The difference is that the chi-square test statistic factors in the degrees of freedom and is therefore less sensitive to differences.

Understanding how this test is useful for comparing variances is helpful for understanding how the chi-square distribution also relates to frequencies of occurrences of categorical data. In the preceding chi-square statistic, we can relate the sample statistic for variance as an observed value whereas the population parameter for variance is what we expect to occur. When determining whether there is statistical significance in the occurrence of categorical factor levels, we can *compare the observed occurrences to expected occurrences*. We will illustrate examples for arguably the two most widely used versions of this test, the **chi-square goodness-of-fit test** and the **chi-square test of independence**, in the following sections. These tests are considered non-parametric as they do not require the assumptions stated in the first section of *Chapter 4, Parametric Tests*.

> **Tailedness of chi-square goodness-of-fit and independence tests**
>
> The chi-square goodness-of-fit test and chi-square test of independence in the following two sections are always right-tailed tests. The null hypothesis in these tests states a difference of zero between the observed and expected frequencies. The alternative hypothesis is that the observed and expected frequencies are not the same. The closer to 0 the χ^2 test statistic, the more likely the observed and expected frequencies approximate each other.

Chi-square goodness-of-fit

The **chi-square goodness-of-fit test** compares the count of occurrences of multiple factor levels for a single variable (factor) to determine whether the levels are statistically equal. For example, a vendor offers three models of phones – three levels (brands) of the single factor (phone) – to customers, who purchase in total an average of 90 phones per week. We can say the expected frequency is 1/3 – so, 30 phones of each model are sold per week, on average. Pearson's chi-square test statistic, which is calculated by measuring the observed frequencies against expected frequencies, is the test statistic used for the chi-square goodness-of-fit test. The linear equation for this test statistic is as follows:

$$\chi^2 = \sum \frac{(O_i - E_i)^2}{E_i}, \text{degrees of freedom} = k-1$$

Where O_i is the observed frequency, E_i is the expected frequency, and k is the number of factor levels. Using our phone example, we learn from the vendor the expected frequency of 1/3 is not actually what is observed; the vendor sells an average of 45, 30, and 15 phones of models A, B, and C, respectively. Suppose we want to know whether this observed frequency differs with statistical significance from the expected frequency. The null hypothesis is that the frequencies are equal. We formulate the **Pearson's chi-square test statistic** as follows:

$$\chi^2 = \frac{(45-30)^2}{30} + \frac{(30-30)^2}{30} + \frac{(15-30)^2}{30} = 15$$

Suppose we want to use a hypothesis test (right-tailed) with a 0.05 level of significance, thus forming the alternative hypothesis that the observed and expected frequencies are not equal. The chi-square critical value table shows that for a level of significance of 0.05 with *df*:3 – 1 = 2 degrees of freedom, the critical value is 5.9915. Because 15 > 5.9915, we may conclude to reject the null hypothesis based on the critical value test.

To perform this operation in Python, we can use the `statsmodels.stats.gof` module's `chisquare` function. Each group volume needs to be passed into a list or array, both for the observed frequencies and the expected frequencies. The `statsmodels chisquare` test will automatically calculate the k-1 degrees of freedom based on the counts of values in the observed (`f_obs`) list, but the degrees of freedom must be provided for the `scipy chi2.ppf` function, which provides the critical value, bypassing the need for a manual table lookup:

```
from statsmodels.stats.gof import chisquare
from scipy.stats import chi2

chi_square_stat, p_value = chisquare(f_obs=[45, 30, 15],
    f_exp=[30, 30, 30])
chi_square_critical_value = chi2.ppf(1-.05, df=2)
print('Chi-Square Test Statistic: %.4f'%chi_square_stat)
print('Chi-Square Critical Value: %.4f'%chi_square_critical_value)
print('P-Value: %.4f'%p_value)
```

As noted previously, since the p-value is below the 0.05 level of significance and the test statistic is larger than the critical value, we can assume we have a reasonable amount of evidence to reject the null hypothesis and conclude we have statistical significance indicating the phone models are not purchased in equal quantities:

```
Chi-Square Test Statistic: 15.0000

Chi-Square Critical Value: 5.9915

P-Value: 0.0006
```

We could then rerun the test for two frequencies at that point, comparing 45 to 30 or 45 to 15, for example, with expected frequencies of 37.5 and 37.5 or 30 and 30, respectively, to identify which phone may be the largest influencer on sales imbalance.

Chi-square test of independence

Suppose we have a dataset of observed vehicle crashes in the state of Texas in 2021, *Restraint Use by Injury Severity and Seat Position* (https://www.txdot.gov/data-maps/crash-reports-records/motor-vehicle-crash-statistics.html), and want to know whether using a seat belt resulted in a statistically significant difference in fatalities. We have the table of observed values as follows:

	Restrained	Unrestrained	Total
Fatal	1,429	1,235	2,664
Not Fatal	1,216,934	22,663	1,239,597
Total	1,218,363	23,898	1,242,261

Figure 5.6 – Chi-square test of independence "observed" table

Let us now create a table of expected values using this equation:

$$E_{ij} = \frac{T_i T_j}{N}$$

Where T_i is the total in the *i*th row, T_j is the total in the *j*th column, and N is the total number of observations. This yields the following:

	Restrained	Unrestrained	Total
Fatal	(2,664 *1,218,363)/1,242,261 = 2,612.75	(2,664 *23,898)/1,242,261 = 51.25	2,664
Not Fatal	(1,216,934* 1,239,597) / 1,242,261 = 1,214,324.31	(1,239,597 * 23,898) / 1,242,261 = 23,846.75	1,239,597
Total	1,218,363	23,898	1,242,261

Figure 5.7 – Chi-square test of independence "expected" table

The modified version of the Pearson's chi-square test statistic we used in the goodness-of-fit test that extends to the chi-square test of independence follows:

$$\chi^2 = \sum \frac{(O_{ij} - E_{ij})^2}{E_{ij}}, \text{degrees of freedom} = (r-1)(c-1)$$

The subscript *j* corresponds to columnar data and the subscript *i* corresponds to the row data for the observed and expected values, denoted as *O* and *E*, respectively. The values *r* and *c* in the degrees of freedom calculation correspond to the number of rows and the number of columns in the table (this is 2x2, so the degrees of freedom will equal (2-1)(2-1)=1.

Using the values from the observed and expected tables, we find the chi-square test statistic to equal the following:

$$\chi^2 = \frac{(1{,}429 - 2{,}612.75)^2}{2{,}612.75} + \frac{(1{,}216{,}934 - 1{,}214{,}324.31)^2}{1{,}214{,}324.31} + \frac{(1{,}235 - 51.25)^2}{51.25} +$$

$$\frac{(22{,}663 - 23{,}846.75)^2}{23{,}846.75} = 27{,}942.43$$

The table lookup test statistic using the one degree of freedom calculated here is 3.84. We can reasonably conclude that since the test statistic of 27,942.43 is greater than the critical value of 3.84, we can conclude that in the state of Texas in 2021, there was a statistically significant difference in the rate of fatalities in vehicle crashes where safety restraints were used compared to where they were not used.

> **Chi-square contingency tables**
>
> The test we performed uses two 2x2 tables called contingency tables. The chi-square test of independence is frequently referred to as the **chi-squared contingency test**. The failure to reject the null hypothesis is **contingent** upon the observed table's values matching the expected table's values within a level of statistical confidence. However, the chi-square test of independence can be extended to **tables of any combination of rows and columns**. However, as the rows and columns increase in value, the tables may be less useful to interpret.

To perform this test in Python, we can use the `chi2_contingency` function from the `scipy stats` module. Inputting only the observed frequencies as a 2x2 numpy array, the `chi2_contingency` test provides the expected frequencies should the null hypothesis be true– the p-value, degrees of freedom used, and the chi-square test statistic:

```
from scipy.stats import chi2_contingency
from scipy.stats import chi2
import numpy as np

observed_frequencies = np.array([[1429, 1235], [1216934, 22663]])
chi_Square_test_statistic, p_value, degrees_of_freedom, expected_frequencies = chi2_contingency(observed_frequencies)
chi_square_critical_value = chi2.ppf(1-.05, df=degrees_of_freedom)
```

```
print('Chi-Square Test Statistic: %.4f'%chi_Square_test_statistic)
print('Chi-Square Critical Value: %.4f'%chi_square_critical_value)
print('P-Value: %.4f'%p_value)
```

Here, we see the chi-square test statistic is much larger than the chi-square critical value at the 0.05 level of significance and the p-value is significant at p < 0.0000. Therefore, we can conclude there is strong statistical evidence to suggest a large difference in car crash fatalities when using a safety restraint compared to not using one:

Chi-Square Test Statistic: 27915.1221

Chi-Square Critical Value: 3.8415

P-Value: 0.0000

> ### Yates' continuity correction
> An additional argument can optionally be added to the `chi2_contingency` test with the `correction` = argument and a bool (`True` or `False`) input. This applies Yates' continuity correction, which, per the `scipy.stats` documentation, adjusts each observed value by 0.5 toward the corresponding expected value. The purpose of the adjustment is to avoid incorrectly detecting the presence of statistical significance due to small expected sample sizes. Subjectively, this is useful for less than 10 samples in an expected frequency cell. However, many practitioners and researchers argue against its use.

Chi-square goodness-of-fit test power analysis

Let's use an example where a phone vendor sells four popular models of phones, models A, B, C, and D. We want to determine how many samples are required to produce a power of 0.8 so we can understand whether there is a statistically significant difference between the popularity of different phones so the vendor can more properly invest in phone acquisitions. In this case, the null hypothesis asserts that 25% of phones from each model were sold. In reality, 20% of phones sold were model A, 30% were model B, 19% were model C, and 31% were model D phones.

Testing different values for the `nobs` argument (number of observations), we find that a minimum of 224 samples produces a power just greater than 0.801. Adding more samples will only improve this. If the true distribution were more divergent from the hypothesized 25% even split, fewer samples would be required. However, since the splits are relatively close to 25%, a high volume of samples is needed:

```
from statsmodels.stats.power import GofChisquarePower
from statsmodels.stats.gof import chisquare_effectsize

# probs0 asserts 25% of each brand are sold
# In reality, 12% of Brand A, 25% of Brand B, 33% sold were Brand C,
and 1% were Brand D.
```

```
effect_size = chisquare_effectsize(probs0=[25, 25, 25, 25],
probs1=[20, 30, 19, 31], cohen=True)
alpha = 0.05
n_bins=4 # 4 brands of phones

analysis = GofChisquarePower()
result = analysis.solve_power(effect_size, nobs=224, alpha=alpha, n_
bins=n_bins)
print('Sample Size Required in Sample 1: {:.3f}'.format(
    result))

# Sample Size Required in Sample 1: 0.801
```

Spearman's rank correlation coefficient

In *Chapter 4, Parametric Tests*, we looked at the parametric correlation coefficient, Pearson's correlation, where the coefficient is calculated from independently sampled, continuous data. However, **when we have ranked, ordinal data**, such as that from a satisfaction survey, we would not want to use Pearson's correlation as it cannot be assumed to guarantee the preservation of order. As with Pearson's correlation coefficient, **Spearman's correlation coefficient** results in a coefficient, r, that ranges from -1 to 1, with -1 being a strong inverse correlation and 1 being a strong direct correlation. Spearman's is derived by dividing the covariance of the two variables' ranks by the product of their standard deviations. The equation for the correlation coefficient, r, is as follows:

$$r_s = \frac{S_{xy}}{\sqrt{S_{xx} S_{yy}}}$$

Where

$$S_{xy} = \sum(x_i - \bar{x})(y_i - \bar{y})$$

$$S_{xx} = \sum(x_i - \bar{x})^2$$

$$S_{yy} = \sum(y_i - \bar{y})^2$$

The preceding correlation equation is safe to use under all ranking scenarios. However, when there are no ties or a minimal proportion of ties compared to the overall sample volume, the following equation may be used:

$$r_s = \frac{6\sum d_i^2}{n(n^2 - 1)}, \text{ where } d_i = (x_i - y_i)$$

However, we will perform this test in Python. Therefore, the more complex and less error-prone formula will be applied regardless since computational power enables us to bypass the manual process entirely.

Suppose we have students being judged in a competition by two judges and there is a concern that one of the judges may be biased toward some of the participants based on confounding factors, such as family ties, rather than performance alone. We decide to run a correlation analysis on the scores to test the hypothesis the two judges scored similarly for each contestant:

```
from scipy.stats import spearmanr
import pandas as pd
df_scores = pd.DataFrame({'Judge A':[1, 3, 5, 7, 8, 3, 9],
                          'Judge B':[2, 5, 3, 9, 6, 1, 7]})
```

We have the following table of contestants:

	Student 1	Student 2	Student 3	Student 4	Student 5	Student 6	Student 7
Judge A	1	3	5	7	8	3	9
Judge B	2	5	3	9	6	1	7

Figure 5.8 – Judge rankings of student contestants for Spearman's correlation analysis

```
correlation, p_value = spearmanr(df_scores['Judge A'],
    df_scores['Judge B'])
print('Spearman Correlation Coefficient: %.4f'%correlation)
print('P-Value: %.4f'%p_value)
```

Based on the p-value of 0.04 here, which is less than a significance level of 0.05, we can say the correlation produced is significant beyond a degree of random chance. Had the p-value been greater than 0.05, we could say with a 95% level of confidence that the correlation may have been purely spurious and may not have enough statistical evidence to support determinism. Spearman's correlation (r) is 0.7748. r_squared is therefore approximately 0.6003, meaning approximately 60.3% of the variation in the score is determined by the judge's scoring:

```
Spearman Correlation Coefficient: 0.7748

P-Value: 0.0408
```

Since the p-value is significant and the correlation coefficient is 0.77 – and a strong correlation coefficient starts at approximately 0.7 – we may conclude that the judges' scores are directly correlated enough to assume there is no bias in scoring present, assuming a relatively objective method for ranking exists; something more subjective may not be as suitable for correlation analysis.

Summary

In this chapter, we discussed some of the most commonly used non-parametric hypothesis tests performed when required assumptions for parametric hypothesis testing cannot be prudently guaranteed. We discussed two-sample Wilcoxon Rank-Sum – also called Mann-Whitney U – tests to draw inferences from medians when two-sample t-testing cannot be performed. Next, we walked through the Wilcoxon Sign-Rank test's paired comparison of medians when a paired t-test comparison of means cannot be performed. After, we looked at the non-parametric chi-square goodness-of-fit test and the chi-square Test of independence for comparing observed frequencies against expected frequencies, both useful for identifying the presence of statistically significant differences in counts of categorical data. Additionally, we discussed the Kruskal-Wallis test, a non-parametric alternative to the analysis of variance (ANOVA). Finally, we discussed Spearman's correlation coefficient and how to derive correlation based on rank when parametric assumptions cannot safely support using Pearson's correlation. Closing out the chapter, we provided an example of using power analysis for the chi-square test.

In the next chapter, we will begin our discussion of predictive analytics, starting with simple linear regression. There, we will discuss methods for fitting a model, interpreting linear coefficients, understanding the required assumptions of linear regression, assessing model performance, and considering modified versions of linear regression.

Part 2: Regression Models

In this part, we discuss the types of problems that can be solved with regression, coefficients of correlation and determination, multivariate modeling, model selection and variable adjustment with regularization.

It includes the following chapters:

- *Chapter 6, Simple Linear Regression*
- *Chapter 7, Multiple Linear Regression*

6
Simple Linear Regression

In previous chapters, we worked with distributions of single variables. Now we will discuss the relationships between variables. In this chapter and the next chapter, we will investigate the relationship between two or more variables using linear regression. In this chapter, we will discuss simple linear regression within the framework of **Ordinary Least Squares** (**OLS**) regression. Simple linear regression is a very useful tool for estimating continuous values from two linearly related variables. We will provide an overview of the intuitions and calculations behind regression errors. Next, we will provide an overview of the pertinent assumptions of linear regression. After that, we will analyze the output summary of OLS in `statsmodels`, and finally, we will address the scenarios of serial correlation and model validation. As highlighted, our main topics in this chapter follow this framework:

- Simple linear regression using OLS
- Coefficients of correlation and determination
- Required model assumptions
- Test for significance
- Handling model errors
- Validating models

Simple linear regression using OLS

We will study one of the simplest machine learning models – simple linear regression. We will provide its overview within the context of OLS, where the objective is to minimize the sum of the square of errors. It is a straightforward concept related to a dependent variable (quantitative response) y and its independent variable x, where their relationship can be drawn as a straight line, approximately. Mathematically, a simple linear regression model can be written in the following form:

$$y = \beta_0 + \beta_1 x + \epsilon$$

Here, β_0 is the intercept term and β_1 is the slope of the linear model. The error term is denoted as ϵ in the preceding linear model. We can see that in an ideal case where the error term is zero, β_0 represents the value of the dependent variable y at $x = 0$. Within the range of the independent variable x, β_1 represents the increase in the outcome y corresponding to a unit change in x. In literature, the independent variable x can be called the explanatory variable, predictor, input, or feature and the dependent variable y can be called the response variable, output, or target. The question is raised, "If we have a sample from a dataset (x_i, y_i) with $i = 1, 2, \ldots, n$ points), how do we determine a line of best fit using the intercept term and the slope?" By estimating these terms, we will fit the best line through the data to show the linear relationship between the independent variable and the dependent variable.

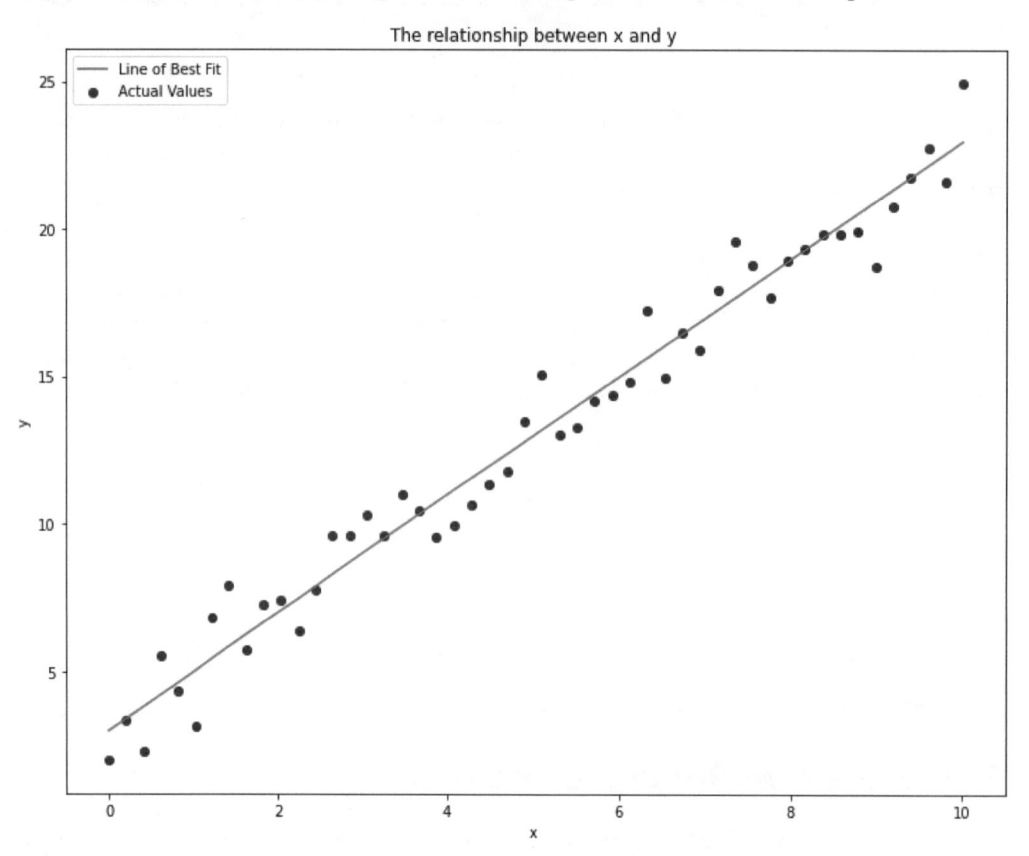

Figure 6.1 – The relationship between x and y – Line of best fit

In the preceding plot example, the blue points are actual values representing the relationship between the independent variable x and the dependent variable y. The red line is the line of best fit through all these data points. Our objective now is to formulate the predicted line (the line of best fit):

$$\hat{y} = \widehat{\beta}_0 + \widehat{\beta}_1 x$$

Take a closer look at the relationship between the predicted values and the true values. In order to find the line of best fit, we will minimize the vertical distances, meaning that we will minimize the errors between the points and the line through the data. In other words, we will minimize the sum of square errors, which is computed by the following:

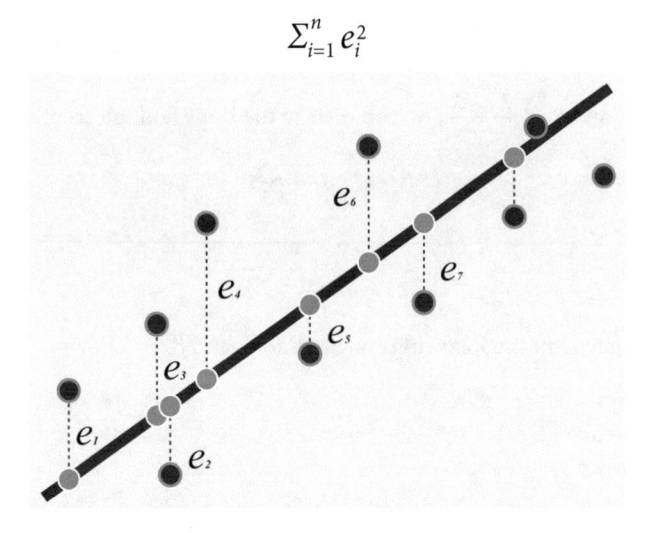

$$\sum_{i=1}^{n} e_i^2$$

Figure 6.2 – Errors between data points and the line through the data

Here,

$$e_i = y_i - \hat{y}_i,$$

where y_i is the observed value and $\hat{y}_i$ is the predicted value of the response variable y for the ith observation. The sum of the square of errors is given by:

$$S = \sum_{i=1}^{n} \left(y_i - \hat{\beta}_0 - \hat{\beta}_1 x_i \right)^2$$

To minimize S, we will take the partial derivatives with respect to $\hat{\beta}_0$ and $\hat{\beta}_1$. Then, we see:

$$\frac{\partial S}{\partial \hat{\beta}_0} = -2\Sigma \left(y_i - \hat{\beta}_0 - \hat{\beta}_1 x_i \right) = 0,$$

$$\frac{\partial S}{\partial \hat{\beta}_1} = -2\Sigma x_i \left(y_i - \hat{\beta}_0 - \hat{\beta}_1 x_i \right) = 0.$$

Therefore, it is easy to see that:

$$n\hat{\beta}_0 + \left(\sum_{i=1}^{n} x_i \right) \hat{\beta}_1 = \sum_{i=1}^{n} y_i,$$

$$\left(\sum_{i=1}^{n} x_i \right) \hat{\beta}_0 + \left(\sum_{i=1}^{n} x_i^2 \right) \hat{\beta}_1 = \sum_{i=1}^{n} x_i y_i,$$

This implies:

$$\hat{\beta}_0 = \frac{\sum_{i=1}^n y_i}{n} - \hat{\beta}_1 \frac{\sum_{i=1}^n x_i}{n},$$

$$\frac{\sum_{i=1}^n x_i \sum_{i=1}^n y_i}{n} - \frac{(\sum_{i=1}^n x_i)^2}{n}\hat{\beta}_1 + \left(\sum_{i=1}^n x_i^2\right)\hat{\beta}_1 = \sum_{i=1}^n x_i y_i.$$

Because $\bar{y} = \frac{\sum_{i=1}^n y_i}{n}$ and $\bar{x} = \frac{\sum_{i=1}^n x_i}{n}$, we can rewrite the slope and intercept as follows:

$$\hat{\beta}_0 = \bar{y} - \hat{\beta}_1 \bar{x},$$

$$\hat{\beta}_1 = \frac{\sum(x_i - \bar{x})(y_i - \bar{y})}{\sum(x_i - \bar{x})^2}.$$

In Python, we can implement the following code to find $\hat{\beta}_0$ and $\hat{\beta}_1$ as follows:

```
def least_squares_method(x,y):
    x_mean=x.mean()
    y_mean=y.mean()
    beta1 = ((x-x_mean)*(y-y_mean)).sum(axis=0)/ ((x-x.mean())**2).sum(axis=0)
    beta0 = y_mean-(beta1*x_mean)
    return beta0, beta1
```

Coefficients of correlation and determination

In this section, we will discuss two related notions – coefficients of correlation and coefficients of determination.

Coefficients of correlation

A coefficient of correlation is a measure of the statistical linear relationship between two variables and can be computed using the following formula:

$$r = \frac{1}{n-1}\sum_{i=1}^n \left(\frac{x_i - \bar{x}}{s_x}\right)\left(\frac{y_i - \bar{y}}{s_y}\right)$$

The reader can go here – https://shiny.rit.albany.edu/stat/corrsim/ – to simulate the correlation relationship between two variables.

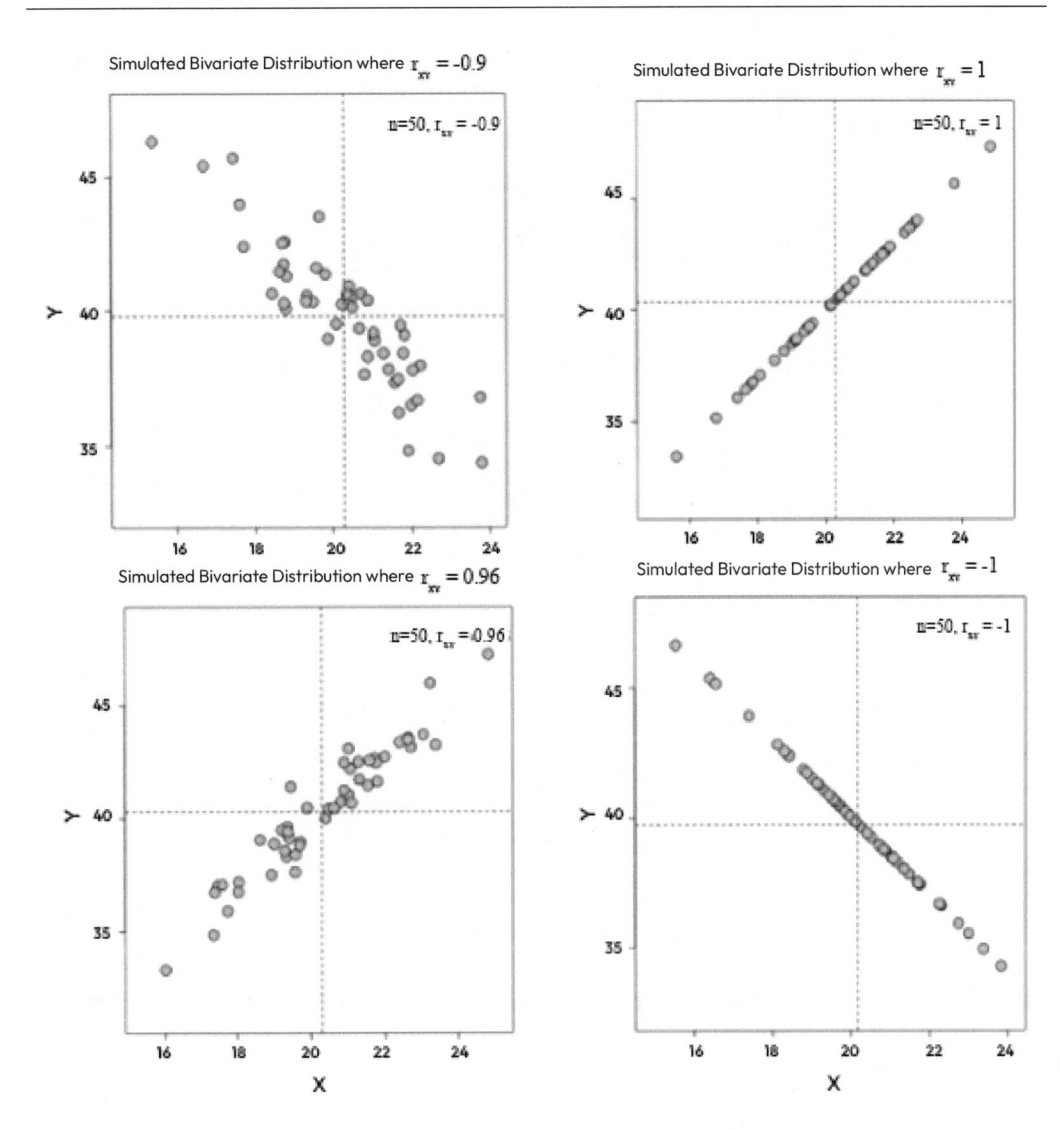

Figure 6.3 – Simulated bivariate distribution

By observing the scatter plots, we can see the direction and the strength of the linear relationship between the two variables and their outliers. If the direction is positive (r>0), then both variables increase or decrease together.

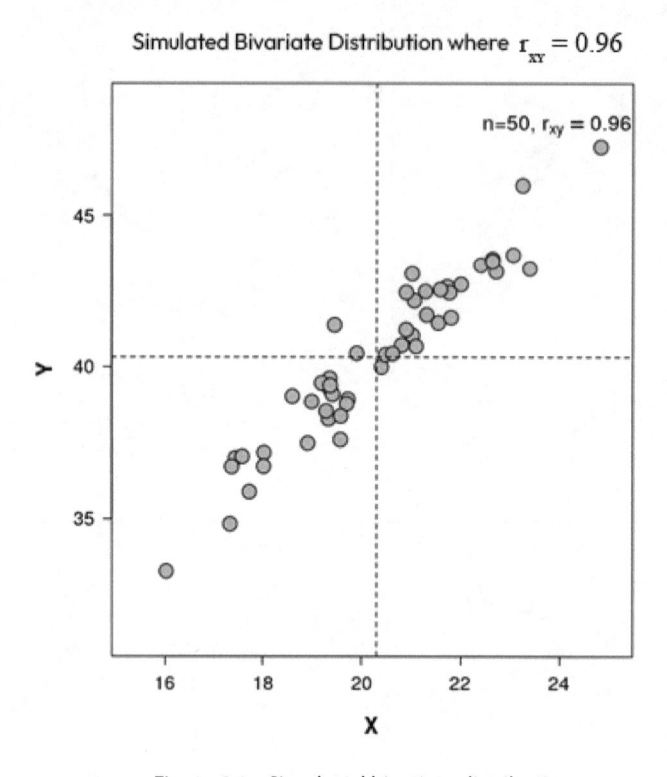

Figure 6.4 – Simulated bivariate distribution

If the direction is negative, then one variable increases while the other decreases. We illustrate this in the following scatter plot:

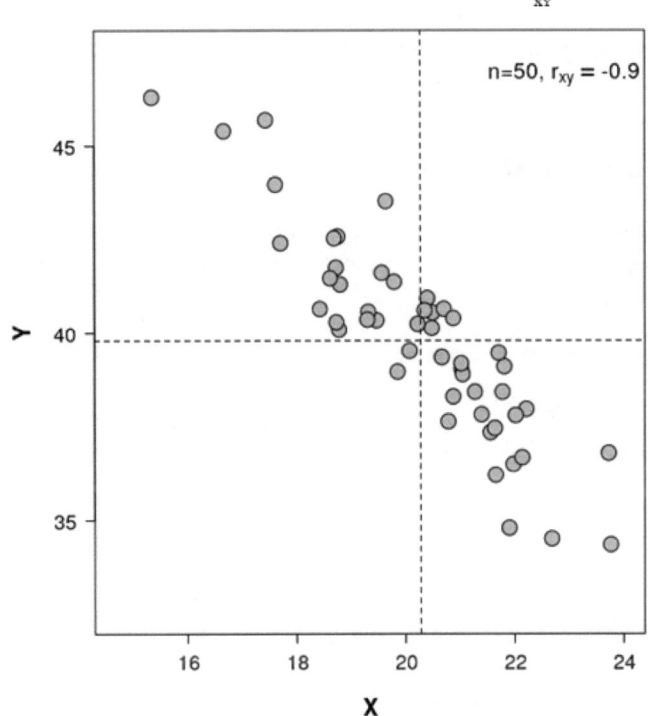

Figure 6.5 – Simulated bivariate distribution

The coefficient of correlation exists as a value between -1 and 1. When r =1, we have a perfect positive correlation, and when r = -1, we have a perfect negative correlation. Observe that the value of the coefficient of correlation does not change if, for example, the variable x and the variable y are switched or if all values of either variable are linearly scaled. In general, correlation does not imply causation.

Coefficients of determination

The coefficient of determination is just r^2, where r is the coefficient of correlation, which is a proportion of the variation in the variable y, explained by the variable x. The value of r^2 is between 0 and 1, with the linear relationship between the two variables becoming stronger as its value approaches 1. The value of r^2 can also be computed using the following formula:

$$r^2 = 1 - \frac{SS_{Res}}{SS_{Tot}}$$

Here, SS_{Res} is the sum of the square of errors and SS_{Tot} is the total sum of the square of the difference between the actual value and the average value. In `statsmodels`, from the output of the OLS regression results, the value of determination can be obtained. This output is discussed in the next sections of this chapter.

Required model assumptions

Like the parametric tests we discussed in *Chapter 4, Parametric Tests*, linear regression is a parametric method and requires certain assumptions to be met for the results to be valid. For linear regression, there are four assumptions:

- A linear relationship between variables
- The normality of the residuals
- The homoscedasticity of the residuals
- Independent samples

Let's discuss each of these assumptions individually.

A linear relationship between the variables

When thinking about fitting a linear model to data, our first consideration should be whether the model is appropriate for the data. When working with two variables, the relationship between the variables should be assessed with a scatter plot. Let's look at an example. Three scatter plots are shown in *Figure 6.6*. The data is plotted, and the actual function used to generate the data is drawn over the data points. The leftmost plot shows data exhibiting a linear relationship. The middle plot shows data exhibiting a quadratic relationship. The rightmost plot shows two uncorrelated variables.

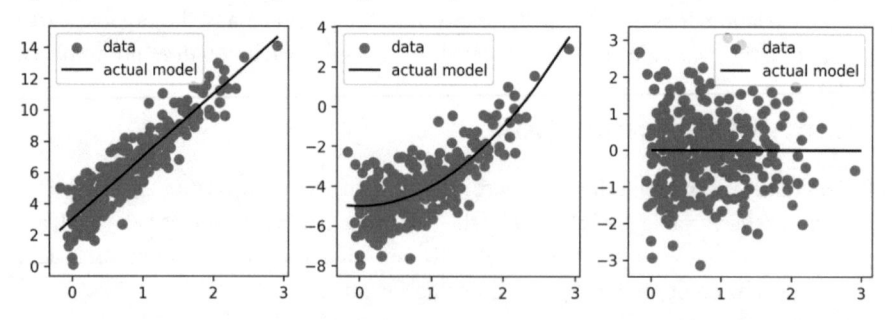

Figure 6.6 – Three scatter plots depicting three distinct relationships between two variables

Of the example data in *Figure 6.6*, a linear model would only be appropriate for the leftmost data. In this example, we have the benefit of knowing the actual relationship of the two variables, but, in real problems, you will have to determine whether a linear model is appropriate. Once we have assessed whether a linear model is appropriate and fitted a linear model, we can assess the two assumptions that depend on the model residuals.

Normality of the residuals

In the previous two chapters, we discussed normality at length and looked at several examples of data that are normally distributed and several examples that were not normally distributed. In this case, we are making the same assessment, but it is for the model residuals rather than for the variables. In fact, unlike the previous parametric methods, the *variables themselves do not need to be normally distributed, only the residuals*. Let's look at an example. We will fit a linear model on two variables, each of which exhibits a skewed distribution. As can be seen in *Figure 6.7*, even though each variable has a skewed distribution, the variables are linearly correlated.

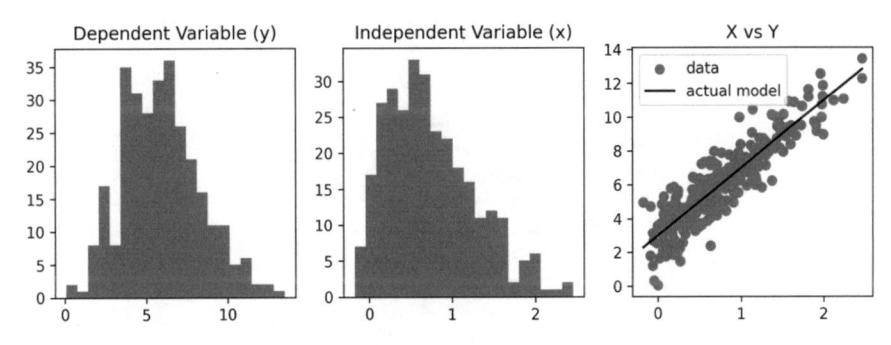

Figure 6.7 – Distributions of X, Y, and their scatter plots

Now, let's look at the residuals from the model fit. The residuals from the model fit are shown in *Figure 6.8*. As we expect, the residuals from this model fit appear to be normally distributed. This model appears to meet the assumption of normally distributed residuals.

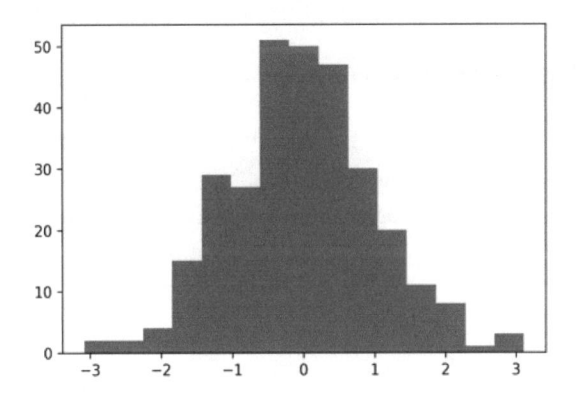

Figure 6.8 – Residuals from model fit

With the assumption of normally distributed residuals verified, let's move on to the assumption of homoscedasticity of the residuals.

Homoscedasticity of the residuals

Not only should the residuals be normally distributed, but the residuals should also have constant variance over the linear model. **Homoscedasticity** refers to the residuals having equal scatter. Conversely, residuals that do exhibit equal spread are said to be **heteroscedastic**. Homoscedasticity is generally assessed by plotting the residuals against the model predictions with a scatter plot. The residuals should appear to be randomly distributed with a mean of 0 and equally dispersed along the x axis. *Figure 6.9* shows a scatter plot of the model residuals against the predicted value of Y. These residuals appear to exhibit homoscedasticity.

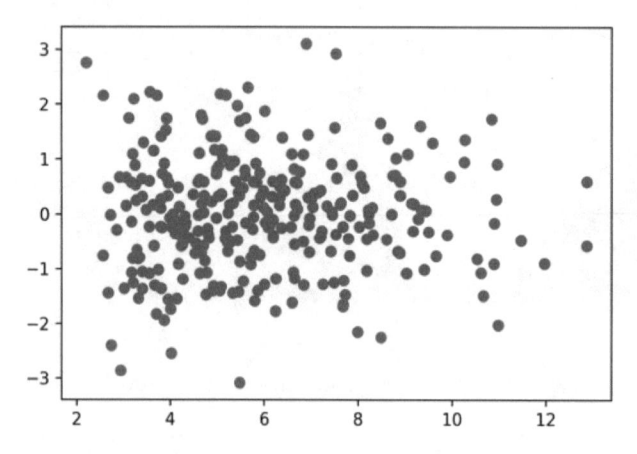

Figure 6.9 – Scatter plot of model residuals versus prediction

There are a few common ways the homoscedasticity of the residuals could be violated:

- The residuals show a systematic change in variance
- There is an extreme outlier

Two examples of these violations are shown in *Figure 6.10*. The left plot shows an extreme outlier, and the right plot shows non-constant variance in the residuals. In both cases, the model assumptions are violated.

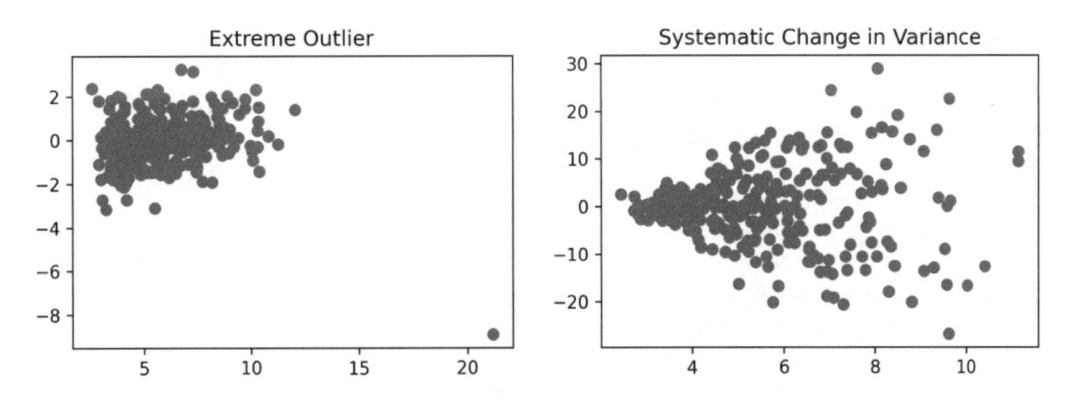

Figure 6.10 – Example residuals that exhibit poor behavior

When the residuals show a systematic pattern, it may be an indication that another type of model would be more appropriate, or it may be helpful to apply a transformation to one or both variables. When an extreme outlier is present, it may be worth verifying and investigating the data point to ensure it should be included in the analysis. In the next section, we will discuss how extreme outliers impact the model fit.

Sample independence

We have discussed sample independence in previous chapters. There are no graphics or statistics that can be used to determine whether samples are independent. *Assessing sample independence requires careful analysis of the sampling method and the populations from which the samples are drawn*. For example, a common type of data that violates the independence assumption is time-series data. Time-series data is a type of data that is sampled over time, making it serially correlated. We will discuss analysis methods for time-series data in later chapters.

In this section, we discussed the linear regression model assumptions. In the next section, we will discuss how to validate a linear model, which will again utilize the model residuals.

Testing for significance and validating models

Up to this point in the chapter, we have discussed the concepts of the OLS approach to linear regression modeling; the coefficients in a linear model; the coefficients of correlation and determination; and the assumptions required for modeling with linear regression. We will now begin our discussion on testing for significance and model validation.

Simple Linear Regression

Testing for significance

To test for significance, let us load `statsmodels` macrodata data set so we can build a model that tests the relationship between real gross private domestic investment, `realinv`, and real private disposable income, `realdpi`:

```
import numpy as np
import pandas as pd
import seaborn as sns
import statsmodels.api as sm
import matplotlib.pyplot as plt
from statsmodels.nonparametric.smoothers_lowess import lowess
df = sm.datasets.macrodata.load().data
```

Least squares regression requires a constant coefficient in order to derive the **intercept**. In the least squares equation, $X^T X \beta = X^T y$, the first column of the **design matrix**, X, requires a set of ones. However, this **constant** must be added in directly, which we do with `statsmodels`' `add_constant` function here:

```
df = sm.add_constant(df, prepend=False)
df_mod = df[['realinv','realdpi','const']]
```

The input data does not need to be normally distributed for least squares regression. However, it is assumed that the residuals of the model are normally distributed. However, it is useful to see the spread of the data to get an understanding of their statistics, such as the mean, median, and range, and whether there are outliers present. If the residual analysis after the modeling suggests there may be some issues, having inspected the model variables will help the analyst understand potential root causes.

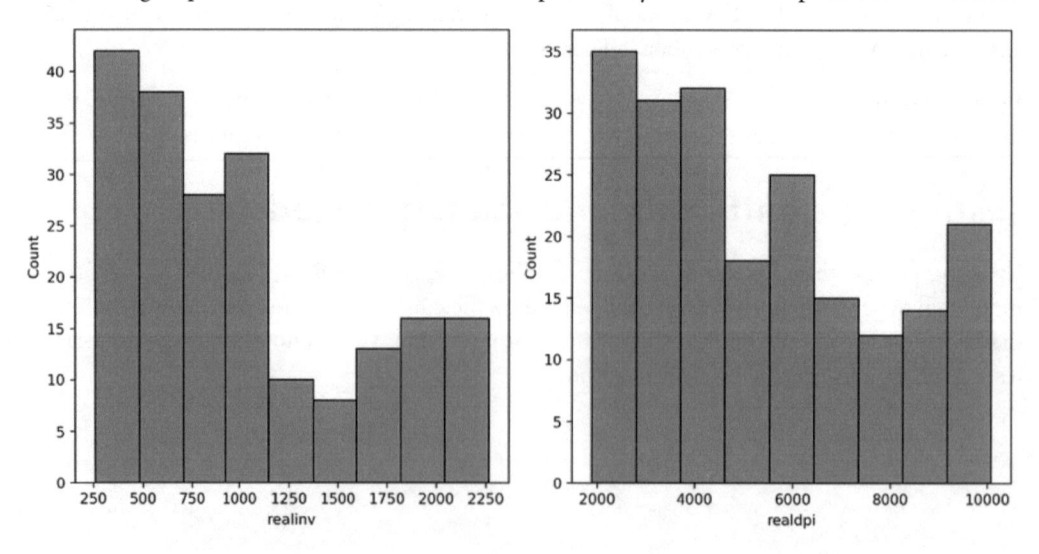

Figure 6.11 – Visualizing the distributions of the *realinv* and *realdpi* variables

When visualizing the relationship between `realinv` and `realdpi`, we can see a strong linear relationship, but also a potential serial correlation in the data as there appears to be a somewhat **cyclical oscillation** in the data, which becomes stronger toward the extreme values, seen in the upper right of the plot.

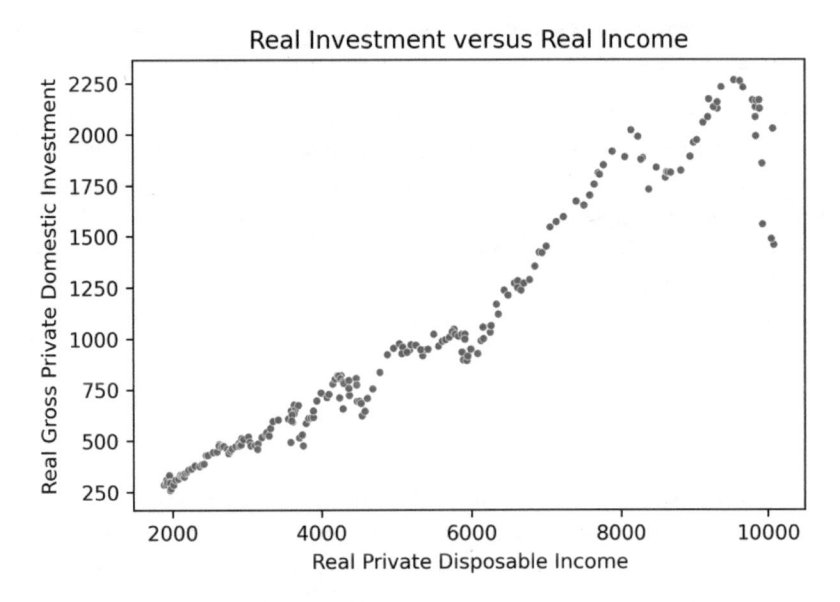

Figure 6.12 – Visualizing the relationship between the *realinv* and *realdpi* variables

We will analyze the potential serial correlation seen in the preceding plot after a few more steps. First, let us fit the input variable and coefficient to the model:

```
ols_model = sm.OLS(df_mod['realdpi'], df_mod[['const','realinv']])
compiled_model = ols_model.fit()
```

Next, we will want to print out a summary of the model performance so we can begin to gauge the significance of the terms involved:

```
print(compiled_model.summary())
```

Referring to the results in *Figure 6.13*, we can see that the **95% confidence interval** for neither the intercept, `const`, nor the input variable, `realinv`, contains zero. Therefore, in addition to the **significant p-values** for these variables, we can conclude they are significant contributors to the target at the coefficients provided. Now that we have confirmed there is statistical significance for both the intercept and the input variable and their coefficients, we want to know the level of correlation to the target they represent. We can see the **R-squared** statistic is 0.944. Because the two variables are the constant and the input variable, this R-squared statistic explains only the input variable.

> **R-squared versus adjusted r-squared**
>
> For multivariate regression, we would look at the adjusted r-squared value, which we can also see is the same – in the univariate/simple regression case – as the r-squared value.

The **correlation of determination** – also called goodness of fit – of 0.944 means real gross private domestic investment explains 94.4% of the variance in real private disposable income.

We can see based on the table that the intercept and the `realinv` variable are both significant in predicting the target with very low p-values. We can also see that neither of the 95% confidence intervals contains 0, which backs up the relevancy of their p-values. However, the confidence interval for the constant (intercept) suggests there may be a risk of uncertainty in the model and that more variables may be needed to reduce that uncertainty and improve model performance. Stated differently, we could say a higher level of uncertainty in the constant indicates the model has variance left to be explained and that the current model is overly biased.

Validating models

The Durbin-Watson test

It is important to acknowledge that based on the fact that data was observed over a 50-year period and random sampling was not performed, it is possible the residuals will have a **serial correlation**. We observe a possibility of serial correlation in the **Durbin-Watson test** statistic of 0.095, which suggests the residuals exhibit positive autocorrelation. The Durbin-Watson statistic tests whether there is autocorrelation at the first time lag. This means the Durbin-Watson tests whether the change in values between each data point is correlated with the value of the previous data point. If that correlation exists, it is considered **lag-one autocorrelation**.

> **Durbin-Watson test statistic**
>
> As a rule of thumb, a Durbin-Watson test statistic below roughly 2.0 (many consider 1.5 to 2.0 to be reasonable) is considered a flag for positive autocorrelation, while a value above roughly 2.0 (many consider 2.0 to 2.5 to be reasonable) indicates possible negative autocorrelation. Procedurally, a table lookup is required to identify the critical values against which the test statistic should be measured. The process follows the same steps for any hypothesis test with respect to comparing the test statistic against the critical value. The Durbin-Watson test is typically interpreted as a one-tailed test as inference from a two-tailed test is not particularly useful. A version of the Durbin-Watson table can be found at https://www.real-statistics.com/statistics-tables/durbin-watson-table/. This table produces the critical values for **positive autocorrelation**. To find the critical values for **negative autocorrelation**, the positive critical values can be subtracted from 4.

Although we have a Durbin-Watson statistic that suggests serial correlation may be present in the data, we will confirm significance by comparing the Durbin-Watson statistic to its corresponding positive autocorrelation **table lookup** using the table found at the preceding link. We have n=203 samples in our dataset but will use n=200 on the table lookup since this value is likely accurate enough to assess, and 203 does not exist in the table. At the 0.05 level of significance, which is used by default with the Durbin-Watson statistic produced in `statsmodels`' OLS regression, we find the bounds for the `k=1` input variable to be `[1.758, 1.779]`. A value within this range should be interpreted as **inconclusive**. A value less than the lower bound indicates the presence of enough evidence to reject the null hypothesis and thus conclude there is statistically significant positive autocorrelation. A value greater than the upper bound indicates there is not enough evidence to reject the null hypothesis and thus there is no autocorrelation. Based on the results of our model, we can conclude with a 95% level of confidence that there is evidence of positive serial correlation in our residuals.

OLS Regression Results

Dep. Variable:	realdpi
Model:	OLS
Method:	Least Squares
Date:	Sat, 01 Oct 2022
Time:	15:53:00
No. Observations:	203
Df Residuals:	201
Df Model:	1
Covariance Type:	nonrobust

R-squared:	0.944
Adj. R-squared:	0.944
F-statistic:	3384.
Prob (F-statistic):	1.01e-127
Log-Likelihood:	-1577.1
AIC:	3158.
BIC:	3165.

	coef	std err	t	P>|t|	[0.025	0.975]
const	1234.5164	80.866	15.266	0.000	1075.062	1393.971
realinv	4.0243	0.069	58.175	0.000	3.888	4.161

Omnibus:	90.643	Durbin-Watson:	0.095
Prob(Omnibus):	0.000	Jarque-Bera (JB):	434.654
Skew:	1.696	Prob(JB):	4.13e-95
Kurtosis:	9.316	Cond. No.	2.34e+03

Notes:

[1] Standard Errors assume that the covariance matrix of the errors is correctly specified.

[2] The condition number is large, 2.34e+03. This might indicate that there are strong multicollinearity or other numerical problems

Figure 6.13 – OLS regression model output for *realdpi*

At this point, an analyst may want to consider addressing the serial correlation before moving forward. However, for the purpose of following the process of end-to-end model validation, we will move forward to the next steps.

Diagnostic plots for analyzing model errors

As mentioned, $y_i = \beta_0 + \beta_1 x_i + e_i$, where in regression, $\beta_0 + \beta_1 x_i$ is the predicted mean for y_i given x_i and e_i is the error term – also called the residual term, which is calculated as the observed value for a data point minus the predicted value for that data point. The residuals in least squares linear regression must approximate a normal distribution centered around the mean with a standard deviation. The method for deriving the residuals for a fitted model is to subtract the predicted value from the actual value for each observation. Four common visualizations for inspecting the fit of a least squares regression model are as follows:

- **Residuals versus fitted plots**, used to inspect the requirement of a linear relationship between the input and target variables
- **Quantile plots**, used to inspect the assumption of residuals being normally distributed
- **Scale-location plots**, used to inspect the assumption of homoscedasticity, or the homogenous distribution of variation
- **Residuals versus leverage influence plots**, which help identify the impact outliers may have on model fit

To produce these plots, we can use the following Python code implementation:

```
model_residuals = compiled_model.resid
fitted_value = compiled_model.fittedvalues
standardized_residuals = compiled_model.resid_pearson # Residuals, normalized to have unit variance.
sqrt_standardized_residuals = np.sqrt(np.abs(compiled_model.get_influence().resid_studentized_internal))
influence = compiled_model.get_influence()
leverage = influence.hat_matrix_diag
cooks_distance = compiled_model.get_influence().cooks_distance[0]

fig, ax = plt.subplots(2, 2, figsize=(10,8))
# Residuals vs. Fitted
ax[0, 0].set_xlabel('Fitted Values')
ax[0, 0].set_ylabel('Residuals')
ax[0, 0].set_title('Residuals vs. Fitted')
locally_weighted_line1 = lowess(model_residuals, fitted_value)
sns.scatterplot(x=fitted_value, y=model_residuals, ax=ax[0, 0])
ax[0, 0].axhline(y=0, color='grey', linestyle='--')
ax[0,0].plot(locally_weighted_line1[:,0], locally_weighted_line1[:,1],
```

```python
color = 'red')
# Normal Q-Q
ax[0, 1].set_title('Normal Q-Q')
sm.qqplot(model_residuals, fit=True, line='45',ax=ax[0, 1], c='blue')

# Scale-Location
ax[1, 0].set_xlabel('Fitted Values')
ax[1, 0].set_ylabel('Square Root of Standardized Residuals')
ax[1, 0].set_title('Scale-Location')
locally_weighted_line2 = lowess(sqrt_standardized_residuals, fitted_value)
sns.scatterplot(x=fitted_value, y=sqrt_standardized_residuals, ax=ax[1, 0])
ax[1,0].plot(locally_weighted_line2[:,0], locally_weighted_line2[:,1], color = 'red')
# Residual vs. Leverage Influence
ax[1, 1].set_xlabel('Leverage')
ax[1, 1].set_ylabel('Standardized Residuals')
ax[1, 1].set_title('Residuals vs. Leverage Influence')
locally_weighted_line3 = lowess(standardized_residuals, leverage)
sns.scatterplot(x=leverage, y=standardized_residuals, ax=ax[1, 1])
ax[1, 1].plot(locally_weighted_line3[:,0], locally_weighted_line3[:,1], color = 'red')
ax[1, 1].axhline(y=0, color='grey', linestyle='--')
ax[1, 1].axhline(3, color='orange', linestyle='--', label='Outlier Demarkation')
ax[1, 1].axhline(-3, color='orange', linestyle='--')
ax[1, 1].legend(loc='upper right')
leverages = []
for i in range(len(cooks_distance)):
    if cooks_distance[i] > 0.5:
        leverages.append(leverage[i])
        ax[1, 1].annotate(str(i) + " Cook's D > 0.5",xy=(leverage[i], standardized_residuals[i]))
if leverages:
    ax[1, 1].axvline(min(leverages), color='red', linestyle='--', label="Cook's Distance")
for i in range(len(standardized_residuals)):
    if standardized_residuals[i] > 3 or standardized_residuals[i] < -3:
        ax[1, 1].annotate(i,xy=(leverage[i], standardized_residuals[i]))

fig.tight_layout()
```

The following output contains four common diagnostic plots.

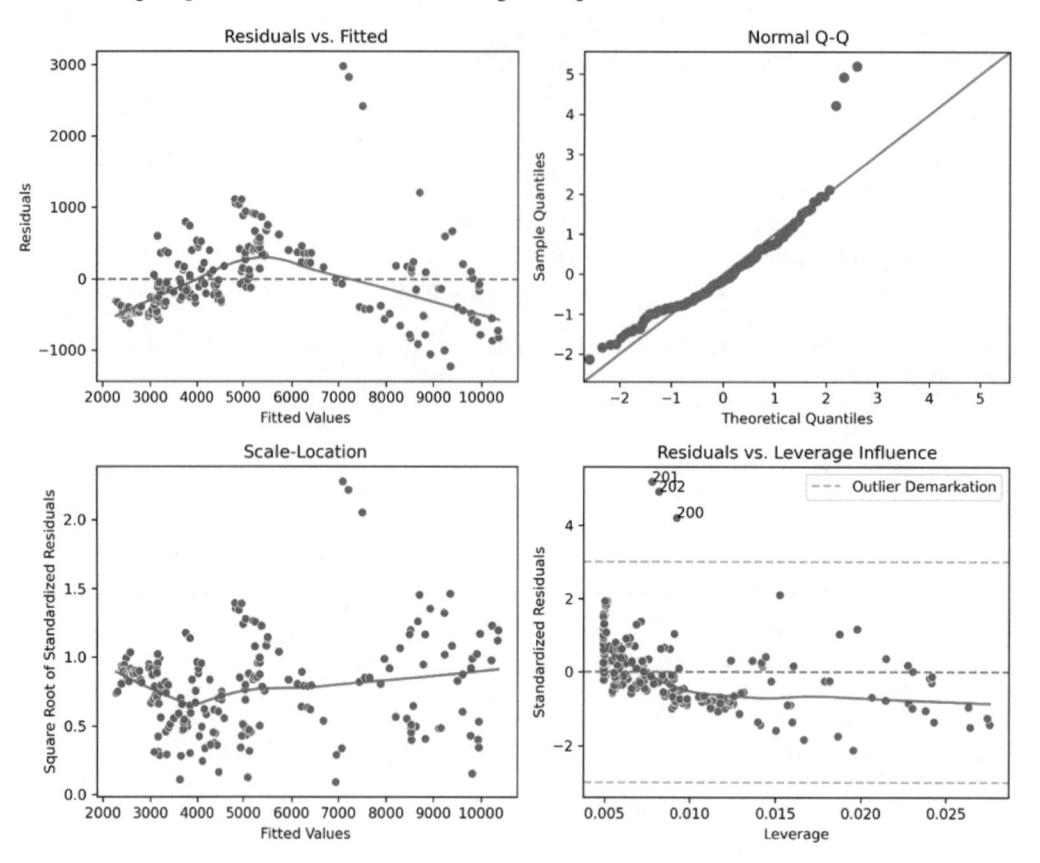

Figure 6.14 – Linear regression diagnostic plots

Residuals versus fitted

One requirement of least squares regression is a **linear relationship** between the input and target variables. Suffice it to say a strongly correlated relationship exists between the input and target variables. This plot helps the analyst assess the linearity between the two.

We should expect the fitted line – representing the model – to be horizontal, representing the strength of the linear relationship between the input and target variable across all data points. The residuals should be roughly evenly spread around the line in reasonable proximity. If these attributes cannot be confirmed, a non-linear relationship may exist between the two variables. A transformation of at least one of the variables – if the variable is skewed – may help resolve this type of issue. However, it could also be that a non-parametric model is more appropriate, which is often the case when categorical features – or encoded categorical features – are present.

Our model produced a plot that does not display evenly spread residuals forming around a horizontal line. Rather, the line is concaved down, which suggests a polynomial or non-parametric regression may be more appropriate. Additionally, the residuals form a pattern toward the right end of the line that suggests there may be some serial correlation in the data as it exhibits some sinusoidal behavior. We were able to identify the presence of serial correlation using the Durbin-Watson statistic. This is likely due to the fact the data was not randomly sampled and was collected over a long period of time.

QQ plot of residuals

Assessment of normality for the QQ plot follows the same procedure as assessing normality using the QQ plot in *Chapter 4, Parametric Tests*. The primary purpose of this plot is to assess the required assumption of the **normal distribution of errors**. Minor deviation of the residuals from the 45-degree line in the plot is normal, but extreme skewness or deviation from the line may indicate a poor fit. Such a scenario could indicate poor statistical power. In the event of skewness, data transformation – such as a log transformation – could be useful to resolve the issue. Dropping outliers may be another suitable solution. However, rather than dropping outliers outright, it is advisable to ensure there was not an error in data handling that could be resolved to make the outliers conform.

In the QQ plot for our model's residuals, we can see some moderate left-side skewness and extreme right-side skewness, which suggests the three values in the right tail could be extreme outliers. Aside from this, we can assume that overall, the residual error is approximately normally distributed.

Scale-location

The scale-location plot of the square root of the standardized residuals helps analysts use visual inspection to determine whether there is **homoscedasticity** – or *non-constant* variance – of the residuals, which is required for a least squares regression model. The plot visualizes the square root of the absolute value of the standardized residuals. Analyzing homoscedasticity can sometimes be useful for identifying potential issues related to **sample independence**, another requirement of least-squares regression. However, the Durbin-Watson test should be given more weight for testing this assumption.

In assessing our model, we can see the line has some deviation from being perfectly horizontal. Additionally, the residuals, while seeming to have a large amount of constant variance, nonetheless at times exhibit patterns, which conflicts with the behavior of homoscedasticity. Furthermore, there appear to be three very notable outliers. Based on this information, we can reasonably assume a risk of heteroscedasticity – or *constant* variance – exists within the model's residuals.

Residuals versus leverage influence

Assessing the residuals versus leverage influence is another approach to assessing model fit. In addition to gauging leverage, this plot also shows when **Cook's distance** is beyond a reasonable level.

Leverage is used to identify the distance of an individual point from all other points. High leverage for a residual likely means the corresponding data point is an outlier strongly influencing the model to fit less approximately to the overall data and instead give more weight to that specific value. Residuals should be between -2 and 2 to not be considered potential outliers. Values between +/-2.5 to 3 suggest data points are extreme outliers. Overall, this plot is useful for separating outliers that have no significant negative impact on the model from the ones that do.

Cook's distance is a measurement that uses both an observation's leverage and its residual value. As the leverage is higher, its calculated Cook's distance is higher. In general, a Cook's distance value greater than 0.5 means a residual has leverage that is negatively impacting the model through over-representation – essentially, an outlier. Cook's distance is included in the code and flags the residuals that have distances greater than 0.5. Cook's distance is particularly useful because removing outliers may not have a significant impact on the model. However, if the outlier has a high Cook's distance value, that is a strong indication that resolving the outlier **will benefit** the model. Cook's distance is calculated for each data point by removing it from the model and calculating the difference in error divided by the mean squared error multiplied by the number of model coefficients plus one. Its formulation follows:

$$D_i = \frac{\sum_{j=1}^{n}\left(\hat{y}_j - \hat{y}_{j(i)}\right)^2}{(p+1)\hat{\sigma}^2}$$

where $\hat{y}_j$ is the jth estimate of the response and $\hat{y}_{j(i)}$ is the value of the jth fitted value with the ith value removed. p is the number of coefficients in the model and $\hat{\sigma}^2$ is the variance of the residuals for all observations, including those removed, also called the squared error.

In the plot for our model, we can observe three extreme outliers. However, we can also see no value of Cook's distance is greater than 0.5. As with the scale-location plot, the standardized residuals should follow even distribution around a smoothed line that is horizontal, which indicates a constant variance. We can see from our model's output that this is not the case.

Addressing issues with residuals

Alone, each plot can be considered enough to negate the validity of a model. However, it is useful to consider all plots since a model may still be useful if there is only some deviance from expectations in the diagnostic plots. Some approaches to resolving issues with residuals are as follows:

- Investigating and resolving any data collection issues that may produce outliers or data inconsistencies
- Removing outliers outright
- Improving the sampling approach using methods such as using random or stratified sampling
- Performing data transformations, such as a log or square root transformation
- Including additional input variables that can help explain the target variable or any serial correlation that may be present

Handling serial correlation

If uncertain of the presence of serial correlation in the residuals, a useful next step is to analyze a **Partial Autocorrelation Function** (**PACF**) plot to assess whether serial correlation exists in the model at a significant level, which could explain issues with the model's residuals.

> ### ACF and PACF
>
> The **PACF** provides a partial correlation between the value of a point and previous points, called lags, by controlling for the lags between. Controlling the lags between is where the term "partial" comes from. If a specific point at lag zero in a dataset is correlated strongly with the value of the third lagging point, there would be a significant correlation with a lag of 3, but not a lag of 2 or 1. In time-series modeling, the PACF is used to directly derive the autoregressive orders, denoted as $AR(n)$ where n is the lag. The **Autocorrelation Function** (**ACF**), which is used for determining moving average orders, denoted as $MA(n)$, considers the correlation between data points at lag zero and all previous lagging points without controlling for the relationships between them. Stated alternatively, autocorrelation in an ACF plot at lag 3 will provide autocorrelation across points 0 through 3 whereas correlation in a PACF plot at lag 3 will provide autocorrelation only between point 0 and point 3, excluding the impact of values at lags 1 and 2.

We can execute the following code to generate the PACF plot with a 95% confidence interval using the most recent 50 data points, ordered by index positioning – in this case, by date:

```
from statsmodels.graphics.tsaplots import plot_pacf
plot_pacf(compiled_model.resid, alpha=0.05, lags=50);
```

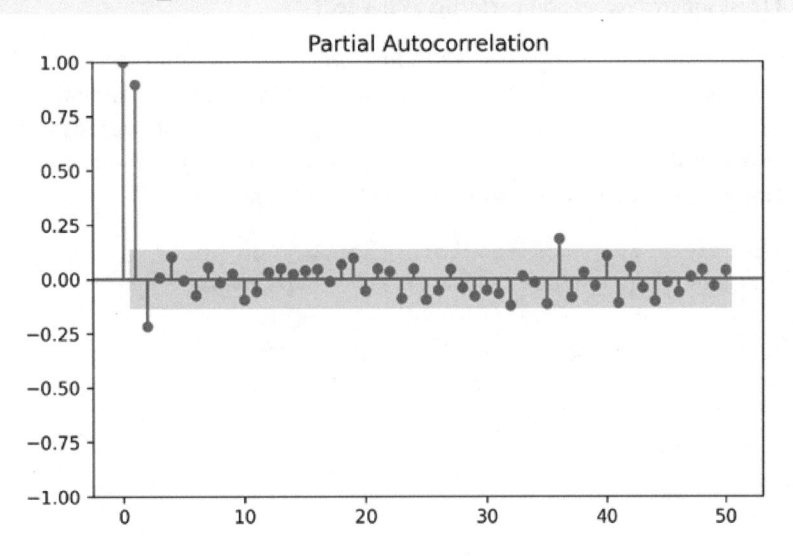

Figure 6.15 – Partial autocorrelation plot for the model residuals

In the PACF plot in *Figure 6.15*, we see very strong and significant correlations at lags 0 and 1. We observe a low level of significant correlation (approximately -0.25) for lags 2 and 36, extending beyond the 95% confidence interval. What this tells us is we might be better off modeling the dataset with a first-order autoregressive (**AR(1)**) time-series model than a linear regression model.

Answering regression questions with a least squares method when there is a presence of significant autocorrelation creates inherit risk. There are methods to adjust the results of a model when there is autocorrelation detected, such as taking a **first-order difference** – or another type of low-pass filter - or applying the Cochrane-Orcutt, Hildreth-Lu, or Yule-Walker adjustments. However, it is ideal to perform a time-series analysis, as the results of that type of modeling are more robust to time-dependent patterns. We will introduce time series in *Chapter 10, Introduction to Time Series*, where we discuss simple and complex methods for approximating time series to produce more useful results.

Let us consider a first-order difference of the data. Regarding *Figure 6.16*, we first look at the ACF plot and original data for the input variable, `realdpi`. We can see an exponentially growing line (original data) that displays an exponentially dampening ACF plot. This ACF behavior is typical for **trended** data that often benefits from **first-order differencing**. The low autoregressive ordering identified in the PACF indicates this may be sufficient for a regression model to proceed. Had the ACF shown different behavior, such as sinusoidal (seasonal) autocorrelation, or the PACF shown more partial autocorrelation, we would have evidence against a first-order difference being a sufficient solution for resolving serial correlation and proceeding with least squares regression. ACFs and PACFs for least squares regression should ideally have no significantly correlated lags. However, when the ACF exhibits this behavior and the PACF shows no significance, or when the level of correlation is very low, regression may perform well. It is preferable to see no patterns or peaks in either plot. Regardless, in either case, model error should be assessed, in addition to the autocorrelation of the residuals, to determine if least squares regression performs as needed.

Using the `diff()` numpy function, we take a first-order, low-pass difference and plot the ACFs, PACFs, and line plots for the original and differenced data. We can see a significant reduction in autocorrelation as a result. As an aside, the post-differencing autocorrelation suggests an **autoregressive moving average** (**ARMA**) model of AR order 1 and MA order 1 may be better suited than least-squares regression. However, we will ignore this for the chapter and continue:

```
from statsmodels.graphics.tsaplots import plot_acf
from statsmodels.graphics.tsaplots import plot_pacf
fig, ax = plt.subplots(2,3, figsize=(15,10))
plot_acf(df_mod['realdpi'], alpha=0.05, lags=50, ax=ax[0,0])
ax[0,0].set_title('Original ACF')
plot_pacf(df_mod['realdpi'], alpha=0.05, lags=50, ax=ax[0,1])
ax[0,1].set_title('Original PACF')
ax[0,2].set_title('Original Data')
ax[0,2].plot(df_mod['realdpi'])
plot_acf(np.diff(df_mod['realdpi'], n=1), alpha=0.05, lags=50, ax=ax[1,0])
```

```
ax[1,0].set_title('Once-Differenced ACF')
plot_pacf(np.diff(df_mod['realdpi'], n=1), alpha=0.05, lags=50,
ax=ax[1,1])
ax[1,1].set_title('Once-Differenced PACF')
ax[1,2].set_title('Once-Differenced Data')
ax[1,2].plot(np.diff(df_mod['realdpi'], n=1))
```

We get the following result:

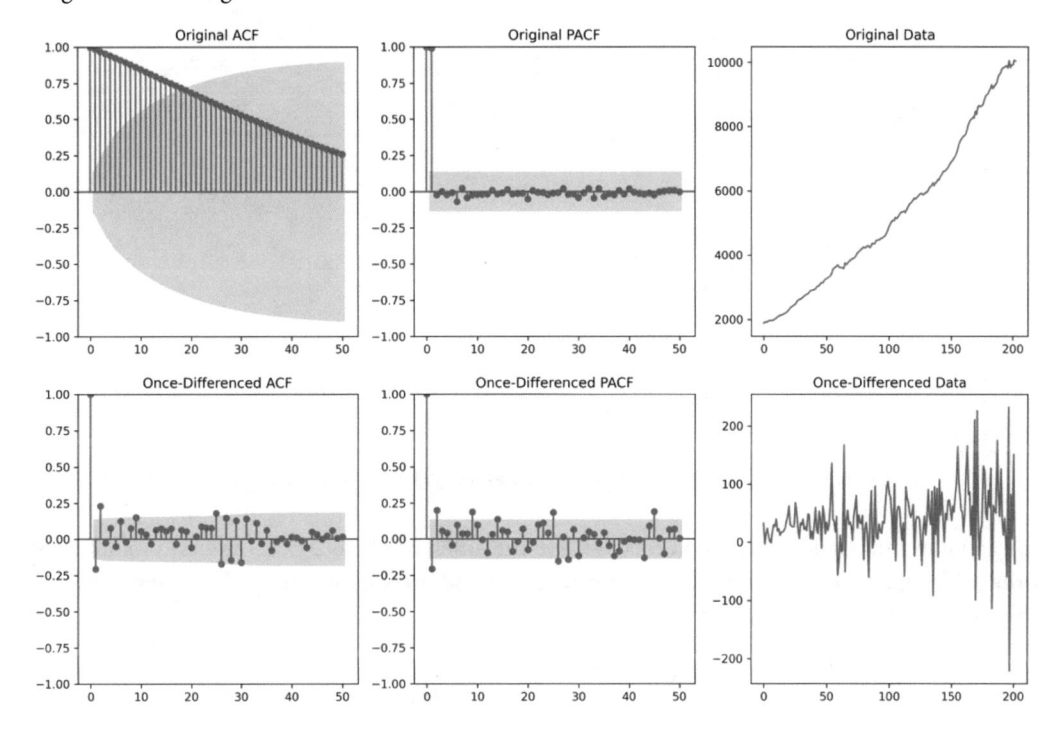

Figure 6.16 – Plots for the *realinv* variable before and after first-order differencing

This gives us data that we may more suitably model using regression. However, when using differencing as a method to resolve serial correlation in a least squares regression model, we must also differentiate the input variable. Observe the input variable, `realinv`, before and after transformation. Here in *Figure 6.17*, we can see similar behavior as with `realdpi`:

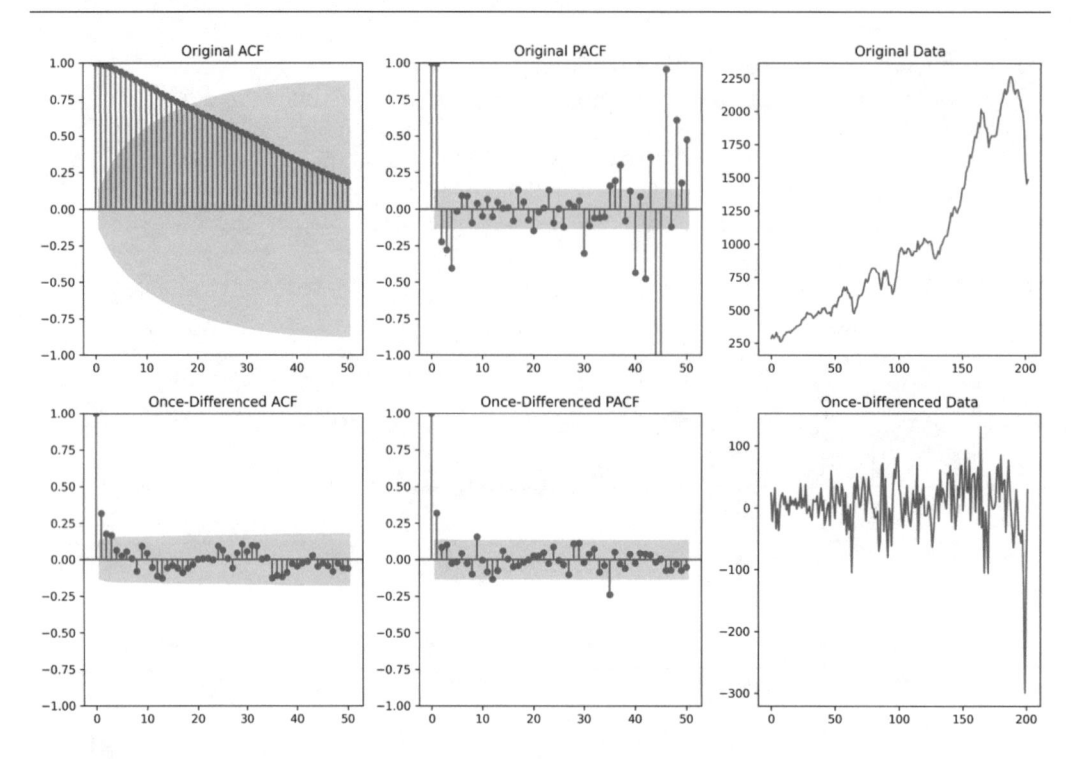

Figure 6.17 – Plots for the *realdpi* variable before and after first-order differencing

We can now build a new regression model using the differenced data. It is important to note first-order differencing removed one data point from the dataset so one value must also be removed from the constant in the design matrix:

```
ols_model_1diff = sm.OLS(np.diff(df_mod['realdpi'], n=1),
pd.concat([df_mod['const'].iloc[:-1], pd.Series(np.diff(df_
mod['realinv'], n=1))], axis=1))
compiled_model_1diff = ols_model_1diff.fit()
```

The output of this model is notably different. We now see an r-squared value of 0.045. Looking at the differenced data, we can see there is a different variance in the differenced data for disposable income than there is for investment. This means that although the realinv variable is useful for predicting realdpi as it appears significant, there is a lot more influencing real disposable gross personal income. We observe that the new Durbin-Watson test statistic is now 2.487. Using the critical values from the table lookup earlier, [1.758, 1.779], we see the Durbin-Watson statistic is now above the upper limit. **To check whether there is negative autocorrelation**, we subtract these values from 4 to find the new limit range [2.221, 2.242]. Because the new statistic, 2.487, is greater than 2.221, we can confirm there is now negative autocorrelation in the model's residuals. We can see based on the PACF plot, as well as the proximity to the critical value range, the autocorrelation is much less significant than before, but still present. It could be argued the autocorrelation of the residuals is now small enough to be considered resolved, however; as noted earlier, a Durbin-Watson statistic less than 2.5 can be considered normal. Nonetheless, least-squares regression remains risky compared to time-series modeling to generate our predictions. We begin an in-depth overview of this topic in *Chapter 10, Introduction to Time Series*.

Let us assume, however, that for the sake of moving through the steps of regression model validation, we have no autocorrelation in our data. The next step will be to test our model.

Model validation

Assuming an analyst has developed a model that produces strong results and appears useful based on the previously mentioned plots, the next step is to test the model. One conventional approach is to perform a train-and-test split on the data used for the original model where we fit all data. In the following code, we run a split with the train dataset having 75% of the data and the test dataset having 25%. The purpose is to assess whether there is a significant difference in performance between the two, as well as compared to the original model. The process follows the same steps as earlier. Acceptable differences are up to the analyst to decide, but in addition to assessing the differences between the metrics already discussed, the analyst should also note the coefficients for the slope and input variable, as these will be used to predict the target.

Here, we split the data. We use the `shuffle` argument to *randomly* shuffle the data so that if there is some order to the data, such as it being ordered by time, the data will be split randomly:

```
from sklearn.model_selection import train_test_split
train, test = train_test_split(df_mod, train_size=0.75, shuffle=True)
```

Now, we build a model using the training data:

```
ols_model_train = sm.OLS(train['realdpi'], train[['const','realinv']])
compiled_model_train = ols_model_train.fit()
print(compiled_model_train.summary())
```

Finally, we build a model using the testing data:

```
ols_model_test = sm.OLS(test['realdpi'], test[['const','realinv']])
compiled_model_test = ols_model_test.fit()
print(compiled_model_test.summary())
```

The idea here is that the two models should produce similar results on two different partitions of the data.

Another method for validating the model is to compare a metric against a naïve model. Let us use the **mean absolute error** (**MAE**). With this model metric, we could compare a naïve linear model's errors where the error is each data point minus the average to the model's predictions. A useful model MAE would need to be lower than the naïve model's MAE. Here, we use the fit model to predict the inputs, then compare the model's predictions to the naïve model's predictions:

Below we can observe the MAE from the *trained* model:

```
from sklearn.metrics import mean_absolute_error
mae = mean_absolute_error(train['realdpi'], compiled_model_train.
predict(train[['const','realinv']]))
mae
#438.5872081797031
```

Here we can see the MAE from the *naïve* model, which makes no assumptions other than that the mean will continue to hold the true values:

```
import numpy as np
errors = []
for i in range(len(train)):
    errors.append(abs(train['realdpi'].iloc[i] - train['realdpi'].
mean()))
np.mean(errors)
#2065.440235457064
```

We can see the model compares favorably against the naïve model.

Another method that is popular to use is testing on a holdout dataset. For this, a model is constructed and trained on the training data. Then, using that model, we apply it to the test data. We would then compare the metric to that of the training data. Using the training model's MAE of 438, we compare it to the following:

```
mae_test = mean_absolute_error(test['realdpi'], compiled_model_train.
predict(test[['const','realinv']]))
mae_test
#408.0253171549187
```

We can see that the error is lower for the test data than for the train data. However, since the train and test data are assumed to have different values and thus means, we should reassess that the naïve model still produces a higher MAE for the test data:

```
errors = []
for i in range(len(test)):
    errors.append(abs(test['realdpi'].iloc[i] - test['realdpi'].
mean()))
np.mean(errors)
#1945.5873125720873
```

Here we can see the model does provide better MAE than the naïve model at approximately the same rate on the test data as on the train data. Therefore, we expect that the linear regression model is better than the naïve model.

Summary

In this chapter, we discussed an overview of simple linear regression between one explanatory variable and one response variable. The topics we covered include the following:

- The OLS method for simple linear regression
- Coefficients of correlation and determination and their calculations and significance
- The assumptions required for least squares regression
- Methods of analysis for model and parameter significance
- Model validation

We looked closely at the concept of the square of error and how the sum of squared errors is meaningful for building and validating linear regression models. Then, we walked through the four pertinent assumptions required to make linear regression a stable solution. After, we provided an overview of four diagnostic plots and their interpretations with respect to assessing the presence of various issues related to heteroscedasticity, linearity, outliers, and serial correlation. We then walked through an example of using the ACF and PACF to assess serial correlation and an example of using first-order differencing to remove serial correlation constraints from data and build an OLS model using the differenced data. Finally, we provided methods for testing and validating least square regression models.

In the next chapter, we will extend this concept to include more than one explanatory variable, a technique called multiple linear regression. We will also discuss various topics related to multiple linear regression, such as variable selection and regularization.

7
Multiple Linear Regression

In the last chapter, we discussed **simple linear regression** (**SLR**) using one variable to explain a target variable. In this chapter, we will discuss **multiple linear regression** (**MLR**), which is a model that leverages multiple explanatory variables to model a response variable. Two of the major conundrums facing multivariate modeling are multicollinearity and the bias-variance trade-off. Following an overview of MLR, we will provide an induction into the methodologies used for evaluating and minimizing multicollinearity. We will then discuss methods for leveraging the bias-variance trade-off to our benefit as analysts. Finally, we will discuss handling multicollinearity using **Principal Component Regression** (**PCR**) to minimize overfitting without removing features but rather transforming them instead.

In this chapter, we're going to cover the following main topics:

- Multiple linear regression
- Feature selection
- Shrinkage methods
- Dimension reduction

Multiple linear regression

In the previous chapter, we discussed SLR. With SLR, we were able to predict the value of a variable (commonly called the response variable, denoted as y) using another variable (commonly called the explanatory variable, denoted as x). The SLR model is expressed by the following equation where β_0 is the intercept term and β_1 is the slope of the linear model.

$$y = \beta_0 + \beta_1 x + \epsilon$$

While this is a useful model, in many problems, multiple explanatory variables could be used to predict the response variable. For example, if we wanted to predict home prices, we might want to consider many variables, which may include lot size, the number of bedrooms, the number of bathrooms, and overall size. In this situation, we can expand the previous model to include these additional variables. This is called MLR. The MLR model can be expressed with the following equation.

$$y = \beta_0 + \beta_1 x_1 + \beta_2 x_2 + \ldots + \beta_p x_p + \epsilon$$

Like the previous equation, β_0 represents an intercept. In this equation, in addition to the intercept value, we have a β parameter for each explanatory variable. Each explanatory variable is denoted as x with a numerical subscript. We can include as many explanatory variables as desired in the model, which is why the equation shows a final subscript of p.

> **Impacts of data sizes on MLR**
>
> In general, we can include as many explanatory variables as desired in the model. However, there are some realistic limits. One of these potential limits is the number of samples relative to the number of explanatory variables. MLR tends to work best when the number of samples (N) is much less than the number of explanatory variables (P). As P approaches N, the model becomes difficult to estimate. If P is large compared to N, it would be wise to consider dimension reduction, which will be discussed later in this chapter.

Let's look at an example. The Python package `scikit-learn` contains several datasets for practice with modeling. We will utilize the `diabetes` dataset, which includes vital measurements and quantitative measures of the disease progression after one year. More information about this dataset can be found at this link: `https://scikit-learn.org/stable/datasets/toy_dataset.html#diabetes-dataset`. We will attempt to model the relationship between the vital statistics and the disease progression.

The explanatory variables in this dataset are `age`, `bmi`, `bp`, `s1`, `s2`, `s3`, `s4`, `s5`, and `s6`. The response variable is a measurement of disease progression. This is how we would express this model mathematically:

$$progression = \beta_0 + \beta_1 x_{age} + \beta_2 x_{bmi} + \beta_3 x_{bp} + \beta_4 x_{s1} + \beta_5 x_{s2} \\ + \beta_6 x_{s3} + \beta_7 x_{s4} + \beta_8 x_{s5} + \beta_9 x_{s6} + \epsilon$$

As discussed previously, we have a β for each variable in the model.

> **Transformations on the diabetes dataset**
>
> The explanatory variables in this dataset have been transformed from the original measurements. Referring to the dataset user's guide, we can see that each variable has been mean-centered and scaled. In addition, the `sex` variable has been converted from a categorical variable into a numerical variable.

Adding categorical variables

Up until this point, we have only considered continuous variables in our model, that is, variables that can have any value on the number line. Thinking back to *Chapter 2, Distributions of Data*, the variables we have considered before now have been ratio data. However, the linear regression model is not limited to using continuous variables. We can include categorical variables in the model. Recall from *Chapter 2, Distributions of Data*, categorical variables are associated with groups of items. In statistical learning, the groups are generally called **levels** of the variable. For example, if we had five students, three with MS degrees, one with a PhD degree, and one with a BS degree, we would say that the student categorical variable has three levels: BS, MS, and PhD. Let's look at how we include categorical variables in the model.

We will include the categorical variable in the model by including additional β terms. Specifically, we will add the number of levels (L) minus one additional β term. For the student example with three levels, we would add two additional β terms. Using one less than the number of levels will allow us to choose one level as a reference (or baseline) and compare the other levels to that reference level. We include the β terms in the model just like the other β terms. Let's say in this example, we are modeling income level, and in addition to the education levels, we have the experience of the individuals. Then, we would construct our model like this:

$$income = \beta_0 + \beta_1 x_{experience} + \beta_2 x_{ms} + \beta_3 x_{phd}$$

With the addition of the beta terms, we also must make a data transformation. We will have to create **dummy variables** for each of the *non-reference* levels in the categorical variable. This process is called the dummification of the categorical variable. In dummification, we create a column for each non-reference level and insert ones where the level occurs in the original variable and zeros where it does not appear in the original variable. This process is demonstrated in *Figure 7.1*.

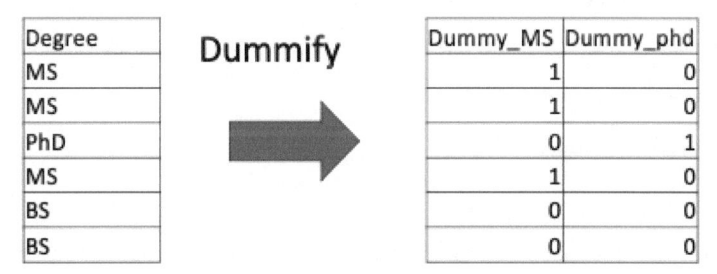

Degree		Dummy_MS	Dummy_phd
MS	Dummify →	1	0
MS		1	0
PhD		0	1
MS		1	0
BS		0	0
BS		0	0

Figure 7.1 – Dummification of a categorical variable

The original variable contains the BS, MS, and PhD levels. BS is chosen as the reference level and two dummy variables are created for the levels MS and PhD. This process of creating additional columns for categorical variables is called **encoding**. There are many types of variable encodings and encoding is a widely used technique in statistical learning and machine learning.

> **Categorical levels are shifts by a constant**
>
> Let's look a little deeper at the impact of categorical variables on the linear regression model. Earlier, we discussed how categorical variables are encoded with dummy variables that take on values of zero and one. Essentially, that means that we are switching the associated β terms, depending on the level, which will shift the response by a constant. For example, in the income model, when we want to calculate the income of an individual with a BS degree, the model resolves to the following:
>
> $$income = \beta_0 + \beta_1 x_{experience}$$
>
> And when we want to calculate the income of an individual with an MS degree, the model resolves to the following:
>
> $$income = \beta_0 + \beta_1 x_{experience} + \beta_2$$
>
> This means that levels of categories only shift the output of the model from the reference level by a constant, the beta associated with the level.

Returning to our previous example, our dataset includes a categorical variable: the sex of the patient. The dataset includes two levels for sex. We will choose one level to serve as a reference and we will create a dummy variable for the other level. With this knowledge, let's fit this data to a linear regression model and discuss the assumptions of multiple linear regression.

Evaluating model fit

Whenever we fit a parametric model, we should verify that the model assumptions are met. As discussed in the previous chapter, if the model assumptions are not met, the model may provide misleading results. In the previous chapter, we discussed the four assumptions for simple linear regression:

- A linear relationship between the response variable and explanatory variables
- Normality of the residuals
- Homoscedasticity of the residuals
- Independent samples

These assumptions also apply here, but in this new modeling context, we have an additional assumption: no or little **multicollinearity**. Multicollinearity occurs when two or more of the explanatory variables in an MLR model are highly correlated. The presence of multicollinearity impacts the statistical significance of the independent variables. Ideally, all the explanatory variables will be uncorrelated (or linearly independent). However, we can accept a small amount of multicollinearity without much impact on the model.

Let's evaluate the model on each of these assumptions.

Linear relationships

To check for linear relationships between the response variable and the explanatory variable, we can look at scatter plots of each variable against the response variable. Based on the plots in *Figure 7.2*, several of the variables, including `bmi`, `bp`, `s3`, `s4`, and `s5`, possibly exhibit a linear relationship with the response variable.

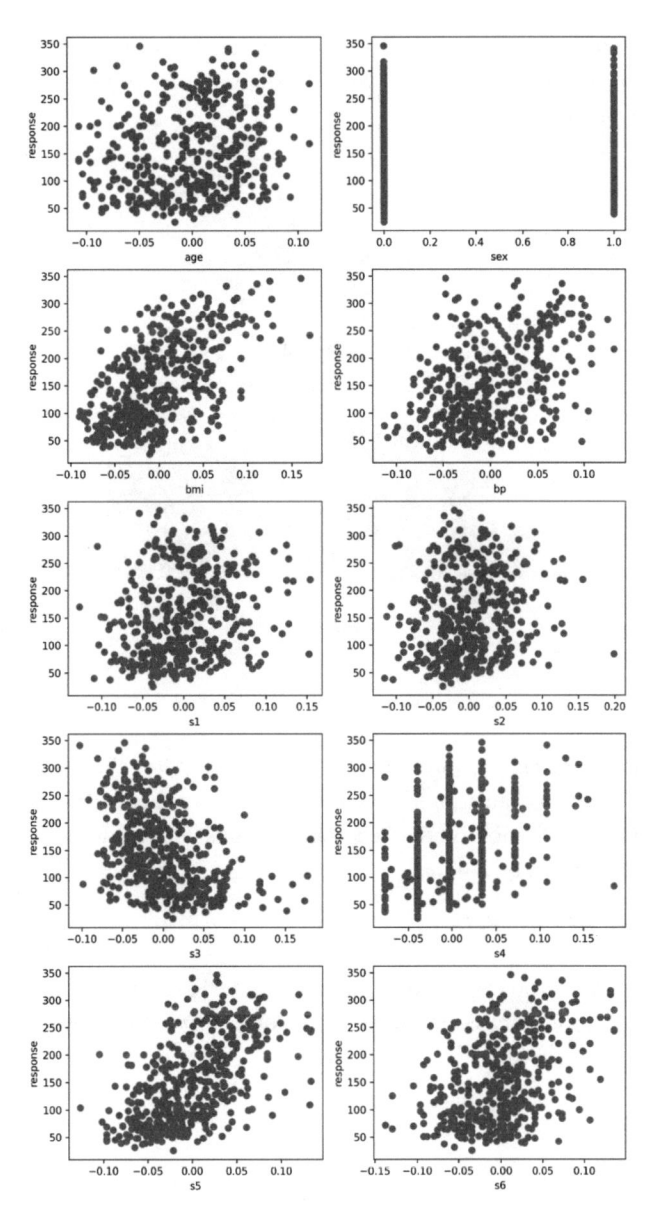

Figure 7.2 – Scatter plots of the response variable against the explanatory variables

While it would be ideal if any of the variables showed a strong linear relationship with the response variable, in an MLR model, we have the benefit of the combination of variables. We can add variables that appear to be more useful and remove variables that do not appear to be useful. This is called **feature selection**, which we will cover in the next section.

Normality of the residuals

Recall from the previous chapter that we expect the residuals from a well-fitted model to appear randomly distributed. We could look at this with a histogram or a QQ plot. The residuals from our model are shown in *Figure 7.3*.

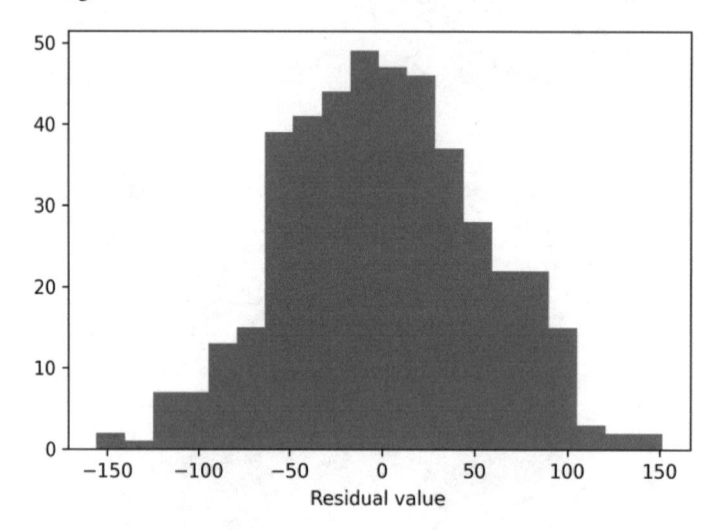

Figure 7.3 – Residual value

Based on the histogram in *Figure 7.3*, we cannot reject the possibility that the residuals are normally distributed. In general, when evaluating this assumption, we are checking for *egregious* violations. It turns out that the linear regression is relatively robust against this assumption. However, that does not mean this assumption can be ignored.

Homoscedasticity of the residuals

In the previous chapter, we also discussed homoscedasticity. For a well-fitted model, we expect that the residuals should exhibit homoscedasticity. The plot in *Figure 7.4* shows the scatter plot of the residuals against the values predicted by the model.

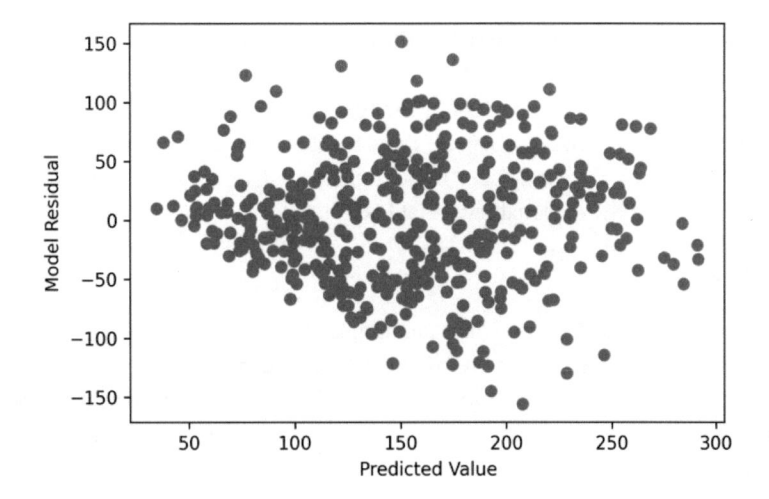

Figure 7.4 – Scatter plot of model residuals for the predicted versus actual values

There does not appear to be a clear pattern of changing variance or a significant outlier that would violate the assumption of homoscedasticity. If there had been a pattern, that would have been a sign that one or more of the variables may need to be transformed, or a sign of a non-linear relationship between the response and the explanatory variables.

Independent samples

In the last chapter, we discussed independent sampling and its impact on this type of model. However, we cannot make a certain determination on whether the samples are independent without knowing the sampling methodology. Since we do not know the sampling strategy for this dataset, we will assume this assumption is met and proceed with the model. In a real modeling setting, this assumption should never be taken for granted.

Multicollinearity

The new assumption for MLR is that there is little or no multicollinearity in the explanatory variables. Multicollinearity is a situation that occurs when two or more variables are strongly linearly correlated. We commonly use the **variance inflation factor** (**VIF**) to detect multicollinearity. The VIF is a measurement of how much the coefficient of an explanatory variable is influenced by other explanatory variables. A lower VIF is better where the minimum value is 1, meaning there is no correlation. We generally consider a VIF of 5 or more to be too high. When a high VIF is detected in a set of explanatory variables, we repeatedly remove the variable with the highest VIF until the VIF values for each variable are below 5. Let's look at an example with our current data. The process of removing variables with high VIFs is shown in *Figure 7.5*.

VIF of Initial Variable Set

	vif
const	2.125790
age	1.217307
sex	1.278071
bmi	1.509437
bp	1.459428
s1	59.202510
s2	39.193370
s3	15.402156
s4	8.890986
s5	10.075967
s6	1.484623

Remove S1 and Recalculate VIF →

	vif
const	2.123128
age	1.216892
sex	1.275049
bmi	1.502320
bp	1.457413
s2	2.926535
s3	3.736890
s4	7.818670
s5	2.172865
s6	1.484410

Remove S4 and Recalculate VIF →

VIF of Final Variable Set

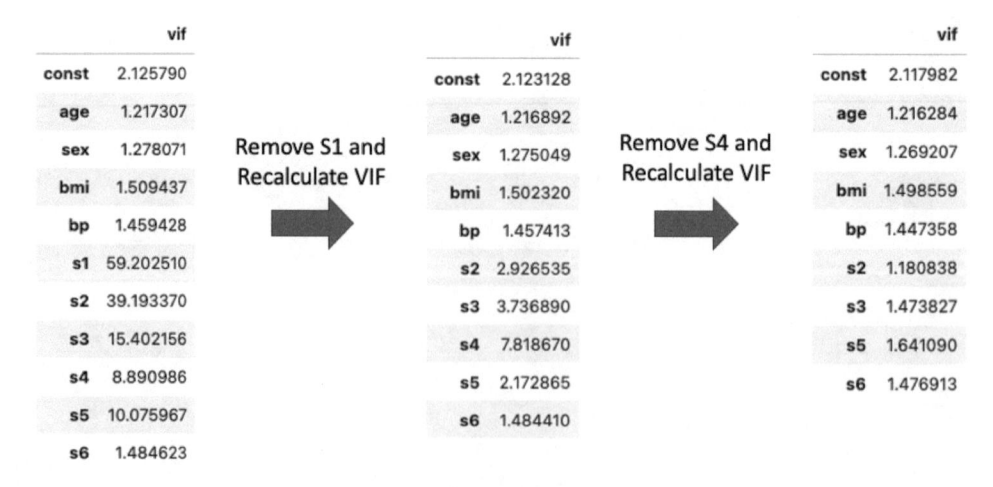

	vif
const	2.117982
age	1.216284
sex	1.269207
bmi	1.498559
bp	1.447358
s2	1.180838
s3	1.473827
s5	1.641090
s6	1.476913

Figure 7.5 – Removing high VIF variables from a dataset

Figure 7.5 shows the process of removing variables with a high VIF from the diabetes dataset. The leftmost table in the figure shows the original dataset where the highest VIF is 59.2, which corresponds to variable S1. Then, we remove this variable from the dataset and recalculate the VIF. Now we see that the highest VIF is 7.8, corresponding to variable `s4`. We remove this variable and recalculate the VIF. Now, all VIFs are below 5, indicating that there is a low correlation between the remaining variables. With these variables removed, we need to fit the model again.

With the model fit and the assumptions of the model verified, let's look at the fit results and discuss how to interpret the results.

Interpreting the results

Fitting the model, we get the following results from `statsmodels`. The output is divided into three sections:

- The top section contains high-level statistics about the model
- The middle section contains details about the model coefficients
- The bottom section contains diagnostic tests about the data and the residuals

Let's walk through each section of this model.

```
                            OLS Regression Results
==============================================================================
Dep. Variable:                      y   R-squared:                       0.513
Model:                            OLS   Adj. R-squared:                  0.504
Method:                 Least Squares   F-statistic:                     57.12
Date:                Tue, 25 Oct 2022   Prob (F-statistic):           4.35e-63
Time:                        21:24:25   Log-Likelihood:                -2387.9
No. Observations:                 442   AIC:                             4794.
Df Residuals:                     433   BIC:                             4831.
Df Model:                           8
Covariance Type:            nonrobust
==============================================================================
                 coef    std err          t      P>|t|      [0.025      0.975]
------------------------------------------------------------------------------
const         162.5006      3.757     43.257      0.000     155.117     169.884
age            -8.5365     59.850     -0.143      0.887    -126.170     109.097
sex           -22.1366      5.828     -3.798      0.000     -33.591     -10.682
bmi           526.9864     66.433      7.933      0.000     396.415     657.558
bp            317.2370     65.288      4.859      0.000     188.915     445.559
s2           -109.9931     58.972     -1.865      0.063    -225.900       5.913
s3           -288.4990     65.883     -4.379      0.000    -417.989    -159.009
s5            480.2121     69.521      6.907      0.000     343.572     616.852
s6             71.3085     65.952      1.081      0.280     -58.317     200.934
==============================================================================
Omnibus:                        1.795   Durbin-Watson:                   2.016
Prob(Omnibus):                  0.408   Jarque-Bera (JB):                1.629
Skew:                           0.038   Prob(JB):                        0.443
Kurtosis:                       2.713   Cond. No.                         35.4
==============================================================================

Notes:
[1] Standard Errors assume that the covariance matrix of the errors is correctly speci
fied.
```

Figure 7.6 – Results from statsmodels OLS regression

High-level statistics and metrics (top section)

In the top section of the fit results, we have model-level information. The left side of the top section contains information about the model such as the **degrees of freedom** (**df**) and the number of observations. The right side of the top section contains model metrics. Model metrics are useful for comparing models. We will discuss more about model metrics in the section on feature selection.

Model coefficient details

The middle section contains details about the model coefficients (the β terms in the equations listed previously). For the purposes of this section, we will focus on two columns in the middle section: coef and P>|t|. The coef column is the model coefficient estimated for the model equation (the estimate of β or termed $\hat{\beta}$). The column labeled P>|t| is the p-value for a significance test for the coefficient. We are interested in both columns for interpreting the model. Let's start with the p-value column.

The null hypothesis for this test is that **the value of the β parameter is equal to zero**. Recall the following from *Chapter 3, Hypothesis Testing*:

- A p-value below the significance threshold means we reject the null hypothesis
- A p-value above the significance threshold means that we fail to reject the null hypothesis

In this example, we would reject the null hypothesis for the bmi variable, but we fail to reject the null hypothesis for the age variable. Once we have determined the significant variables, we can move on to interpreting the meaning of the significant variables. We will not be able to provide an interpretation of the other variables because we cannot reject that their coefficient values might be zero. If the coefficient value is zero, then the variable makes no contribution to the model. Let's look at how to interpret the coefficients.

Often, when we construct a model, we want to understand how the parts of the model affect the output. In an MLR model, this comes down to understanding the coefficients of the model. We have two types of variables, continuous and categorical, and the associated coefficients have different meanings.

Interpreting continuous variable coefficients

For the continuous variables, such as BMI, as the value of the variable increases, so does its significance to the output of the model. A unit increase in the value of the variable is associated with an increase in the mean of the dependent variable by the size of the coefficient with all other variables held constant. Let's take the bmi variable as an example. Since the coefficient of the bmi variable is approximately 526, we would say, "A unit increase in bmi would be associated with a 526 increase in the mean of the diabetes measurement with all other variables held constant." Of course, the coefficients can also take on negative values.

Interpreting categorical variable coefficients

For the categorical variables, such as sex, recall that, unlike the continuous variables, the values of the categorical are dummy-encoded, and therefore can only take on two values: zero and one. Also, recall that one level was chosen to be the reference level. In this case, sex level 0 is the reference level, and we can compare this reference to sex level 1. When we use the reference level, the coefficient will not affect the output of the model. Thus, the categorical-level change is associated with a change in the mean of the dependent variable by the size of the coefficient with all other variables held constant. Since the coefficient of the sex variable is approximately -22, we would say, "The level of sex is associated with a decrease of 22 in the mean of the diabetes measurement compared to the reference level with all other variables held constant."

Diagnostic tests

The bottom section of the fit results contains diagnostic statistics for the data and residuals. Glancing over the list, several should be from *Chapter 6, Simple Linear Regression*. The Durbin-Watson test is a test for serial correlation (data sampled sequentially over time). A result around 2 is not indicative of serial correlation. The skew and kurtosis are measurements of the shape of the distribution of the residuals. These results indicate almost no skew, but possibly some kurtosis. These are likely small deviations from the normal distribution and are not cause for concern as we saw in the plots earlier.

In this section, we looked at our first model that can use multiple explanatory variables to predict a response variable. However, we noticed that several of the variables included in the model were not statistically significant. For this model, we only selected features by removing any features that had high VIF scores, but there are other methods to consider when choosing features. In the next section, we will discuss comparing models and feature selection.

Feature selection

The are many factors that influence the success or failure of a model, such as sampling, data quality, feature creation, and model selection, several of which we have not covered. One of those critical factors is **feature selection**. Feature selection is simply the process of choosing or systematically determining the best features for a model from an existing set of features. We have done some simple feature selection already. In the previous section, we removed features that had high VIFs. In this section, we will look at some methods for feature selection. The methods presented in this section fall into two categories: statistical methods for feature selection and performance-based methods for feature selection. Let's start with statistical methods.

Statistical methods for feature selection

Statistical methods for feature selection rely on the primary tool that we have used throughout the previous chapters: statistical significance. The methods presented in this sub-section will be based on the statistical properties of the features themselves. We will cover two statistical methods for feature selection: correlation and statistical significance.

Correlation

The first statistical method we will discuss is **correlation**. We have discussed correlation in this chapter and in previous chapters; recall that correlation is a description of the relationship between two variables. Variables can be positively correlated, uncorrelated, or negatively correlated. In terms of feature selection, we want to *remove features that are uncorrelated with the response variable*. A feature that is uncorrelated with the response variable does not have a relationship with the response variable. Thus, an uncorrelated feature would not be a good predictor of the response variable.

Recall from *Chapter 4, Parametric Tests*, that we can use Pearson's correlation coefficient to measure the linear correlation between two variables. In fact, we can calculate the correlation coefficient between all features and the target variable. After performing those calculations, we can construct a correlation ranking as shown in *Figure 7.7*.

Features Correlated with Response

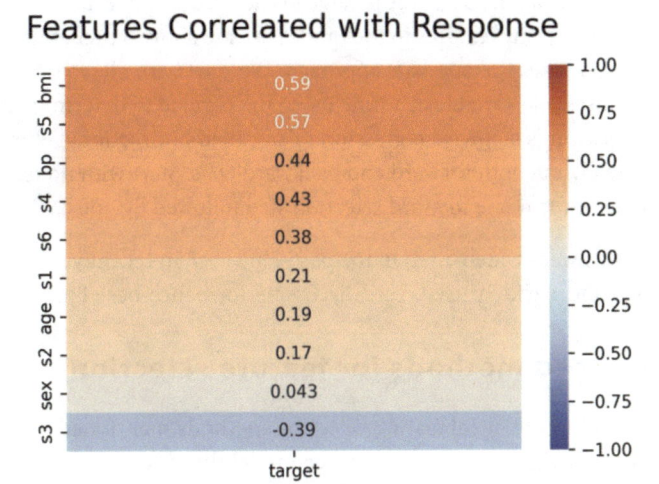

Feature	Correlation
bmi	0.59
s5	0.57
bp	0.44
s4	0.43
s6	0.38
s1	0.21
age	0.19
s2	0.17
sex	0.043
s3	-0.39

Figure 7.7 – Feature correlation ranking

When evaluating features based on correlation, we are most interested in features with a high *absolute* correlation. For example, the correlation ranking in *Figure 7.7* shows the following:

- `bmi` and `s5` exhibit a strong correlation with the response variable
- `bp`, `s4`, `s6`, and `s3` exhibit moderate correlation with the response variable
- `s1`, `age`, and `s2` exhibit a weak correlation with the response variable

While `sex` may appear to show no correlation with the response variable, Pearson's correlation coefficient cannot be used with categorical features. From this correlation ranking, we can see that, at least, `bmi`, `s5`, and `bp`, are likely among the best features in this dataset for predicting the response. In fact, these features were considered statistically significant in our model. Now let's discuss selection using statistical significance.

Statistical significance

Assessing features using correlation is generally a good first step for feature selection. We can easily eliminate features that are uncorrelated with the response variable. However, depending on the problem, we could still be left with many features that are correlated with the response. We can further select features using the statistical significance of the feature in the context of a model. However, in recent years, these methods for feature selection have somewhat fallen out of favor in the community. For this reason, we will not focus on these methods, but will only describe them for understanding.

Recall when we fit the MLR model, the results included a test for statistical significance for each feature in the model. We can use that test to select features. There are three well-known algorithms for selecting features based on statistical significance: **forward selection**, **backward selection**, and **stepwise regression**. In forward selection, we start without any variables in the model and then iteratively add one variable at a time using the p-value to choose the best feature to add at each iteration. We stop once the p-values of any features in the model size are above a predefined threshold, such as 0.05. Backward selection takes the opposite approach. We start with all features in the model, then iteratively remove features one variable at a time using the p-value to determine the least important feature. The final algorithm is stepwise regression (also called bidirectional elimination). Stepwise regression is performed using both forward and backward tests. Start with no features in the model, then in each iteration, perform one forward selection step followed by one backward selection pass.

These selection methods were widely used in the past. However, in recent years, performance-based methods have become more widely used. Let's discuss performance-based methods now.

Performance-based methods for feature selection

The primary issue with the statistical feature selection methods mentioned previously is that they tend to create **overfit** models. An overfit model is a model that fits the given data exactly and fails to generalize to new data. Performance-based methods overcome overfitting using a method called **cross-validation**. In cross-validation, we have two datasets: a **training dataset** used to fit the model and a **test dataset** for evaluating the model. We can build models from multiple sets of features, fit all those potential models on the training set, and finally rank them based on the performance of the models on the testing set with a given metric.

Comparing models

Before we get into feature selection methods, let's first discuss how to compare models. We use metrics to compare models. On a basic level, we can use metrics to help us determine whether one set of features is better than another set of features based on model performance. Many metrics can be used for comparing models. We will discuss two metrics, **mean square error** (**MSE**) and **mean absolute percentage error** (**MAPE**), which are, by far, two of the most commonly used metrics for regression models.

MSE

The MSE is given by the following formula, where N is the number of samples, y is the response variable, and $\hat{y}$ is the predicted value of the response variable.

$$MSE = \frac{1}{N} \sum_{i=1}^{N} (y_i - \hat{y}_i)^2$$

In other words, take the differences between the response values and the predicted response values, square the differences, and finally take the mean of the squared differences. A small extension of this metric commonly used is the **root mean squared error** (**RMSE**), which is simply the square root of the MSE. The RMSE is used when it is desirable for the metric to have the same units as the response variable.

MAPE

The MAPE is given by the following formula, where N is the number of samples, y is the response variable, and $\hat{y}$ is the predicted value of the response variable:

$$MAPE = \frac{100\%}{N} \sum_{i=1}^{N} \left| \frac{y_i - \hat{y}_i}{y_i} \right|$$

This formula is like the formula for the MSE, but instead of taking the mean of the squared error, we take the mean of the percent error. This makes the MAPE easier to interpret than the MSE, which is a distinct advantage over the MSE.

Now that we have discussed model validation and metrics, let's put these concepts together to perform feature selection using model performance as an indicator.

Recursive feature elimination

Recursive feature elimination (**RFE**) is a method for selecting an optimal number of features in a model using a metric. Much like the backward selection method mentioned previously, the RFE algorithm starts with all features in the model, then removes features with the least influence on the model. At each step, cross-validation is performed. When RFE is completed, we will be able to see the cross-validation performance of the model over the various sets of features.

In this example, we use the linear regression implementation and RFE implementation from `scikit-learn` (`sklearn`), which is a primary package used for machine learning in the Python ecosystem. In the following code example, we set up RFE to use the MAPE as the scoring metric (`make_scorer(mape ,greater_is_better=False)`), remove one feature at each step (`step=1`), and indicate with `cv=2` that it should score the model using two cross-validation sets:

```
linear_model = LinearRegression()
linear_model.fit(X, y)
rfecv = RFECV(
    estimator=linear_model,
    step=1,
    cv=2,
    scoring=make_scorer(mape ,greater_is_better=False),
    min_features_to_select=1
)
rfecv.fit(X,y)
```

Once we fit the RFE object, we can look at the results to see how to model scored over the various sets of features. The performance of the model using the MAPE is shown in *Figure 7.8*.

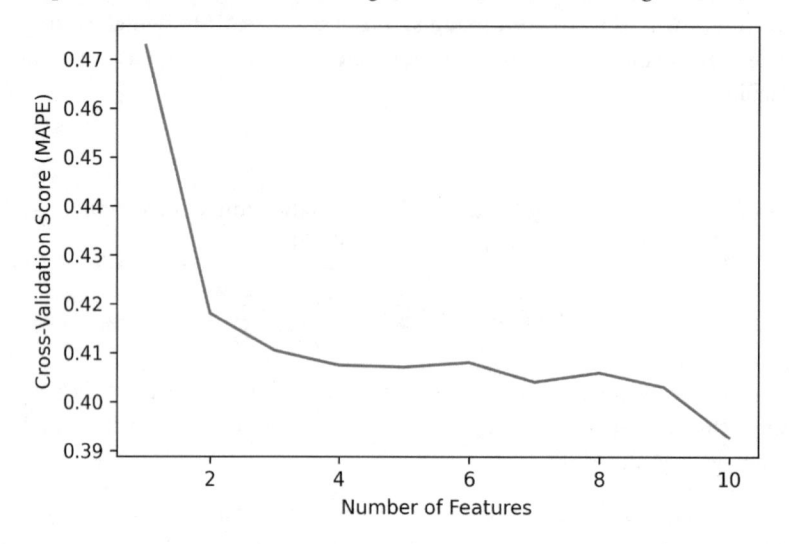

Figure 7.8 – Linear regression performance over RFE steps, scoring with the MAPE

It's clear that including all features produces the best-performing model but including all 10 features only provides a small increase over the performance of just five features. While the best-performing model contains all 10 features, we would still need to consider model assumptions and verify that the model is well fit.

In this section, we have looked at several methods for feature selection, including statistical methods and performance-based methods. We also discussed metrics and how to compare models, including the reason for splitting the data into training and validation sets. In the next section, we will look at linear regression shrinkage methods. These types of models use a method called regularization, which in some ways acts like model-based feature selection.

Shrinkage methods

The **bias-variance trade-off** is a decision point all statistics and machine learning practitioners must balance when performing modeling. Too much of either renders results useless. To catch these when they become issues, we look at test results and the residuals. For example, assuming a useful set of features and the appropriate model have been selected, a model that performs well on validation, but poorly on a test set could be indicative of too much variance and conversely, a model that fails to perform well at all could have too much bias. In either case, both models fail to generalize well. However, while bias in a model can be identified in poor model performance from the start, high variance can be notoriously deceptive as it has the potential to perform very well during training and even during validation, depending on the data. High-variance models frequently use values of coefficients that are unnecessarily high when very similar results can be obtained from coefficients that are not. Further, in using coefficients that are not unnecessarily high, the model is more likely to be in bias-variance equilibrium, which provides it a better chance of generalizing well on future data. Additionally, more reliable insights into the influence of current factors on a given target are provided when model coefficients are not exaggerated, which aids in more useful descriptive analysis. This brings us to the concept of shrinkage.

Shrinkage is a method that reduces model variance by shrinking model parameter coefficients toward zero. The coefficients are derived by applying least squares regression to all variables considered for a model. The amount of shrinkage is based on the parameters' contribution to least squares estimates; parameters that contribute to a high level of squared error will have their coefficients pushed toward zero or zeroed out altogether. In this case, shrinkage can be used for variable elimination. Variables that do not contribute to high squared error across the model fit will have a minimal reduction in coefficient value, thus being useful for model fitting, assuming the practical purpose of their inclusion is vetted. The reason shrinkage – also called **regularization** – is important is because it helps models include useful variables while preventing them from introducing too much variance and thus overfitting. Preventing excess variance is particularly useful for ensuring that models generalize well over time. Let's look at some of the most common shrinkage techniques, **ridge regression** and **Least Absolute Shrinkage and Selection Operator (LASSO) Regression**.

Ridge regression

Recall the formulation for the **residual sum of squared errors** (RSS) for multiple linear regression using least squares regression is as follows:

$$RSS = \sum_{i=1}^{n} \left(y_i - \hat{\beta}_0 - \sum_{j=1}^{p} \hat{\beta}_j x_{ij} \right)^2$$

where n is the number of samples and p is the number of parameters. Ridge regression adds a scalar, λ, called a **tuning parameter** – which must be greater than or equal to 0 – that gets multiplied by the model parameter coefficient estimates, $\hat{\beta}_j^2$, to create a **shrinkage penalty**. This gets added back to the RSS equation such that the new least squares regression's fitting procedure is now defined as follows:

$$RSS + \lambda \sum_{j=1}^{p} \hat{\beta}_j^2$$

When $\lambda=0$, the respective ridge regression **penalty term** is 0. However, as $\lambda \longrightarrow \infty$, the model coefficients shrink toward zero. The new fitting procedure can be rewritten as follows:

$$\|y - X\beta\|_2^2 + \lambda \|\beta\|_2^2$$

where RSS = $\|y - X\beta\|_2^2$, $\|\beta\|_2^2$ is the squared L^2 **norm** (Euclidean norm) of the regression coefficients array, and X is the design matrix. Further simplification brings this to the following:

$$\|y - X\beta\|_2^2 + \lambda \beta^T \beta$$

which can be rewritten in the closed form by using the tuning parameter's scalar multiple of the identity matrix to derive the **ridge regression coefficient estimates**, $\hat{\beta}^R$, as follows:

$$\hat{\beta}^R = (X^T X + \lambda I)^{-1} X^T y$$

Simply stated, the ridge regression coefficient estimates are the set of coefficients that minimizes the least squares regression output as follows:

$$\hat{\beta}^R = \text{argmin} \left\{ \sum_{i=1}^{n} \left(y_i - \hat{\beta}_0 - \sum_{j=1}^{p} \hat{\beta}_j x_{ij} \right)^2 + \lambda \sum_{j=1}^{p} \hat{\beta}_j^2 \right\}$$

> **Standardizing coefficients prior to applying ridge regression**
>
> Because ridge regression seeks to minimize the error for the entire dataset and the tuning parameter for doing so is applied within the L^2 norm normalization process of taking the root of summed squared terms, it is important to apply a standard scaler to each variable in the model so that all variables are on the same scale, prior to applying ridge regression. If this is not performed, ridge regression will almost certainly fail to be useful for helping the model generalize across datasets.

In summary, ridge regression reduces variance in a model by using the **L^2 penalty**, which penalizes the sum of squared coefficients. However, it is important to note the **L^2 norm**, and consequently, ridge regression, **will never produce a zero-valued coefficient**. Therefore, ridge regression is an excellent tool for reducing variance when the analyst seeks to use all terms in the model. However, ridge regression cannot be used for variable elimination. For variable elimination, we can use the LASSO regression shrinkage method, which uses the **L^1 penalty** to regularize coefficients with the **L^1 norm**, which uses the absolute value to shrink values to and including zero.

Let's walk through an example of a ridge regression implementation in Python.

First, let's load the Boston home prices dataset from `scikit-learn`. In the final line, we add the constant for the intercept. Note that this process can be repeated with comparable results, albeit with different input variables, using the California housing data set by running `from sklearn.datasets import fetch_california_housing` in place of `from sklearn.datasets import load_boston`:

```
from sklearn.metrics import mean_squared_error as MSE
from sklearn.model_selection import train_test_split
from sklearn.preprocessing import StandardScaler
```

```
from sklearn.datasets import load_boston
import statsmodels.api as sm
import pandas as pd

boston_housing = load_boston()
df_boston = pd.DataFrame(boston_housing.data, columns = boston_
housing.feature_names)
df_boston['PRICE'] = boston_housing.target
df_boston = sm.add_constant(df_boston, prepend=False
```

The first three records are given here:

CRIM	ZN	INDUS	CHAS	NOX	RM	AGE	DIS	RAD	TAX	PTRATIO	B	LSTAT	PRICE	const
0.00632	18	2.31	0	0.538	6.575	65.2	4.09	1	296	15.3	396.9	4.98	24	1
0.02731	0	7.07	0	0.469	6.421	78.9	4.9671	2	242	17.8	396.9	9.14	21.6	1
0.02729	0	7.07	0	0.469	7.185	61.1	4.9671	2	242	17.8	392.83	4.03	34.7	1

Figure 7.9 – First three records of the Boston housing data

We set `PRICE` as our target. Recall, ridge regression is most useful when there are either more parameters than samples or there is excessive variance. Let's assume both of these assumptions are met:

```
X = df_boston.drop('PRICE', axis=1)
y = df_boston['PRICE']
```

As noted previously, **data must be standardized** for ingestion into the L^2 norm, so let's apply that with scikit-learn's `StandardScaler` function:

```
sc = StandardScaler()
X_scaled = sc.fit_transform(X)
```

Next, let's take a 75/25 train/test split of the data for the model. We use `shuffle=True` to randomly shuffle the data so we test with a random sample, which is more likely to be representative of the population:

```
X_train, X_test, y_train, y_test = train_test_split(
    X_scaled, y, test_size=0.25, shuffle=True)
```

Now we can compare **ordinary least squares** (**OLS**) regression to OLS regression using coefficients regularized through ridge regression. In the imported function here, `fit_regularized`, we set the required argument for `method` to `'elastic_net'`. We will discuss Elastic Net shortly, but for now, note the `L1_wt` argument applies ridge regression **when set to 0**. Alpha is the tuning parameter, λ, in the ridge regression penalty term. A small alpha allows for large coefficients and a large alpha pushes the coefficients toward zero. Here, we fit the training data to derive the mean squared error:

```
ols_model = sm.OLS(y_train, X_train)
compiled_model = ols_model.fit()
```

```
compiled_model_ridge = ols_model.fit_regularized(method = 'elastic_
net', L1_wt=0, alpha=0.1,refit=True)
print('OLS Error: ', MSE(y_train,
    compiled_model.predict(X_train)) )
print('Ridge Regression Error: ', MSE(y_train,
    compiled_model_ridge.predict(X_train)))
```

We can see ridge regression has a slightly higher amount of error

`OLS Error:    530.7235449265926`

`Ridge Regression Error:    533.2278083730833`

Next, we fit the test data to see how the model generalizes on unseen data. We measure again with the mean squared error:

```
print('OLS Error: ', MSE(y_test, compiled_model.predict(X_test)) )
print('Ridge Regression Error: ', MSE(y_test, compiled_model_ridge.
predict(X_test)))
```

We can see that the number of errors increased for both on the test data. However, the OLS regression produced a slightly higher error in proportion to the ridge regression approach:

`OLS Error:    580.8138216493896`

`Ridge Regression Error:    575.5186673728349`

Here, we can observe the OLS regression coefficients and the regularized coefficients in *Figure 7.10*:

	CRIM	ZN	INDUS	CHAS	NOX	RM	AGE	DIS	RAD	TAX	PTRATIO	B	LSTAT	const
Before ridge regression	-0.43	-0.84	0.14	1.33	-1.07	2.86	-0.72	-1.86	-0.05	0.48	-1.82	1.36	-4.69	0.00
After ridge regression	-0.47	-0.69	0.04	1.29	-0.76	2.85	-0.62	-1.41	-0.01	0.10	-1.62	1.28	-4.08	0.00

Figure 7.10 – OLS regression coefficients before and after ridge regression regularization

LASSO regression

Earlier in this section, we mentioned how variable coefficients with unnecessarily high values contribute to model variance, and in so doing, take away a model's ability to generalize as well as it could were the coefficients reasonable. We demonstrated applying ridge regression to enact this. A very popular alternative to ridge regression, however, is LASSO regression. LASSO regression follows a similar procedure of adding a penalty term to the model's residual sum of squared error and seeks to minimize the resulting value (error). However, LASSO uses the L^1 norm rather than L^2. Consequently, it is possible to obtain absolute zero coefficient values as the tuning parameter reaches a sufficient size, thus functioning as a feature selection and shrinkage tool.

The LASSO equation seeks to minimize overall model error by shrinking each variable's coefficient using the following method:

$$\widehat{\beta}^L = \operatorname{argmin}\left\{\sum_{i=1}^{n}\left(y_i - \widehat{\beta}_0 - \sum_{j=1}^{p}\widehat{\beta}_j x_{ij}\right)^2 + \lambda \sum_{j=1}^{p}|\widehat{\beta}_j|\right\}$$

The only difference between LASSO and ridge regression is the penalty term $|\widehat{\beta}_j|$.

> **Choosing λ**
>
> The value of the tuning parameter, λ, is best selected using cross-validation. While the tuning parameter can go to infinity, in theory, it is typical to start with values less than 1, such as at 0.1 increasing by increments of tenths up to 1. After, typically integer values are used.

Using the same data we used for ridge regression, we apply LASSO regression. Here, we set `L1_wt=1`, indicating the L^1 norm will be applied:

```
ols_model = sm.OLS(y_train, X_train)
compiled_model = ols_model.fit()
compiled_model_lasso = ols_model.fit_regularized(method='elastic_net',
L1_wt=1, alpha=0.1,refit=True)
```

We will follow the same steps for LASSO as with ridge regression in that we first check the errors on the training data, then again on the test data to see how the models generalize:

```
print('OLS Error: ', MSE(y_train, compiled_model.predict(X_train)) )
print('LASSO Regression Error: ', MSE(y_train, compiled_model_lasso.
predict(X_train)))
```

The output shows OLS regression slightly outperforming LASSO. This could be due to higher variance. However, practically speaking, the results are the same:

OLS Error: 530.7235449265926

LASSO Regression Error: 531.2440812254207

Now, we need to check the model's performance on the holdout data:

```
print('OLS Error: ', MSE(y_test, compiled_model.predict(X_test)) )
print('LASSO Regression Error: ', MSE(y_test, compiled_model_lasso.
predict(X_test)))
```

We can see the two models again have essentially the same error. However, with fewer features, the LASSO model may be easier to trust to generalize on future data. That may depend on the researcher's level of subject knowledge, however:

```
OLS Error: 546.1338399374557

LASSO Regression Error: 546.716239805892
```

As with ridge regression, some of the coefficients have been minimized in the L^1 norm regularization process. However, we also see, using the same alpha value, four variables have been minimized to zero, thus eliminating them from the model altogether. Three of the features were comparatively shrunk almost to zero by ridge regression, but not ZN, which we see in *Figure 7.11* has been reduced to 0. The model error has been slightly improved, which may not appear significant, but when considering the elimination of four variables, we can consider the model to have more generalization with less dependence on exogenous variables.

	CRIM	ZN	INDUS	CHAS	NOX	RM	AGE	DIS	RAD	TAX	PTRATIO	B	LSTAT	const
Before LASSO regression	-0.434	-0.845	0.144	1.326	-1.078	2.86	-0.729	-1.869	-0.059	0.484	-1.821	1.359	-4.692	0.000
After LASSO regression	-0.428	0.000	0.000	1.321	-0.826	2.731	-0.643	-2.382	0.000	0.000	-1.473	1.327	-4.722	0.000

Figure 7.11 – OLS regression coefficients before and after LASSO regression regularization

Elastic Net

Elastic Net is another common shrinkage method that can be applied to manage a bias-variance trade-off. This method applies the tuning parameter to a combination of ridge and LASSO regression where the proportion of influence from either is determined by the hyperparameter α. The equation Elastic Net minimizes is as follows:

$$(\hat{\beta}_0, \hat{\beta}) = \operatorname{argmin}\left\{\frac{1}{2n}\sum_{i=1}^{n}\left(y_i - \hat{\beta}_0 - \sum_{j=1}^{p}\hat{\beta}_j x_{ij}\right)^2 + \lambda\left(\frac{1-\alpha}{2}\sum_{j=1}^{p}\hat{\beta}_j^2 + \alpha\sum_{j=1}^{p}|\hat{\beta}_j|\right)\right\}$$

Naturally, depending on the values of α, Elastic Net can also generate absolute zero-valued coefficient parameter estimates where it cancels out the ridge regression penalty term. When all input variables are needed – for example, if they have already been pruned following procedures such as those outlined in the *Feature selection* section of this chapter – Elastic Net is most likely to outperform both ridge and LASSO regression, especially when there are highly correlated features in the dataset that must be included to capture necessary variance. In the next section, we will discuss dimension reduction. Specifically, we p rovide an in-depth overview of PCR, which uses **Principal Component Analysis (PCA)** to extract useful information from systems that contain correlated features required to evaluate the target.

First, let's walk through a regression example using Elastic Net using the same data we used for ridge and LASSO regression. Following the previous Elastic Net minimizing equation, we set `Lt_wt=0.5`, meaning an equal, 50/50 balance between ridge and LASSO regression. Differently, however, we applied `alpha=8` instead of `0.1` as we used in ridge and LASSO regression to gain an improvement over the OLS regression coefficients. Recall that as the tuning parameter approaches infinity, coefficients approach 0. Therefore, we can conclude based on the Elastic Net coefficients that 8 is a very high tuning parameter:

```
ols_model = sm.OLS(y_train, X_train)
compiled_model = ols_model.fit()
compiled_model_elastic = ols_model.fit_regularized(method='elastic_
net', L1_wt=0.5, alpha=8,refit=True)
```

Let's test the model on the training data:

```
print('OLS Error: ', MSE(y_train, compiled_model.predict(X_train)) )
print('Elastic Net Regression Error: ', MSE(y_train, compiled_model_
elastic.predict(X_train)))
```

Here, we see Elastic Net has added errors into the model compared to OLS regression:

```
OLS Error: 530.7235449265926

Elastic Net Regression Error: 542.678919923863
```

Now let's check the model errors for the holdout data:

```
print('OLS Error: ', MSE(y_test, compiled_model.predict(X_test)) )
print('Elastic Net Regression Error: ', MSE(y_test, compiled_model_
elastic.predict(X_test)))
```

Observing the results in *Figure 7.12*, we can see how Elastic Net has traded variance – which increased the error on training – for bias, which has enabled the model to generalize better on holdout. We can see better results with Elastic Net than OLS regression on holdout. However, the improved error suggests the added bias provided a better chance of lower error, but is not something to be expected:

```
OLS Error: 546.1338399374557

Elastic Net Regression Error: 514.8301731640446
```

Here we can see the coefficients before and after Elastic Net's implementation:

	CRIM	ZN	INDUS	CHAS	NOX	RM	AGE	DIS	RAD	TAX	PTRATIO	B	LSTAT	const
Before Elastic Net	-0.43	-0.84	0.14	1.33	-1.07	2.86	-0.72	-1.86	-0.05	0.48	-1.82	1.36	-4.69	0.00
After Elastic Net	0.00	0.00	0.00	0.00	0.00	3.05	0.00	0.00	0.00	0.00	0.00	0.00	-5.45	0.00

Figure 7.12 – OLS regression coefficients before and after Elastic Net regularization

We can see with a sufficiently large tuning parameter value (`alpha=8`), most of the coefficients with the balanced Elastic Net have been forced to absolute zero. The only coefficients remaining had comparatively large values with OLS regression. Notably, for the variables that remain with coefficients (`RM` and `LSTAT`), Elastic Net increased both coefficients where ridge regression and LASSO either reduced them slightly.

Dimension reduction

In this section, we will use a specific technique – **PCR** – to study MLR. This technique is useful when we need to deal with a multicollinearity data issue. Multicollinearity occurs when an independent variable is highly correlated with another independent variable, or an independent variable can be predicted from another independent variable in a regression model. A high correlation can affect the result poorly when fitting a model.

The PCR technique is based on PCA as used in unsupervised machine learning for data compression and exploratory analysis. The idea behind it is to use the dimension reduction technique, PCA, on these original variables to create new uncorrelated variables. The information obtained on these new variables helps us to understand the relationship and then apply the MLR algorithm to these new variables. The PCA technique can also be used in a classification problem, which we will discuss in the next chapter.

PCA – a hands-on introduction

PCA is a dimension reduction technique based on linear algebra by linearly transforming data using linear combinations of original variables into a new coordinate system. These new linear combinations are called **principal components** (**PCs**). The difference between the PCA technique and the feature selection technique or shrinkage methods mentioned in previous sections is that the original independent variables are maintained but the new PC variables are transformed into a new coordinate space. In other words, PCA uses the original data to arrive at a new representation or a new structure. The number of PC variables is the same as the number of original independent variables, but these new PC variables are uncorrelated with each other. The PC variables are created and ordered from the most to the least amount of variability. The sum of variances is the same between the original independent variables and the newly transformed PC variables.

Before conducting PCA, we perform pre-processing for the dataset by subtracting the mean from each data point and normalizing the standard deviation of each independent variable. In a high-level structure, the goal of the PCA technique is to find vectors $v_1, v_2, ..., v_k$ such that any data point, x, in the dataset can be approximately represented as a linear combination:

$$x = \sum_{i=1}^{k} a_i v_i$$

for some constants a_i with $i = \overline{1, k}$. Assume that we have these k vectors; then, each data point can be written as a vector in R^k corresponding the projections:

$$x = \langle x, v_1 \rangle \cdot v_1 + \langle x, v_2 \rangle \cdot v_2 + ... + \langle x, v_k \rangle \cdot v_k$$

In other words, if we have d original independent variables, we will construct a $d \times k$–dimensional transformation matrix that can map any data point onto a new k-dimensional variable subspace with k smaller than d. It means that we have performed a dimension reduction by a linear transformation of the original independent variables. There are several applications of PCA, but here, we will cite a paper, *Gene mirror geography within Europe*, published in *Nature (2008)* (https://pubmed.ncbi.nlm.nih.gov/18758442/). Authors considered a sample of 3,000 European individuals genotyped at over half a million variable DNA sites in the human genome; then, each individual was represented using more than half a million genetic markers. This means that it produced a matrix with dimensions larger than 3,000 x 500,000. They performed a PCA on the dataset to find the most meaningful vectors, v_1 and v_2 (the first and second components), where each person only corresponds to two numbers. The authors plotted each person based on two numbers in a two-dimensional plane and then colored each point according to the country they came from.

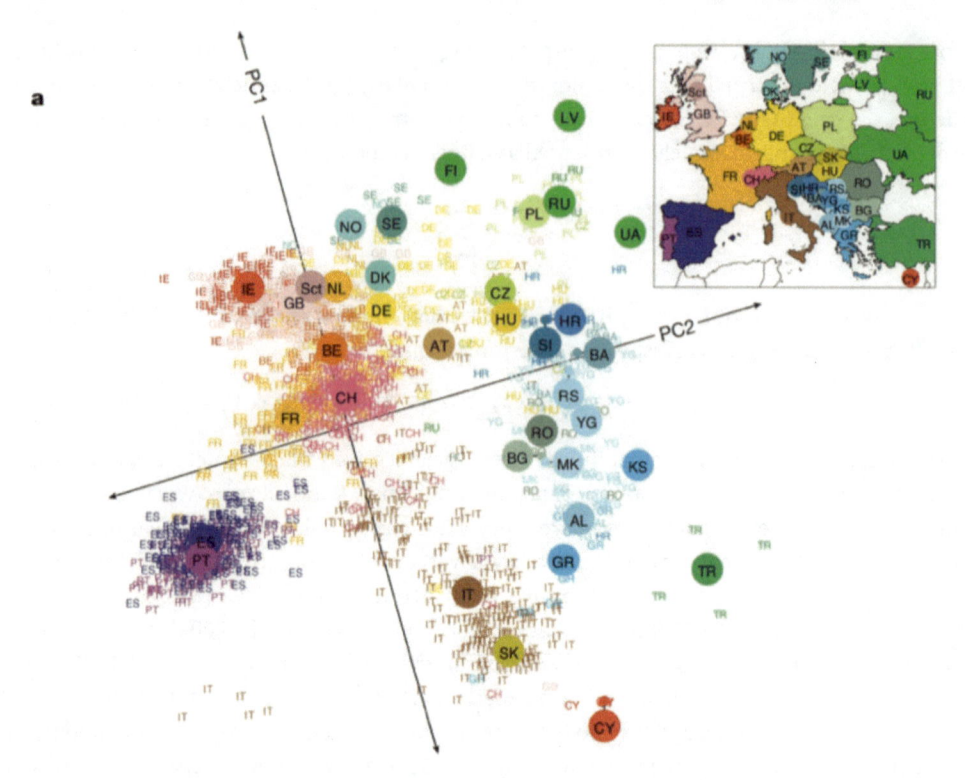

Figure 7.13 – PCA analysis for gene mirror geography within Europe

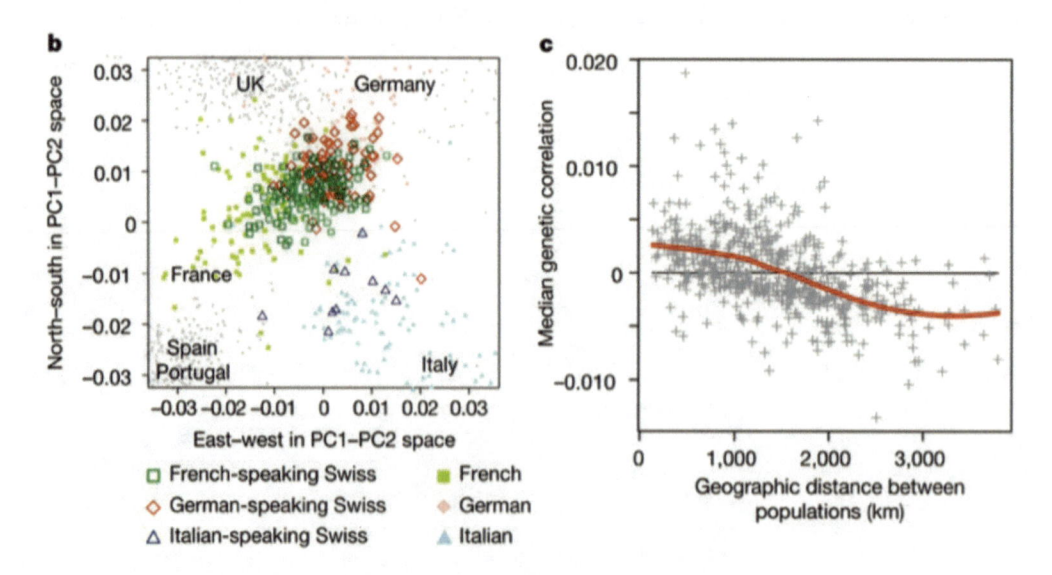

Figure 7.14 – PCA analysis for gene mirror geography within Europe

Surprisingly, the PCA technique performed well in the plots by showing genetic similarities that are similar to the European map.

In the next part, we will discuss how to use PCA to conduct a PCR analysis in practice.

PCR – a hands-on salary prediction study

To conduct a PCR analysis, we first perform a PCA to obtain the PCs and then decide to keep the first k PCs that contain the most explainable amount of variability. Here, k is the dimensionality of the new PC variable space. Finally, we fit MLR on these new variables.

We will consider a hands-on salary prediction task from the open source Kaggle data – https://www.kaggle.com/datasets/floser/hitters – to illustrate the PCR method. If following along, please download the dataset from the Kaggle URL:

1. **Setting up and loading the data**

    Import the necessary libraries to be used in this study and loading the Hitters data. For simplicity, we will drop all missing values in the dataset. There are 19 independent variables (16 numerical and 3 categorical) with the target 'Salary'. The categorical independent variables 'League', 'Division', and 'NewLeague' are converted into dummy variables. We preprocess and standardize the features, and create a train and a test set before conducting the PCA step:

    ```
    # Import libraries
    import numpy as np
    import pandas as pd
    import matplotlib.pyplot as plt
    from sklearn.preprocessing import scale
    from sklearn.linear_model import LinearRegression
    from sklearn.model_selection import KFold, cross_val_score, train_test_split
    from sklearn.metrics import mean_squared_error
    from sklearn.decomposition import PCA
    #location of dataset
    url = "/content/Hitters.csv"
    #read in data
    data = pd.read_csv(url).dropna() # to simply the analysis, we drop all missing values
    # create dummies variables
    dummies_variables = pd.get_dummies(data[['League', 'Division', 'NewLeague']])
    # create features and target
    target = data['Salary']
    feature_to_drop = data.drop(['Salary', 'League', 'Division', 'NewLeague'],axis=1).astype('float64')
    X = pd.concat([feature_to_drop, dummies_variables[['League_N',
    ```

```
                'Division_W', 'NewLeague_N']]], axis=1)
#scaled data - preprocessing
X_scaled = scale(X)
# train test split
X_train, X_test, y_train, y_test = train_test_split(X_scaled,
target, test_size=0.2, random_state=42)
```

2. **Generating all PCs**

   As the next step, we generate all the PCs for the training set. This performance produces 19 new PC variables because there are 19 original independent variables:

   ```
   # First generate all the principal components
   pca = PCA()
   X_pc_train = pca.fit_transform(X_train)
   X_pc_train.shape
   ```

3. **Determining the best number of PCs to be used**

   The next step is to perform a 10-fold cross-validation MLR and choose the best number of PCs to use by using the RMSE:

   ```
   # Define cross-validation folds
   cv = KFold(n_splits=10, shuffle=True, random_state=42)
   model = LinearRegression()
   rmse_score = []
   # Calculate MSE score - based on 19 PCs
   for i in range(1, X_pc_train.shape[1]+1):
       rmse = -cross_val_score(model, X_pc_train[:,:i], y_train,
   cv=cv, scoring='neg_root_mean_squared_error').mean()
       rmse_score.append(rmse)
   # Plot results
   plt.plot(rmse_score, '-o')
   plt.xlabel('Number of principal components in regression')
   plt.ylabel('RMSE')
   plt.title('Salary')
   plt.xlim(xmin=-1)
   plt.xticks(np.arange(X_pc_train.shape[1]), np.arange(1, X_pc_
   train.shape[1]+1))
   plt.show()
   ```

Here's the plot produced:

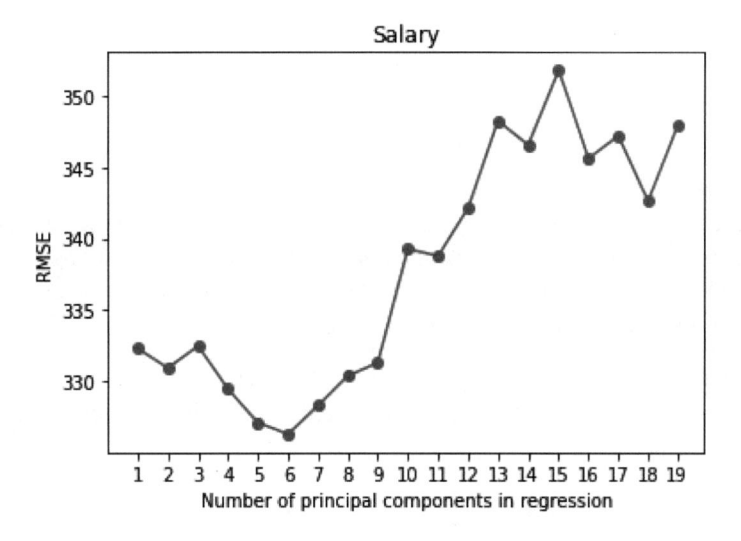

Figure 7.15 – Number of PCs

From this, we see that the best number of PCs is 6, corresponding with the lowest cross-validation RMSE.

4. **Retraining the model and performing prediction**

   We will use this number to train a regression model on the training data and make predictions on the test data:

   ```
   # Train regression model on training data
   model = LinearRegression()
   model.fit(X_pc_train[:,:6], y_train)

   pcr_score_train = -cross_val_score(model, X_pc_train[:,:6], y_train, cv=cv, scoring='neg_root_mean_squared_error').mean()

   # Prediction with test data
   X_pc_test = pca.fit_transform(X_test)[:,:6]
   pred = model.predict(X_pc_test)
   pcr_score_test = mean_squared_error(y_test, pred, squared=False)
   ```

Remark that PCR analysis is more difficult to interpret the results of than feature selection or shrinkage methods, and this analysis performance is better if the few first PCs capture the most explainable amount of variability.

Summary

In this chapter, we discussed the concept of MLR and topics aiding in its implementation. These topics included feature selection methods, shrinkage methods, and PCR. Using these tools, we were able to demonstrate approaches to reduce the risk of modeling excess variance. In doing so, we were able to also induce model bias so that models can have a better chance of generalizing on unseen data with minimal complications as frequently faced when overfitting.

In the next chapter, we will begin a discussion on classification with the introduction of logistic regression, which fits a sigmoid to a linear regression model to derive probabilities of binary class membership.

Part 3: Classification Models

In this part, we discuss the types of problems that can be solved with classification, coefficients of correlation and determination, multivariate modeling, model selection and variable adjustment with regularization.

It includes the following chapters:

- *Chapter 8, Discrete Models*
- *Chapter 9, Discriminant Analysis*

8
Discrete Models

In the previous two chapters, we discussed models for predicting a continuous response variable. In this chapter, we will begin discussing models for predicting discrete response variables. We will start by discussing the probit and logit models for predicting binary outcome variables (categorical variables with two levels). Then, we will extend this idea to predicting categorical variables with multiple levels. Finally, we will look at predicting count variables, which are like categorical variables but only take values of integers and have an infinite number of levels.

In this chapter, we're going to cover the following main topics:

- Probit and logit models
- Multinomial logit model
- Poisson model
- The negative binomial regression model

Probit and logit models

Previously, we discussed different types of problems that can be solved with regression models. In particular, the dependent variable is continuous, such as house prices, salaries, and so on. A natural question is if dependent variables are not continuous – in other words, if they are categorical – how would we adapt our regression equation to predict a categorical response variable? For instance, a human resources department in a company wants to conduct an attrition study to predict whether an employee will stay with the company or a car dealership wants to know if one car can be sold or not based on prices, car models, colors, and so on.

First, we will study **binary classification**. Here, the outcome (dependent variable) is a binary response such as yes/no or to do/not to do. Let's look back at the simple linear regression model:

$$y = \beta_0 + \beta_1 x + \epsilon$$

Here, the predicted outcome is a line crossing data points. With this model, we could build a model to predict house prices (the dependent variable) based on different independent variables, such as districts, units, stories, or distances. The same idea cannot be applied to a binary classification problem. Let's consider a simple example that we want to predict if a house is sold only based on its price. Let's build a model based on linear regression:

$$Sold = \beta_0 + \beta_1 * Price + \epsilon \quad (1)$$

Here, *Sold* = 1 if the house is sold and *Sold*=0 if it was not sold. By performing visualization using linear regression, we can draw a line of best fit through the data points. It looks like when the price increases, the chance of selling a house decreases:

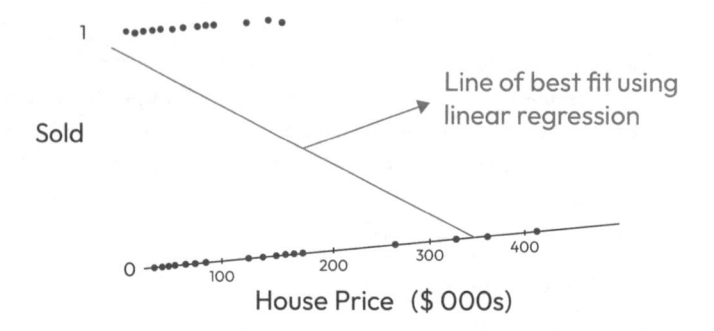

Figure 8.1 – Line of best fit using linear regression

So, instead of considering the equation model (*1*), we will be considering the following:

$$Prob(Sold) = \beta_0 + \beta_1 * Price + \epsilon$$

Here, *Prob(Sold)* is the probability of selling a house. The probability value will be between 0 and 1; the midpoint of its range is 0.5 when the probability of selling a house is equal to the probability of not selling a house. However, when the price of a house is very high, then the value of *Prob(Sold)* can be negative. To avoid this problem, we can model the problem with the following equation:

$$\frac{P}{1-P} = \beta_0 + \beta_1 * Price + \epsilon \quad (2)$$

Here, P is the probability of selling a house and $1 - P$ is the probability of not selling a house. The value of $\frac{P}{1-P}$ is in the range of $[0, \infty)$ and when the probability of selling a car is the same as the probability of not selling a car, (P=0.5), then $\frac{P}{1-P}$ = 1, or the midpoint of its range is 1. This is also called the **odds ratio**. To interpret the odds ratio, when $\frac{P}{1-P}$ = 1, for every house that is sold, there is a house that is not sold. When P = 0.75, then the odds ratio is 3. In this case, the interpretation is that the odds of being able to sell the house are three times higher than the odds of not being able to sell the house. Going back to the equation (*2*), the range of the odds ratio is $[0, \infty)$ and its lower limit is 0, but as we discussed previously for equation (*1*), when the price of a house is really high, then the estimate of the odds ratio could be negative, which contradicts the probability properties. In addition, because the midpoint is 1 but the range of the odds ratio is from 0 to infinity, then the distribution is very skewed with a long right tail. However, we expected it to be normally distributed per one of the assumptions of linear regression. So, instead of using equation *(2)*, we will consider the following:

$$log(\tfrac{P}{1-P}) = \beta_0 + \beta_1 * Price + \epsilon$$

The value of $log(\tfrac{P}{1-P})$ is in the range of $(-\infty, \infty)$ and the midpoint is 0. Similar to linear regression, we can use the following formula:

$$log(odds) = log(\tfrac{P}{1-P}) = \beta_0 + \beta_1 X_1 + \beta_2 X_2 + \ldots = z \quad (3)$$

We can use this to interpret how the explanatory variables are associated with the categorical outcome. We can rewrite the previous equation *(3)* as follows:

$$F(z) = P = \frac{e^z}{1+e^z},$$

The preceding formula is bounded below by 0 and above by 1 (the predicted probability, *P*, of the outcome takes place) and is called the logit model. It is the cumulative distribution function of the **logistic distribution**.

Another approach to model the probability of a binary dependent variable is **probit regression**. Instead of using the cumulative distribution function of logistic regression, we use the cumulative distribution function of the standard normal distribution:

$$F(z) = \int_{-\infty}^{z} \phi(u) du = \phi(z).$$

The preceding formula is bounded below by 0 and above by 1 (the predicted probability, *P*, of the outcome takes place), and is called the **probit model**. Remember the following:

$$\phi(z) = P(Z \leq z), \quad Z \sim \mathcal{N}(0,1).$$

Both the logit and probit models are estimated by using the **maximum likelihood method**. In `statsmodels`, we have classes for the probit and logit models. The documentation for these classes can be found at https://www.statsmodels.org/dev/generated/statsmodels.discrete.discrete_model.Probit.html and https://www.statsmodels.org/dev/generated/statsmodels.discrete.discrete_model.Logit.html.

Now, let's create a training dataset to illustrate how to use the `Logit` class from `statsmodels` to conduct a logit study:

```
import pandas as pd
# create gpa train data
train = pd.DataFrame({'Admitted': [1, 1, 1,1, 1, 0, 1, 1, 0, 1,1,1,
1,1,0, 1, 0, 0, 0, 0, 0, 0, 0, 0 ,0 ,0, 1,1,1,1, 0],
                'GPA': [2.8, 3.3, 3.7, 3.7, 3.7, 3.3, 3.7, 3, 1.7,
3.6, 3.3, 4, 3.2, 3.4, 2.8, 4, 1.5, 2.7, 2.3, 2.3, 2.7, 2.2, 3.3,3.3,
```

```
              4, 2.3, 3.6, 3.4, 4, 3.7, 2.3],
                      'Exp': [8, 6, 5, 5, 6, 3, 4, 2, 1, 5, 5, 3, 6,5,
              4, 4, 4, 1, 1, 2, 2, 2, 1, 4, 4, 4, 5, 2, 4, 6, 3]})
              train.head()
```

Here, `Admitted` is the dependent variable and has two possibilities (1 – admitted and 0 – not admitted). `GPA` is the GPA grade and `Exp` is the number of years of experience. The output of the preceding code is as follows:

	Admitted	GPA	Exp
0	1	2.8	8
1	1	3.3	6
2	1	3.7	5
3	1	3.7	5
4	1	3.7	6

Figure 8.2 – Training dataset on GPA grades and years of experience

We must also create a testing dataset for this model:

```
test = pd.DataFrame({'Admitted': [1, 0, 1, 0, 1],
                     'GPA': [2.9, 2.4, 3.8, 3, 3.3],
                     'Exp': [9, 1, 6, 1,4 ]})
test.head()
```

The resulting output is as follows:

	Admitted	GPA	Exp
0	1	2.9	9
1	0	2.4	1
2	1	3.8	6
3	0	3.0	1
4	1	3.3	4

Figure 8.3 – Testing dataset on GPA grades and years of experience

We will use `logit` from `statsmodels`:

```
import statsmodels.formula.api as smf
#fit logistic regression
model = smf.logit('Admitted ~ GPA + Exp', data =train).fit()
#summary
model.summary()
```

This will print the following summary:

```
Optimization terminated successfully.
         Current function value: 0.316480
         Iterations 7
```

Logit Regression Results

Dep. Variable:	Admitted	No. Observations:	31
Model:	Logit	Df Residuals:	28
Method:	MLE	Df Model:	2
Date:	Sun, 20 Nov 2022	Pseudo R-squ.:	0.5403
Time:	20:36:04	Log-Likelihood:	-9.8109
converged:	True	LL-Null:	-21.342
Covariance Type:	nonrobust	LLR p-value:	9.818e-06

	coef	std err	z	P>|z|	[0.025	0.975]
Intercept	-11.4485	4.320	-2.650	0.008	-19.915	-2.982
GPA	2.7606	1.291	2.139	0.032	0.231	5.290
Exp	0.7569	0.383	1.977	0.048	0.006	1.507

Figure 8.4 – Logit Regression Output

From this output, we can see that GPA and Exp are significant at the $\alpha = 0.05$ significance level.

The values under the [0.025 0.975] heading are the 95% confidence interval for Intercept, GPA, and Exp, respectively. The next step is to use `confusion_matrix` and `accuracy_score` to compute the accuracy of the model on the test set:

```
from sklearn.metrics import confusion_matrix, accuracy_score, ConfusionMatrixDisplay
# X_test and y_test
X_test = test[['GPA', 'Exp']]
y_test = test['Admitted']
#
y_hat = model.predict(X_test)
pred = list(map(round, y_hat))
# confusion matrix
cm = confusion_matrix(y_test, pred)
```

```
ConfusionMatrixDisplay(cm).plot()

# Accuracy
print('Test accuracy = ', accuracy_score(y_test, pred))
```

The output is as follows:

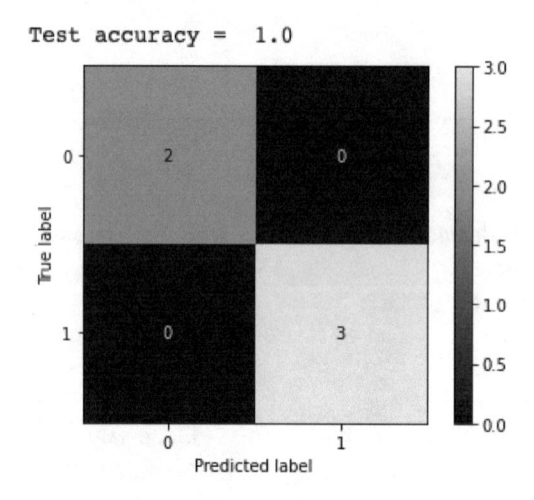

Figure 8.5 – Confusion matrix on testing dataset

By using these train and test datasets, the model can predict the outcome perfectly. In the next section, we will discuss **multi-class regression** using a similar idea as in binary logistic regression.

Multinomial logit model

In practice, there are many situations where the outcomes (dependent variables) are not binary but have more than two possibilities. **Multinomial logistic regression** can be understood as a general case of the logit model, which we studied in the previous section. In this section, we will consider a hands-on study on Iris data by using the `MNLogit` class from `statsmodels`: https://www.statsmodels.org/dev/generated/statsmodels.discrete.discrete_model.MNLogit.html.

Iris data (https://archive.ics.uci.edu/ml/datasets/iris) is one of the best-known statistical and machine learning datasets for education. The independent variables are sepal length (in cm), sepal width (in cm), petal length (in cm), and petal width (in cm). The dependent variable is a categorical variable with three levels: Iris Setosa (0), Iris Versicolor (1), and Iris Virginia (2). The following Python codes illustrate how to conduct this using sklearn and statsmodels:

```python
# import packages
import numpy as np
import pandas as pd
from sklearn import datasets
from sklearn.model_selection import train_test_split
from sklearn.linear_model import LogisticRegression
from sklearn.metrics import confusion_matrix, accuracy_score,  ConfusionMatrixDisplay
import statsmodels.discrete.discrete_model as sm

# import Iris data
iris = datasets.load_iris()
print(iris.feature_names)
print(iris.target_names)

#create dataframe
df = pd.DataFrame(iris.data, columns = ['sepal_length', 'sepal_width', 'petal_length', 'petal_width'])
df['target'] = iris.target
df.head()

# check missing values
df.isna().sum()

# create train and test data
X = df.drop('target', axis=1)
y = df['target']
X_train, X_test, y_train, y_test = train_test_split(X, y, test_size =0.2, random_state =1)

# fit the model using sklearn
model_sk = LogisticRegression(solver = 'newton-cg', multi_class = 'multinomial')
model_sk.fit(X_train, y_train)
y_hat_sk = model_sk.predict(X_test)
pred_sk = list(map(round, y_hat_sk))
# confusion matrix
cm_sk = confusion_matrix(y_test, pred_sk)
```

```
ConfusionMatrixDisplay(cm_sk).plot()
# Accuracy
print('Test accuracy = ', accuracy_score(y_test, pred_sk))

#fit the model using statsmodels
model_stat = sm.MNLogit(y_train, X_train).fit(method='bfgs')
model_stat.summary()
y_hat_stat = model_stat.predict(X_test)
pred_stat = np.asarray(y_hat_stat).argmax(1)
# confusion matrix
cm_stat = confusion_matrix(y_test, pred_stat)
ConfusionMatrixDisplay(cm_stat).plot()
# Accuracy
print('Test accuracy = ', accuracy_score(y_test, pred_stat))
```

Both methods give us the same test accuracy value (96.67%), with the confusion matrix produced as follows:

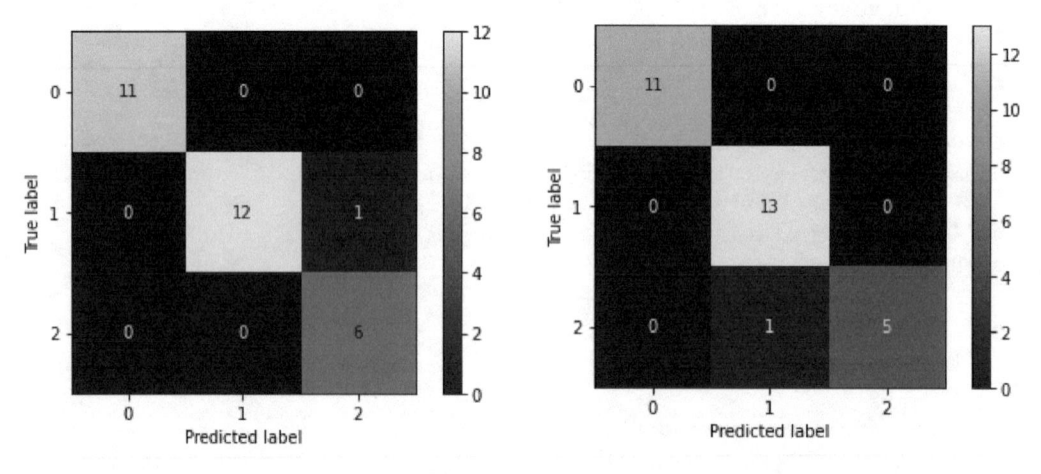

Figure 8.6 – Confusion matrices using sklearn (left) and statsmodels (right)

Poisson model

In the previous section, we discussed models where the response variable was categorical. In this section, we will look at a model for count data. Count data is like categorical data (the categories are integers), but there are an infinite number of levels (0, 1, 2, 3, and so on). We model count data with the **Poisson distribution**. In this section, we will start by examining the Poisson distribution and its properties. Then, we will model a count variable with explanatory variables using the Poisson model.

The Poisson distribution

The Poisson distribution is given by the following formula:

$$P(k) = \frac{\lambda^k e^{-\lambda}}{k!}$$

Here, λ is the average number of events and k is the number of events for which we would like the probability. $P(k)$ is the probability that the k events occur. This distribution is used to calculate the probability of k events occurring in a fixed time interval or a defined space.

The shape of the distribution changes with the value of λ. When λ is greater than 10, the distribution appears approximately normal. However, as λ approaches 0, the distribution becomes right-skewed. This is because count data cannot be negative. Three example Poisson distributions are shown in *Figure 8.7* with means of 12, 5, and 2. Notice that the distribution with the mean of 12 is approximately normally distributed, while the distributions of 5 and 2 are right-skewed:

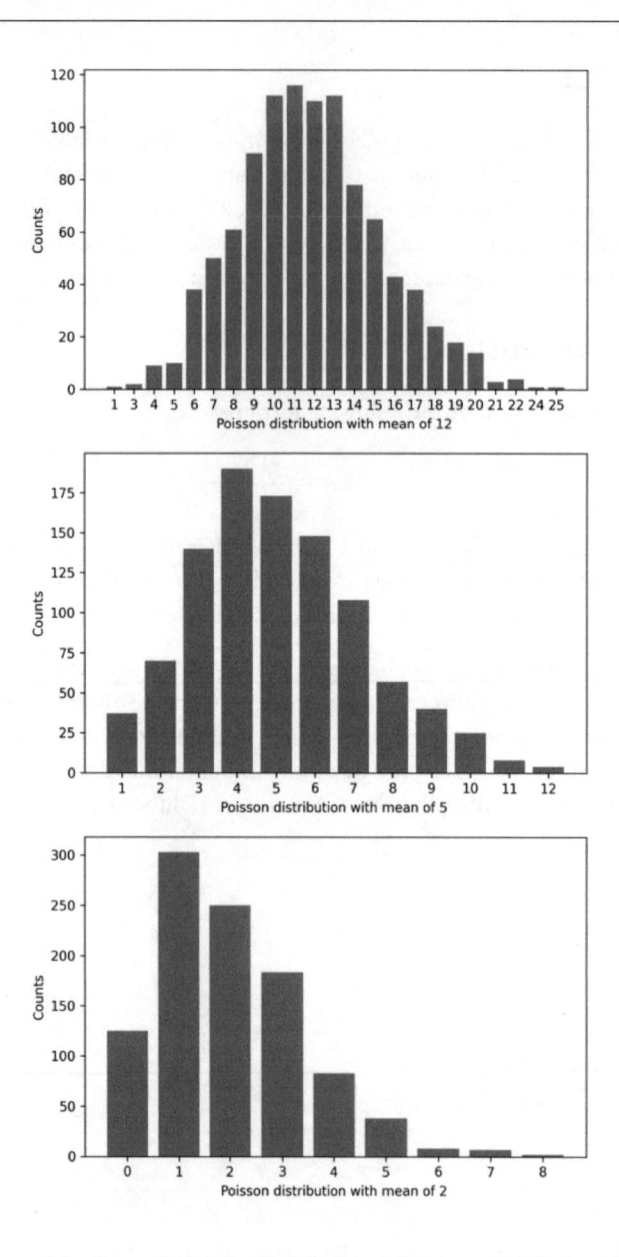

Figure 8.7 – Example Poisson distributions with means of 12, 5, and 2

Another interesting aspect of the distribution is that the mean and variance are equal. This means that as the mean increases, the spread of the distribution also increases. We can see this in action in the examples in *Figure 8.7*. The distribution with a mean of 2 has a small spread with a large peak at the mean, while the distribution with a mean of 12 has a much wider spread with a lower peak at the mean.

Now that we have discussed the Poisson distribution, let's look at how to set up a Poisson model.

Modeling count data

Now, let's look at how to model the response variable of counts with the Poisson model. As mentioned previously, count data often follows a Poisson distribution. The Poisson model is expressed mathematically as follows:

$$y = e^{b_0 + b_1 x_1 + \ldots + b_n x_n}$$

Here, y is the response variable, *b* values are model coefficients, and *x* variables represent explanatory variables. This should appear similar to the equation we used in the previous chapter but with the addition of the exponentiation of the explanatory variables. This type of model is called a **log-linear** model, which is a model where the logarithm of the response variable is modeled by a linear combination of variables. We can rewrite this equation by applying the natural logarithm to both sides of the equation to make it explicit:

$$\ln(y) = b_0 + b_1 x_1 + \ldots + b_n x_n$$

Now, we have the logarithm of the response variable (*ln(y)*) expressed as a linear combination of explanatory variables.

> **The natural logarithm**
>
> The Poisson model uses a special logarithm called the natural logarithm. The natural logarithm of a number is the logarithm of that number using the mathematical constant *e* as the base. The natural logarithm is generally written as *ln*(x), $log_e(x)$, or *log*(x) (the first two options are explicit, but the third option can be ambiguous). The logarithm operation is the inverse of exponentiation. In this case, the natural logarithm is the inverse of the exponential function: $ln(e^x) = x = e^{ln(x)}$. The natural logarithm and exponential function are commonly used in statistics, mathematics, and science.

Let's look at an example. We will be using the Bike Sharing Dataset from UCI (`https://archive.ics.uci.edu/ml/datasets/bike+sharing+dataset`.). In this dataset, we have counts of bikes rented each day. There are two types of rentals: pre-registered (registered) and on-demand at a location (causal). In this example, we will model the weekly mean count of casually rented bikes over a given year. The dataset provides several explanatory variables, including environmental factors such as temperature and calendar information such as whether a holiday occurred.

We will start by setting up the equation for our model and then take a look at the results from fitting the model with statsmodels. The model equation follows the form we discussed previously:

$$\ln(weekly\_mean\_rental\_count) = b_0 + b_1(temperature) + b_1(season)$$
$$+ b_2(weather\_situation) + b_3(humidity) + b_4(wind\_speed) + b_5(holiday)$$

We can fit this model with the given data using statsmodels, similar to how we did in the previous chapter. An excerpt of the code to fit the model is shown here (see the Jupyter Notebook for details on preprocessing):

```
# select variables
X = df.groupby('isoweek').mean()[['atemp', 'season',
'weathersit','hum','windspeed', 'holiday']]
# transform holiday variable as an indicator that a holiday occurs
within that week
X['holiday'] = X['holiday'].apply(lambda x: 1 if x > 0.1 else 0)
# add a constant for the model
X = sm.add_constant(X)
# get the response variable
y = df.groupby('isoweek').mean()['casual']
fit_model = sm.Poisson(y, X).fit()
fit_model.summary()
```

After fitting the model, we can get details on the coefficients using the summary() method. For this model, we get the following output for the coefficients:

	coef	std err	z	P>|z|	[0.025	0.975]
const	4.1252	0.072	57.096	0.000	3.984	4.267
atemp	3.0915	0.042	73.155	0.000	3.009	3.174
season	0.0427	0.007	6.179	0.000	0.029	0.056
weathersit	−0.2387	0.032	−7.422	0.000	−0.302	−0.176
hum	1.2369	0.100	12.345	0.000	1.040	1.433
windspeed	1.2132	0.177	6.856	0.000	0.866	1.560
holiday	0.0955	0.014	6.771	0.000	0.068	0.123

Figure 8.8 – Poisson model summary

Just like the modeling example for linear regression, these coefficient values are estimates of the parameters listed in our model. All the explanatory variables in the model appear to be significant in the model. Interestingly, based on the value of the coefficient estimates, temperature (`atemp`) appears to be the most influential factor, followed by humidity (`hum`) and wind speed. With the model fit and no need to remove insignificant variables, we can assess the model's performance. This model has an MAE of 155, which corresponds to a MAPE of 36%.

The Poisson model depends strongly on the assumption that the response variable has a Poisson distribution. In the next section, we will look at a similar type of model for count data, but with weaker distributional assumptions.

The negative binomial regression model

Another useful approach to **discrete regression** is the **log-linear negative binomial regression** model, which uses the negative binomial probability distribution. At a high level, negative binomial regression is useful with *over-dispersed count data* where the *conditional mean of the count is smaller than the conditional variance of the count*. Model **over-dispersion** is where the variance of the target variable is greater than the mean assumed by the model. In a regression model, the mean is the regression line. We make the determination of using the negative binomial model based on target variable counts analysis (mean versus variance) and supply a measure of model over-dispersion to the negative binomial model to adjust for the over-dispersion, which we will discuss here.

It is important to note that the negative binomial model is not for modeling simply discrete data, but specifically **count data** associated with a fixed number of **random** trials, such as modeling the number of attempts before an event occurs – or failing to occur – in a random sampling scenario. The model operates only on count data, where each count response of the target variable is based on a finite set of outcomes. Additionally, because the count data is the result of repeated binomial trials, the order of count arrangement does not matter.

Negative binomial distribution

The following is an example of the negative binomial distribution of failure counts with a conditional mean of 17 and a conditional variance of 52:

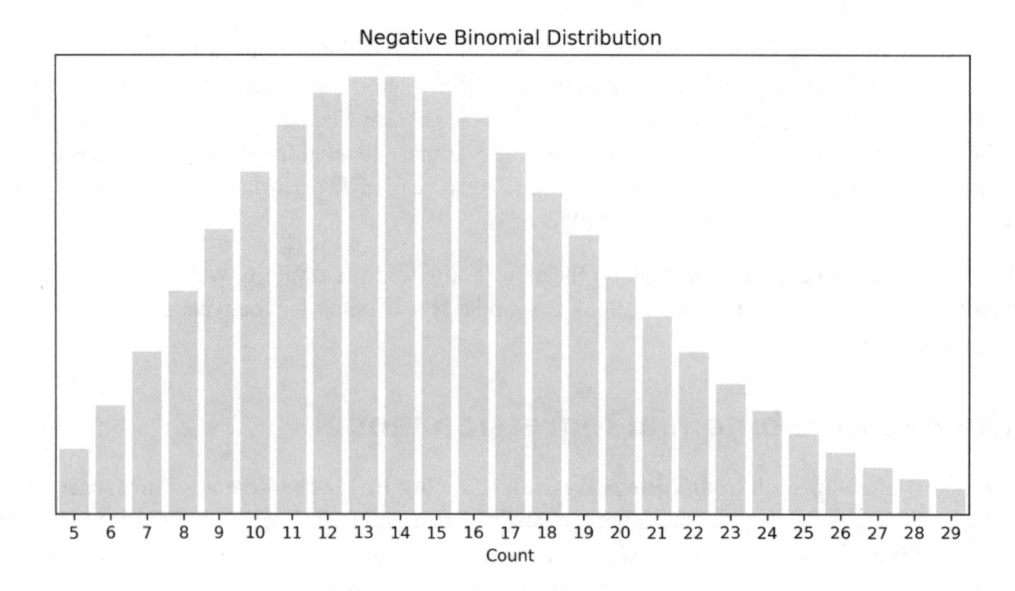

Figure 8.9 – Negative binomial distribution example

The binomial distribution is a construct of counts of successes in a fixed count of random trials ($X = (X_1, X_2, \ldots)$) of a Bernoulli random variable, which is a variable that has one of two outcome values: 0 or 1. The negative binomial distribution is a construct of the count of failures in a fixed count of random draws of the Bernoulli random variable. This distinction is important because a model using binomial regression models a binary outcome across observations, whereas a negative binomial regression model models a count outcome across observations. The negative binomial distribution is the inverse of the binomial distribution. The probability mass function for the negative binomial distribution is as follows:

$$P(X = x) = \binom{x + n - 1}{n - 1} p^n (1 - p)^x$$

Here, there are x failures and $n - 1$ successes in a set of $x + n - 1$ trials reaching success at trial $x + n$.

Concerning the Poisson distribution, the negative binomial *does not require strict adherence to the assumption that conditional variance is equal to the conditional mean* as it includes an additional parameter, α, to explain the extra variance, whereas the Poisson model assumes the variance is equal to the mean, μ. The negative binomial model assumes the variance, here based on the Poisson-gamma mixture distribution, is equal to the following:

$$\mu(1 + \alpha\mu)$$

This reduces to the following:

$$(\mu + \alpha\mu^2)$$

Stated in terms of the probability of success, *p*, the negative binomial variance is calculated as follows:

$$n\frac{1-p}{p^2}$$

It has the following mean:

$$n\frac{1-p}{p}$$

Because the variance is not expected to equal the mean with the negative binomial model, the negative binomial is likely to outperform the Poisson approach *when the counts are large enough that the variance in response exceeds the mean*. Similarly, to Poisson, a negative binomial is a log-linear model with confidence intervals being based on the Wald and drop-in-deviance likelihood ratio. The form of the negative binomial model's regression equation, shown here, is the same as that for Poisson regression:

$$\ln(y) = b_0 + b_1 x_1 + \ldots + b_n x_n$$

It reduces to the following:

$$y = e^{b_0} + e^{b_1 x_1} + \ldots + e^{b_n x_n}$$

The **maximum likelihood estimate** of the probability of success for a given sample of the distribution is as follows:

$$\frac{n}{n + \overline{x}'}$$

> **Maximum likelihood estimation**
>
> **Maximum likelihood estimation (MLE)** is an underpinning of the log-odds (or logit) and log-likelihood approaches to statistical modeling. Likelihood is the probability of the known outcome of a regression model being observed given specific regression coefficient values. By default, a set of variables will have a higher likelihood than another set of variables if the set provides a higher probability of the observed outcome being obtained. The logarithm of the likelihood is taken as a measure of goodness-of-fit for a regression model. Out of a set of potential coefficient values for each variable, the set of coefficients with the maximum log-likelihood values for each variable are referred to as the **maximum likelihood estimates**. These values are obtained through an iterative approach that generates multiple possible values. If the sample size is sufficiently large and an appropriate set of variables has been obtained for the model, the maximum likelihood estimates can be considered unbiased.

Calculating the confidence intervals for negative binomial regression is similar to that for logistic regression. The **Wald** approach to the calculation leverages a **z-ratio**. Where there are *j* variables in the model, the z-ratio is calculated as follows:

$$\frac{(\hat{\beta}_j - \beta_j)}{SE(\hat{\beta}_j)}$$

Here, *SE* is the standard error. The confidence interval is the variable's coefficient estimate plus and minus the half-width confidence interval percentile multiplied by the standard error of the coefficient estimate. The z-ratio can be used because it is assumed the estimates have standard normal sampling distributions. Therefore, we can derive the 95% confidence interval for the variable's estimated coefficient as follows:

Lower 95% confidence limit:

$$\hat{\beta}_j - 0.475 \times SE(\hat{\beta}_j)$$

Upper 95% confidence limit:

$$\hat{\beta}_j + 0.475 \times SE(\hat{\beta}_j)$$

There are three required assumptions specific to negative binomial regression:

1. Independence between samples.
2. A linear relationship between the log of the target variable and input variables (log-linear model).
3. Conditional variance is greater than or equal to the conditional mean.

Independence between samples means there is no serial correlation nor any cluster or other conditional dependence between samples. A linear relationship between the log of the target variable and input variables means that the relationship between the logarithm of the target variable and changes in each input variable scales linearly. Except for requirement 3, which we discussed at the start of this section, the requirements are essentially the same as for the Poisson model.

Let's walk through an example in Python using statsmodels. For this, let's load the statsmodels affairs dataset to model child count (the `children` variable) using the remaining variables. In line three, we must add the constant required to generate an intercept coefficient:

```
import statsmodels.api as sm
data = sm.datasets.fair.load().data
data = sm.add_constant(data, prepend=False)
```

First, let's numerically confirm there is over-dispersion present in the target variable:

```
print('Mean count of children per marriage: ', data['children'].mean())
print('Variance of the count of children per marriage: ', data['children'].var())
```

We can see the variance is greater than the mean:

```
Mean count of children per marriage:    1.3968740182218033
Variance of the count of children per marriage:    2.054838616333698
```

Here, we can see that the conditional mean per marriage is smaller than the conditional variance. While not a massive difference, it is enough to consider using negative binomial regression. Let's visually observe the distribution of the response:

Figure 8.10 – Distribution of children

The first five rows of the data can be seen here. Note that the first column of the design matrix should always contain the constant:

const	age	religious	yrs_married	Educ	occupation	occupation_husb	affairs	rate_marriage	children
1	32	3	9	17	2	5	0.1111111	3	3
1	27	1	13	14	3	4	3.2307692	3	3
1	22	1	2.5	16	3	5	1.3999996	4	0
1	37	3	16.5	16	5	5	0.7272727	4	4
1	27	1	9	14	3	4	4.666666	5	1

Figure 8.11 – First five records of example data set, including the added constant

In our visual inspection of the distribution of children in *Figure 8.10*, we identified that there is a value of 5.5 for children. This may be the result of averaging or an error. A subject matter expert may help determine this, but for our analysis, let's assume it was a mistake and round to a whole number of children since people are not fractional. Let's set up the target array, *y*, and design matrix, *X*:

```
y = round(data['children'])
X = data[['const','age','religious','yrs_
married','educ','occupation','occupation_husb','affairs','rate_
marriage']]
```

Now, let's create a train and test split for regression modeling. Note that `shuffle=True` will provide different results. To obtain a representative sample, the data should be randomly shuffled:

```
from sklearn.model_selection import train_test_split
X_train, X_test, y_train, y_test = train_test_split(X, y, test_size=0.25, shuffle=True)
```

Because the negative binomial model is based on a Poisson-gamma mixture model, we need to estimate the model's measure of over-dispersion using a Poisson model. A method referred to as **auxiliary OLS regression (without constant)** is provided by A. Colin Cameron and Pravin K. Trivedi in *Microeconometrics: Methods and Applications*. The authors propose the creation of an over-dispersion test statistic where the null hypothesis is α=0 and the alternate hypothesis is α ≠ 0, where α is the estimate of over-dispersion. The auxiliary OLS regression formula is as follows:

$$\frac{(y_i - \hat{\mu}_i)^2 - y_i}{\hat{\mu}_i} = \alpha \frac{g(\hat{\mu}_i)}{\hat{\mu}_i} + \mu_i$$

Here, μ_i is an error term and $g(\hat{\mu}_i)$ is $\hat{\mu}_i^2$. Thus, the right-hand operand reduces to $\alpha\hat{\mu}_i + \mu_i$. In terms of negative binomial regression, we consider the error to equal zero, so we can factor α in as the over-dispersion estimate.

In the following code, we have fit our training data to a generalized linear model using the Poisson model for the linkage. Then, we used the regression mean of the model to build the estimated auxiliary target variable. Because the method is "without constant," we subtract 1 to remove the constant from the process:

```
from statsmodels.formula.api import ols as OLS
import statsmodels.api as sm

poisson_model = sm.GLM(y_train, X_train, family=sm.families.Poisson()).fit()

df_aux = pd.DataFrame()
df_aux['y_mu_hat'] = poisson_model.mu
df_aux['children'] = y_train
df_aux['y_auxiliary'] = df_aux.apply(lambda x: ((x['children'] - x['y_mu_hat'])**2 - x['y_mu_hat']) / x['y_mu_hat'], axis=1)

ols_model = OLS('y_auxiliary ~ y_mu_hat - 1', df_aux).fit()
print(ols_model.params)
```

As we can see, the estimated dispersion for the negative binomial model is `0.622034`. Now, we need to assess if the auxiliary estimate is statistically significant. We can do this using the p-value from the OLS model:

```
print(ols_model.summary())
```

This will print the following output:

OLS Regression Results

	coef	std err	z	P>|z|	[0.025	0.975]
y_mu_hat	0.6220	0.070	8.857	0.000	0.484	0.760

Figure 8.12 – OLS regression results

Because the coefficient is significant and greater than 0, we can confirm the model has over-dispersion based on the target. The coefficient can be used as the measure for that over-dispersion in the negative binomial model, which we can use to adjust the variance in the `alpha` argument here:

```
from statsmodels.genmod.families.family import NegativeBinomial

negative_binomial_model = sm.GLM(y_train, X_train,
family=NegativeBinomial(alpha=ols_model.params.values)).fit()
print(negative_binomial_model.summary())
```

The output is generated as follows:

Generalized Linear Model Regression Results

	coef	std err	z	P>|z|	[0.025	0.975]
const	-0.0784	0.183	-0.430	0.667	-0.436	0.279
age	-0.0088	0.006	-1.431	0.153	-0.021	0.003
religious	0.0467	0.021	2.207	0.027	0.005	0.088
yrs_married	0.1137	0.006	19.596	0.000	0.102	0.125
educ	-0.0317	0.009	-3.377	0.001	-0.050	-0.013
occupation	-0.0348	0.021	-1.673	0.094	-0.076	0.006
occupation_husb	0.0162	0.014	1.142	0.253	-0.012	0.044
affairs	-0.0009	0.010	-0.092	0.927	-0.021	0.019
rate_marriage	-0.0634	0.018	-3.434	0.001	-0.100	-0.027

Figure 8.13 – Generalized linear model regression results

Finally, let's use the model we built on the training data to predict the training data, then again to predict on the test data so that we can compare generalizability on unseen data using residual error as a basis of comparison:

```
from sklearn.metrics import mean_squared_error as RMSE

print('Training Root Mean Squared Error: ', RMSE(y_train, negative_binomial_model.predict(X_train)) )
print('Testing Root Mean Squared Error: ', RMSE(y_test, negative_binomial_model.predict(X_test)))
```

We can observe from the root mean squared error that the model's performance is approximately constant across training and test data, indicating a consistent model:

```
Training Root Mean Squared Error: 1.2553439918425695
```

```
Testing Root Mean Squared Error: 1.266620561303553
```

Summary

In this chapter, we explained the issue of encountering negative raw probabilities that are generated by building a binary classification probability model based strictly on linear regression, where probabilities in a range of [0, 1] are expected. We provided an overview of the log-odds ratio and probit and logit modeling using the cumulative distribution function of both the standard normal distribution and logistic distribution, respectively. We also demonstrated methods for applying logistic regression to solve binary and multinomial classification problems. Lastly, we covered count-based regression using the log-linear Poisson and negative binomial models, which can also be logically extended to rate data without modification. We provided examples of their implementations.

In the following chapter, we will introduce conditional probability using Bayes' theorem in addition to dimension reduction and classification modeling using linear discriminant analysis and quadratic discriminant analysis.

9
Discriminant Analysis

In the previous chapter, we discussed discrete regression models, including classification using logistic regression. In this chapter, we will begin with an overview of probability, expanding into conditional and independent probability. We then discuss how these two approaches to understanding the laws of probability form the basis for Bayes' Theorem, which is used directly to expand an approach called Bayesian statistics. Following this topic, we dive into **Linear Discriminant Analysis** (**LDA**) and **Quadratic Discriminant Analysis** (**QDA**), two powerful classifiers that model data using the Bayesian approach to probability modeling.

In this chapter we're going to cover the following main topics:

- Bayes' Theorem
- LDA
- QDA

Bayes' theorem

In this section, we will discuss **Bayes' Theorem**, which is used in the classification models described later in this chapter. We will start the chapter by discussing the basics of probability. Then, we will take a look at dependent events and discuss how Bayes' Theorem is related to dependent events.

Probability

Probability is a measurement of the likelihood that an event occurs or a certain outcome occurs. Generally, we can group events into two types of events: **independent events** and **dependent events**. The distinction between the types of events is in the name. An independent event is an event that is not affected or influenced by the occurrences of other events, while a dependent event is affected or influenced by the occurrences of other events.

Let's think about some examples of these events. For the first example, think about a fair coin toss. A coin toss can result in one of two states: heads and tails. If the coin is fair, there is a one-in-two chance that the toss will result in heads or tails. That means that the probability of heads or tails is 0.5 in a fair coin toss. We can calculate probability using the following formula when events are countable:

$$P(E) = \frac{Count\ of\ Desired\ Outcomes}{Total\ Number\ of\ Outcomes}$$

Here, $P(E)$ is the probability of the desired outcome. In the coin-toss example, the desired outcome is either heads or tails, each of which can only occur once per coin toss. Thus, for the coin toss, the count of desired events is one. The total number of outcomes is two because there are two possible states from a coin toss. Putting these numbers in the preceding equation yields 0.5 for heads and 0.5 for tails. Now, putting your experience to work, does a previous coin toss give you any knowledge about the next toss of the same coin? It does not because *each coin toss is independent of every other coin toss*; it is an independent event.

Now, let's consider another example. Say we have a bag with three red marbles and five blue marbles. Let's calculate the probability of selecting a red marble from the bag: $P(Red)$. There are eight total marbles, which is the total number of outcomes in the equation. We want to select a red marble, which means the count of desired outcomes is three. This means that the probability of selecting red marble from the bag is 3/8. Now, let's calculate the probability of selecting a blue marble from the bag: $P(Blue)$. There are five blue marbles and eight total marbles, so the probability of selecting a blue marble from the bag is 5/8. Notice that the sum of the two probabilities is one: $P(Red) + P(Blue) = 3/8 + 5/8 = 1$. This brings up an important property of probabilities: the total probability must be equal to one. This, of course, means that once we calculated the probability of selecting a red marble, we could have calculated the probability of selecting a blue marble by subtracting the probability of selecting a red marble from one. The property of probabilities is very useful and is used commonly in statistics and **machine learning** (**ML**).

Coming back to the example with red and blue marbles, let's think about the type of event represented by this example. We select a marble from the bag. Now, there are two possible options next: (1) we replace the selected marble and select a marble from the bag again, and (2) we do not replace the marble and select a new marble from the remaining seven marbles in the bag. The first case is called **selection with replacement**. Selection with replacement is an independent event because the bag is reset before each draw from the bag. The second case is called **selection without replacement**. Selection without replacement is where you find dependent events. The first time you select a marble, the probability is the same as described previously (an independent event). However, if the marble is not replaced, the state of the bag has changed, and the probability of the next marble selection depends on which one was selected from the bag in the previous draw. Then, in selection without replacement, each subsequent marble draw is a dependent event. When discussing dependent events, it is useful to describe events using **conditional probability**. Let's discuss conditional probability now.

Conditional probability

Let's start with a formal definition. When we have dependent events, we may be interested in the probability of Event B given Event A, which means we want to know the probability of Event B after Event A occurred. This is called the conditional probability of Event B given Event A and is denoted $P(B \mid A)$. This can be thought of as "the probability of Event B on the condition that Event A has already occurred." This may feel a bit vague, so let's look at conditional probability in terms of our last example.

Let's say Event A is drawn a red marble and Event B is drawn a blue marble. We want to find the conditional probability of drawing a blue marble given that we have already drawn a red marble and can denote this as $P(Blue \mid Red)$. Let's work through this example. After drawing the first red marble from the bag, there are seven marbles remaining: two red marbles and five blue marbles. We calculate the probability of selecting a blue marble from the new set using the same equation, which gives us the probability of 5/7. This means that the conditional probability of drawing a blue marble, given this a red marble was already drawn, is 5/7.

Now that we have discussed conditional probability, let's look at the probability of making two draws from the bag. Formally, we want to calculate the probability of Event A and Event B. Continuing to follow our example, let's calculate the probability of drawing a red marble and then drawing a blue marble. We can describe the probability of two sequential events with the following equation:

$$P(A \text{ and } B) = P(A) * P(B \mid A)$$

In words, the probability of A followed by B is the probability of Event A times the probability of Event B given Event A. In our example, we have calculated $P(Red)$ and $P(Blue \mid Red)$. We can use those values to calculate the probability of drawing a red marble and then a blue marble $P(Red \text{ and } Blue)$, as follows:

$$P(Red \text{ and } Blue) = P(Red) * P(Blue \mid Red) = \frac{\frac{3}{8} * 5}{7} = \frac{15}{56} \cong 0.27$$

Performing the calculation, we find that $P(Red \text{ and } Blue)$ is about 0.27.

Now, out of interest, let's make the calculations again, but *exchanging the order of the colors*; the probability of drawing a blue marble and then drawing a red marble or $P(Blue \text{ and } Red)$. Following the same logic as shown in the previous example, we will find that $P(Blue) = 5/8$ and $P(Red \mid Blue) = 3/7$. Thus, the equation looks like this:

$$P(Blue \text{ and } Red) = P(Blue) * P(Red \mid Blue) = \frac{\frac{5}{8} * 3}{7} = \frac{15}{56} \cong 0.27$$

Notice that $P(Red \text{ and } Blue)$ and $P(Blue \text{ and } Red)$ have the same probability. This is no coincidence. This is known as Bayes' Theorem, which we will discuss next.

Discussing Bayes' Theorem

In the previous section on conditional probability, we happened across Bayes' Theorem in the example of drawing marbles from a bag. Let's discuss Bayes' Theorem in more detail.

Bayes' Theorem is a basis of a field of statistics known as **Bayesian statistics**. The central idea of Bayesian statistics is that given a prior probability, we have used new information to update probability estimates. While a full description of Bayesian statistics is beyond the scope of this book, we will discuss Bayes' Theorem as it relates to the concepts discussed in this book. We generally describe Bayes' Theorem with the following equation:

$$P(B|A) = \frac{P(A|B)P(B)}{p(A)}$$

In Bayesian terminology, $P(B)$ is called the **prior** and $P(B \mid A)$ is called the **posterior**. The posterior gives us an update given another event. Let's make use of the Theorem in a real-world example.

Let's say that we want to determine whether someone has a disease based on the results of a test. We would present this in Bayes' Theorem as follows:

$$P(sick|test\ pos) = \frac{P(test\ pos|sick)P(sick)}{P(test\ pos)}$$

Here, $P(sick)$ is the probability of having the disease, $P(test\ pos)$ is the probability of getting a positive test result, and $P(test\ pos|sick)$ is the conditional probability of a positive test result when the patient tested is sick (also known as the accuracy of the test).

For this example, we will assume the probability of having the disease is 0.2 and the test has an accuracy of 0.95 and a false positive rate of 0.3. The prior probability in this example is $P(sick)=0.2$. We can improve our knowledge of whether someone has the disease with the test. We want to find the posterior: $P(sick|test\ pos)$.

Since we know the accuracy of the test is 0.95 and the prior, we already know the values of the terms in the nominator of the equation. To find $P(test\ pos)$, we need to consider all cases where we can get a positive test result, which could be a true positive or a false positive. We calculate the probability of the positive test result as follows:

$$P(test\ pos) = P(test\ pos|sick)P(sick) + P(test\ pos|not\ sick)P(not\ sick)$$

Recall that the total probability of possible states must equal one: $P(not\ sick) = 1 - P(sick) = 0.8$. With this in mind, we can calculate $P(sick|test\ pos)$:

$$P(sick|test\ pos) = \frac{0.95(0.2)}{0.95(0.2) + 0.3(0.8)} \cong 0.44$$

Given one test result, we see that the likeliness of the patient being sick has increased from 0.2 to 0.44 with the additional information. 0.44 is not a particularly convincing result. This is due to the relatively high false positive rate of 0.3, even though the test has high accuracy. If we are not convinced, we can run the test again using 0.44 as the new prior. Assuming the second test is positive, we would get the following result:

$$P(sick|test\ pos) = \frac{0.95(0.441)}{0.95(0.441) + 0.3(1 - 0.441)} \cong 0.71$$

A second positive test provides a more convincing result of 0.71. We could continue to iteratively improve the probability estimation that a patient is sick by incorporating the results from additional tests. This incremental improvement of probability estimation is the basic idea of Bayesian statistics.

This section contained an introduction to probability. We covered independent and dependent events and then provided more details on dependent events, which led to the introduction of Bayes' Theorem. In the next section, we will discuss a type of classification model that uses Bayes' Theorem.

Linear Discriminant Analysis

In the previous chapter, we discussed logistic regression as a classification model leveraging linear regression to model directly the probability of a target distribution given an input distribution. One alternative to this approach is LDA. LDA models the probability of target distribution class memberships given input variable distributions corresponding to each class using decision boundaries constructed using Bayes' Theorem, which we discussed previously. Where we have k classes, using Bayes' Theorem, we have the probability density function for LDA class membership simply as $P(Y = k|X = x)$ for any discrete random variable, X. This relies on the posterior probability that an observation x in variable X belongs to the kth class.

Before proceeding, we must first make note that LDA makes three pertinent assumptions:

- Each input variable is normally distributed.
- Across all target classes, there is **equal covariance** among the predictors. In other words, the shared variance of all input variables is uniform. For this reason, it is useful to **standardize each input variable** to have a mean of 0 and **scale to unit variance** (standard deviation of 1).
- Samples are assumed to be independent; **random sampling** is highly important to avoid complications resulting from **serial or cluster effects**.

With these assumptions met, we can make the assumption that the class probability density function is Gaussian (normal), and thus, in a one-dimensional form (one input variable), we have the following density function:

$$\frac{1}{\sqrt{2\pi}\ \sigma_k} e^{\frac{-1}{2\sigma_k^2}(x-\mu_k)^2}$$

Here, μ_k is the kth class mean and σ_k^2 is the kth class variance. In the multivariate case, the normal probability density function is this:

$$\frac{1}{2\pi^{p/2}\sqrt{|D|}} e^{-\frac{1}{2}(x-\mu)^T D^{-1}(x-\mu)}$$

Here, D is the covariance matrix of the input variables, p is the number of input variables (or parameters), and π is the prior probability. The univariate calculation of the posterior probability is shown here (where k corresponds to the kth class and K corresponds to the number of classes):

$$\frac{\pi_k \frac{1}{\sqrt{2\pi}\sigma} e^{\frac{-1}{2\sigma^2}(x-\mu_k)^2}}{\sum_{i=1}^{K} \pi_i \frac{1}{\sqrt{2\pi}\sigma} e^{\frac{-1}{2\sigma^2}(x-\mu_i)^2}}$$

And for the multivariate case, the calculation looks like this:

$$\frac{\pi_k \frac{1}{\sqrt{|D|}} e^{-\frac{1}{2}(x-\mu)^T D^{-1}(x-\mu)}}{\sum_{i=1}^{K} \pi_i \frac{1}{\sqrt{|D|}} e^{-\frac{1}{2}(x-\mu)^T D^{-1}(x-\mu)}}.$$

The univariate linear discriminant function, $\delta_k(x)$, can be written as follows:

$$\delta_k(x) = \frac{x^* \mu_k}{\sigma^2} - \frac{\mu_k^2}{2\sigma^2} + \log(\pi_k)$$

Here, the class with the largest value of $\delta_k(x)$ has its label assigned to the given observation. This uses the Bayesian decision boundary for K classes:

$$x = \frac{\mu_1^2 - \mu_2^2 - \ldots - \mu_K^2}{K(\mu_1 - \mu_2)}.$$

For the multivariate case, the linear discriminant function, which follows the same Bayesian decision boundary, looks like this:

$$\delta_k(x) = x^T D^{-1} \mu_k - \frac{1}{2} \mu_k^T D^{-1} \mu_k + \log(\pi_k).$$

Now that we have identified the probability density functions, calculations for class membership posterior probabilities, the Bayesian decision boundary, and linear discriminant functions, we can understand how the required assumptions for LDA are very important. While transformations can be used to make data fit the required assumptions, these transformations must also be adaptable across future datasets. Therefore, if the data changes frequently and heavy transformations are needed to meet the parametric assumptions required of LDA to produce useful results across training and testing, this may not be the proper method of classification for the task (something such as **QDA** may be more useful). However, if these assumptions can be met at the researcher's comfort, and with fair subject-matter knowledge to confirm this, LDA is an outstanding algorithm that is capable of producing highly reliable and stable results across very large datasets, as defined by both large feature space and high observation count. LDA also performs well on smaller datasets.

Let's look at an example in practice. Let's load the `affairs` dataset from `statsmodels` so we can check class imbalance:

```
import statsmodels.api as sm
df_affairs = sm.datasets.fair.load().data
total_count = df_affairs.shape[0]
positive_count = df_affairs.loc[df_affairs['affairs'] > 0].shape[0]
positive_pct = positive_count / total_count
negative_pct = 1 - positive_pct
print("Class 1 Balance: {}%".format(round(positive_pct*100, 2)))
print("Class 2 Balance: {}%".format(round(negative_pct*100, 2)))
```

Here we can see the class imbalance as follows:

Class 1 Balance: 32.25%

Class 2 Balance: 67.75%

We can see the class imbalance is 67.75/32.25. Class imbalance doesn't often become a large concern until it approaches around 90/10, so we'll leave this as-is without any additional work, such as upsampling or downsampling. We'll recode any `affairs` value greater than 0 to 1 to make this a binary classification problem:

```
df_affairs['affairs'] = np.where(df_affairs['affairs'] > 0, 1, 0)
```

Let's select the features we want to use:

```
X=df_affairs[['rate_marriage','age','yrs_married','children','religious','educ','occupation','occupation_husb']]
y=df_affairs['affairs']
```

Discriminant Analysis

Recall that input feature standard deviations must be the same across both classes for LDA to effectively produce a linear separation boundary. Using the method shown next, we can see all input features share the same standard deviation across both target classes with the exception of `rate_marriage`. The standard deviation here has a difference of about 20%. Considering we have eight input features and all features in the model will be scaled, this likely won't present an issue. If the model underperforms, based on this information, we can reasonably assume it isn't the algorithm's fault as the required assumptions are met. Rather, it would more likely be that we don't have enough features or observations to fully explain the target. We exclude `affairs` because that's the target variable and `occupation` and `occupation_husb` because they're categorical encodings that will be one-hot encoded since we are not considering them to be ordinal based on our analysis scope:

```
df_affairs_sd = pd.concat([X, y], axis=1)

for col in df_affairs_sd.columns:
    if col not in ['affairs','occupation','occupation_husb']:
        print('Affairs = 0, Feature = {}, Standard Deviation = {}'.format(col, round(np.std(df_affairs_sd.loc[df_affairs_sd['affairs'] == 0, col]), 2)))
        print('Affairs = 1, Feature = {}, Standard Deviation = {}'.format(col, round(np.std(df_affairs_sd.loc[df_affairs_sd['affairs'] == 1, col]), 2)))
```

```
Affairs = 0, Feature = rate_marriage, Standard Deviation = 0.82
Affairs = 1, Feature = rate_marriage, Standard Deviation = 1.07
Affairs = 0, Feature = age, Standard Deviation = 6.81
Affairs = 1, Feature = age, Standard Deviation = 6.7
Affairs = 0, Feature = yrs_married, Standard Deviation = 7.1
Affairs = 1, Feature = yrs_married, Standard Deviation = 7.18
Affairs = 0, Feature = children, Standard Deviation = 1.42
Affairs = 1, Feature = children, Standard Deviation = 1.41
Affairs = 0, Feature = religious, Standard Deviation = 0.89
Affairs = 1, Feature = religious, Standard Deviation = 0.84
Affairs = 0, Feature = educ, Standard Deviation = 2.21
Affairs = 1, Feature = educ, Standard Deviation = 2.09
```

In the previous chapter, we identified that `occupation` and `occupation_husb` do not capture much difference in explaining affairs so we will drop `occupation_husb` from the dataset to minimize the volume of one-hot encodes, as follows:

```
import pandas as pd
pd.options.mode.chained_assignment = None
X['occupation'] = X['occupation'].map({1:'Occupation_One',
                                       2:'Occupation_Two',
                                       3:'Occupation_Three',
                                       4:'Occupation_Four',
                                       5:'Occupation_Five',
                                       6:'Occupation_Six'})
X = pd.get_dummies(X, columns=['occupation'])
X.drop('occupation_husb', axis=1, inplace=True)
```

Let's build a train/test split with a test size of 33% of the overall data:

```
from sklearn.model_selection import train_test_split
X_train, X_test, y_train, y_test = train_test_split(X, y, test_size=0.33, random_state=42)
```

We want to retain a copy of the original data, but as noted earlier, we also need to center and scale the data for LDA to have an optimal chance of success. Therefore, we'll take copies of the X data and scale them. Note, however, we do not want to scale the one-hot encoded data since one-hot encodes cannot be scaled and retain their meaning. Therefore, we will use scikit-learn's `ColumnTransformer()` pipeline function to apply `StandardScaler()` to all but the one-hot-encoded columns, like so:

```
from sklearn.compose import ColumnTransformer
from sklearn.preprocessing import StandardScaler

X_train_sc = X_train.copy()
X_test_sc = X_test.copy()

ct = ColumnTransformer([
        ('', StandardScaler(), ['rate_marriage','age','yrs_married','children','religious','educ'])], remainder='passthrough')
X_train_sc = ct.fit_transform(X_train_sc)
ct = ColumnTransformer([
        ('', StandardScaler(), ['rate_marriage','age','yrs_married','children','religious','educ'])], remainder='passthrough')
X_test_sc = ct.fit_transform(X_test_sc)
```

Now, let's fit the LDA model to the training data. We fit the model to the training data and then use that to predict the training data to get a performance benchmark. Then, we will use the model to predict the testing data to see how well the model can generalize on unseen data.

Fit the data, like so:

```
from sklearn.discriminant_analysis import LinearDiscriminantAnalysis
lda = LinearDiscriminantAnalysis()
lda.fit(X_train_sc, y_train)
```

Here, we build functions to measure precision and recall. Precision is a measure that basically tells you how well a model is at finding only the positive values of the target (affairs = 1) and none of the others (affairs = 0). Recall is a measure that practically tells you how well the model performs at finding all positive values of the target, regardless of how many of the others it finds. The ideal scenario is both of these metrics are very high. However, this may not be possible. If neither is very high, the use case should determine which metric is more important:

```
def precision_score(true_positives:int, false_positives:int):
    precision = true_positives / (true_positives + false_positives)

    return precision;

def recall_score(true_positives:int, false_negatives:int):
    recall = true_positives / (true_positives + false_negatives)

    return recall;
```

Let's now run the model and verify performance:

```
from sklearn.metrics import confusion_matrix
import seaborn as sns

y_train_pred = lda.predict(X_train_sc)

cf_train = confusion_matrix(y_train, y_train_pred, labels=[0,1])
tn_train, fp_train, fn_train, tp_train = cf_train.ravel()

cf_matrix = sns.heatmap(cf_train, annot=True, fmt='g', cbar=False)
cf_matrix.set(xlabel='Predicted', ylabel='Actual', title='Confusion Matrix - Train');

print('Precision on Train: ', round(precision_score(tp_train, fp_train), 4))
print('Recall on Train: ', round(recall_score(tp_train, fn_train), 4))
```

Here we can see our model results on the training data:

```
Precision on Train:    0.6252

Recall on Train:    0.3444
```

Here, we can visualize performance using a confusion matrix to compare actual target values against the predicted values on the training data:

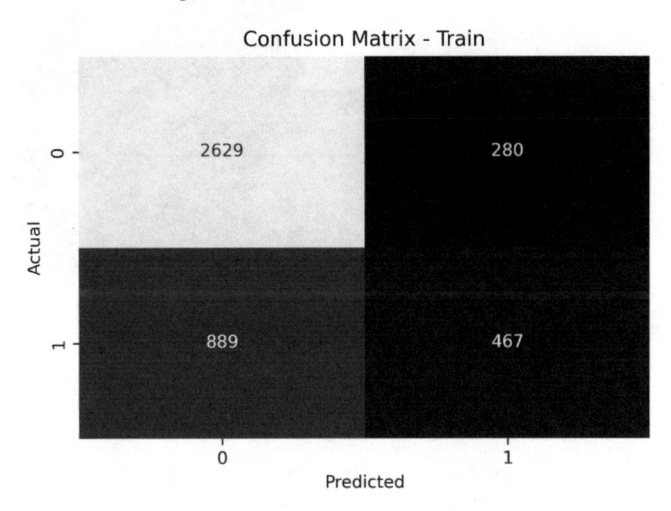

Figure 9.1 – Confusion matrix for LDA training data

Let's repeat the process on the test data:

```
y_test_pred = lda.predict(X_test_sc)

cf_test = confusion_matrix(y_test, y_test_pred, labels=[0,1])
tn_test, fp_test, fn_test, tp_test = cf_test.ravel()

cf_matrix = sns.heatmap(cf_test, annot=True, fmt='g', cbar=False)
cf_matrix.set(xlabel='Predicted', ylabel='Actual', title='Confusion Matrix - Test');

print('Precision on Test: ', round(precision_score(tp_test, fp_test), 4))
print('Recall on Test: ', round(recall_score(tp_test, fn_test), 4))
```

Here we can see our model results on the test data:

```
Precision on Test:    0.6615

Recall on Test:    0.3673
```

Here, we can visualize performance using a confusion matrix to compare actual target values against the predicted values on the test data:

Confusion Matrix - Test

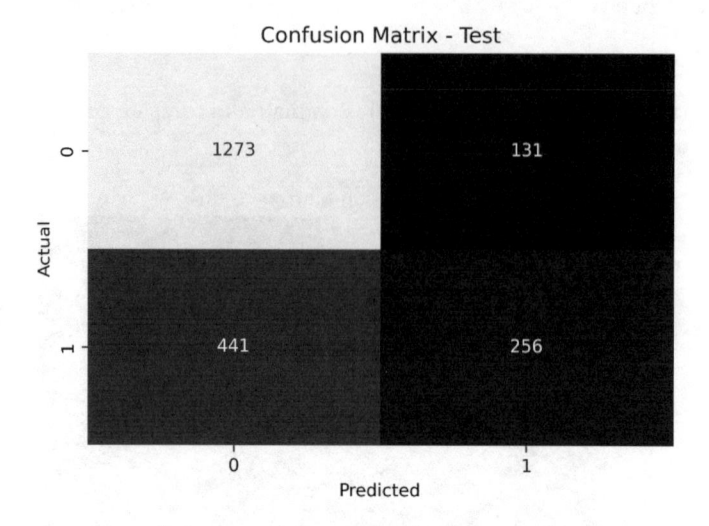

	0	1
0	1273	131
1	441	256

Figure 9.2 – Confusion matrix for LDA test data

From these results, we can see the scores are consistent for precision and recall across both training and test sets using the model we built on the training data, so we can conclude the model generalizes well on unseen data compared to the benchmark (training) performance. However, the performance is not great. The issue may be that our target encoding is not very useful (perhaps the discretization of affairs is not useful and a statistical approach should be taken to discretize the values). However, it is also possible the features do not explain enough of the variance in the `response` variable to build a model that is of much use (it's also possible affairs cannot be predicted at all and there is too much random noise to model). However, for the intent of the chapter, we were able to demonstrate how a model using LDA could be constructed and tested in addition to the preprocessing steps that must be taken before data can be modeled.

Supervised dimension reduction

Unlike PCA, which can be used to perform unsupervised dimension reduction, LDA can be leveraged to perform supervised dimension reduction. That is, upon training the model to learn the input variance in relation to the output target, a derived set of features can be obtained. The differences are illustrated in the following diagram:

PCA:
component axes that maximize the variance

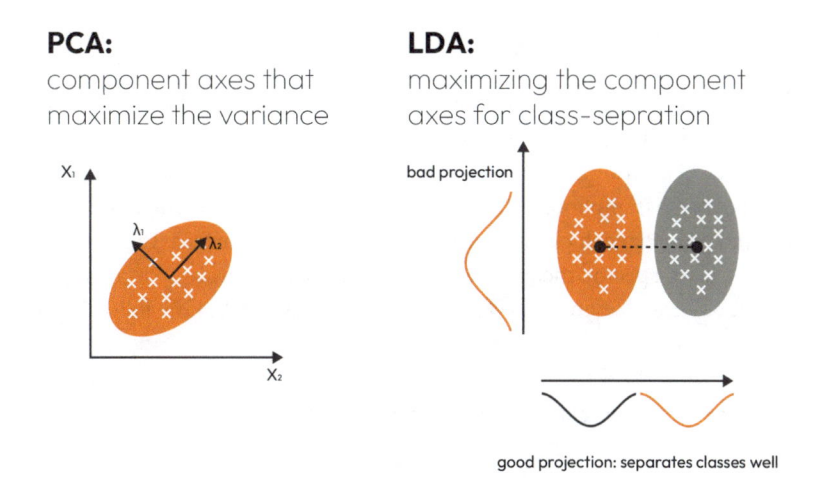

LDA:
maximizing the component axes for class-sepration

bad projection

good projection: separates classes well

Figure 9.3 – Dimension reduction: LDA versus PCA

Suppose our data produced a good result when using LDA for classification. We could then confidently use the same data to build a supervised dimension reduction technique. To perform this, we would run the following code using `fit_transform()` to transform the input data with respect to the target (rather than the `fit()` function we used to fit a classifier earlier). As with the classifier, the data should still conform to the assumptions of the LDA model. *Data is reduced according to the number of classes in the response variable*. Where the number of classes is *C*, the number of reduced features will be *C-1*. Because we have only two classes in our target variable, y, LDA reduces the input features to a dimension of 1:

```
from sklearn.discriminant_analysis import LinearDiscriminantAnalysis
lda = LinearDiscriminantAnalysis()
X_reduced = lda.fit_transform(X_train_sc, y_train)
```

We can see the data has now been reduced to a one-dimensional feature:

```
print('Input data dimensions: ', X_train_sc.shape)
print('Transformed data dimensions: ', X_reduced.shape)
```

Here, we can see our data is reduced from 12 features to 1:

```
Input data dimensions:   (4265, 12)
Transformed data dimensions:   (4265, 1)
```

Quadratic Discriminant Analysis

In the last section, we discussed LDA. The data within each class needs to be drawn from a multivariate Gaussian distribution, and the covariance matrix is the same across different classes. In this section, we consider another type of discriminant analysis called QDA but the assumptions for QDA can be relaxed on the covariance matrix assumption. Here, we do not need the covariance matrix to be identical across different classes but only for each class to have its own covariance matrix. The multivariate Gaussian distribution with a class-specific mean vector within each class for observations is still required to conduct QDA. We assume that an observation from a kth class satisfies the following formula:

$$X \sim N(\mu_k, \Sigma_k)$$

We'll thus consider a generative classifier, as follows:

$$p(X|y = k, \theta) = N(X|\mu_k, \Sigma_k)$$

And then, its corresponding class posterior is this:

$$p(y = k|X, \theta) \propto \pi_k N(X|\mu_k, \Sigma_k).$$

Here, $\pi_k = p(y = k)$ is the prior probability of the class k. In literature, this is called **Gaussian Discriminant Analysis (GDA)**. Then, the discriminant function is computed by taking the log posterior over k classes, as follows:

$$log p(y = k|X, \theta) = log \pi_k - \frac{log|2\pi\Sigma_k|}{2} - \frac{(X - \mu_k)^T \Sigma_k^{-1}(X - \mu_k)}{2} + C$$

Here, C is a constant. Unlike in LDA, the quantity of X appears as a quadratic function in the preceding formula and it is known as QDA. The choice between LDA or QDA is related to the bias-variance trade-off. If the assumption that shares the same covariance matrix over k classes is not good, then a high bias can be significant in LDA. In other words, LDA is used when a boundary between classes is linear, and QDA is performed better in the case of a nonlinear boundary between classes.

Now, we consider an example using the same *Iris* dataset that we studied in the last chapter. To perform QDA, an option is to use `QuadraticDiscriminantAnalysis` in sklearn. As in the last chapter, we use a `train_test_split` function to create a training and a test set (80% versus 20%), and then we fit a QDA model on the training set and use this model for prediction:

```
from sklearn.discriminant_analysis import
QuadraticDiscriminantAnalysis
# fit the model using sklearn
model_qda = QuadraticDiscriminantAnalysis(store_covariance=True)
model_qda.fit(X_train, y_train)
y_hat_qda = model_qda.predict(X_test)
pred_qda = list(map(round, y_hat_qda))
```

The confusion matrix related to the test set is shown here:

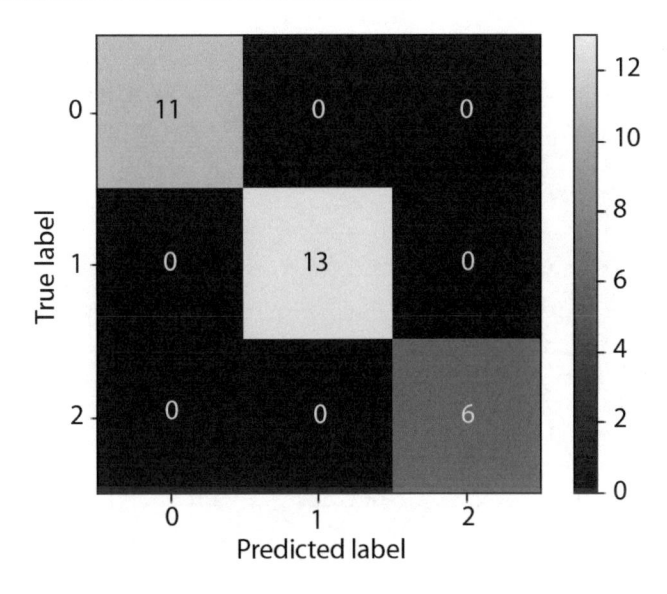

Figure 9.4 – QDA confusion matrix

The accuracy is 100% for this dataset. This analysis is based on all variables (`sepal_length` (cm), `sepal_width` (cm), `petal_length` (cm), `petal_width` (cm)), and for three dependent variables (targets): `setosa`, `versicolor`, and `virginica`. For visualization educational purposes, we will only consider the analysis in two dimensions using `sepal_length` and `sepal_width` as the independent variables:

```
df1 = df[["sepal_length", "sepal_width", "target"]]
df1.head()
```

	sepal_length	sepal_width	target
0	5.1	3.5	0
1	4.9	3.0	0
2	4.7	3.2	0
3	4.6	3.1	0
4	5.0	3.6	0

Figure 9.5 – *sepal_length* and *sepal_width*

The visualization in two dimensions between `sepal_length` and `sepal_width` can be reproduced using the following Python code:

```python
import matplotlib.pyplot as plt
FIGSIZE = (8,8)
Xax= np.array(df1["sepal_length"])
Yax= np.array(df1["sepal_width"])
labels= np.array(df1["target"])
cdict={0:'red',1:'green', 2 :'blue'}
labl={0:'setosa',1:'versicolor', 2: 'virginica'}
marker={0:'*',1:'o', 2:'p'}
alpha={0:.3, 1:.5, 2:.3}
fig,ax=plt.subplots(figsize=FIGSIZE)
fig.patch.set_facecolor('white')
for l in np.unique(labels):
    ix=np.where(labels==l)
    ax.scatter(Xax[ix],Yax[ix],c=cdict[l],s=40,
        label=labl[l],marker=marker[l],alpha=alpha[l])
plt.xlabel("Sepal Length",fontsize=14)
plt.ylabel("Sepal Width",fontsize=14)
plt.legend()
plt.show()
```

The visualization is as follows:

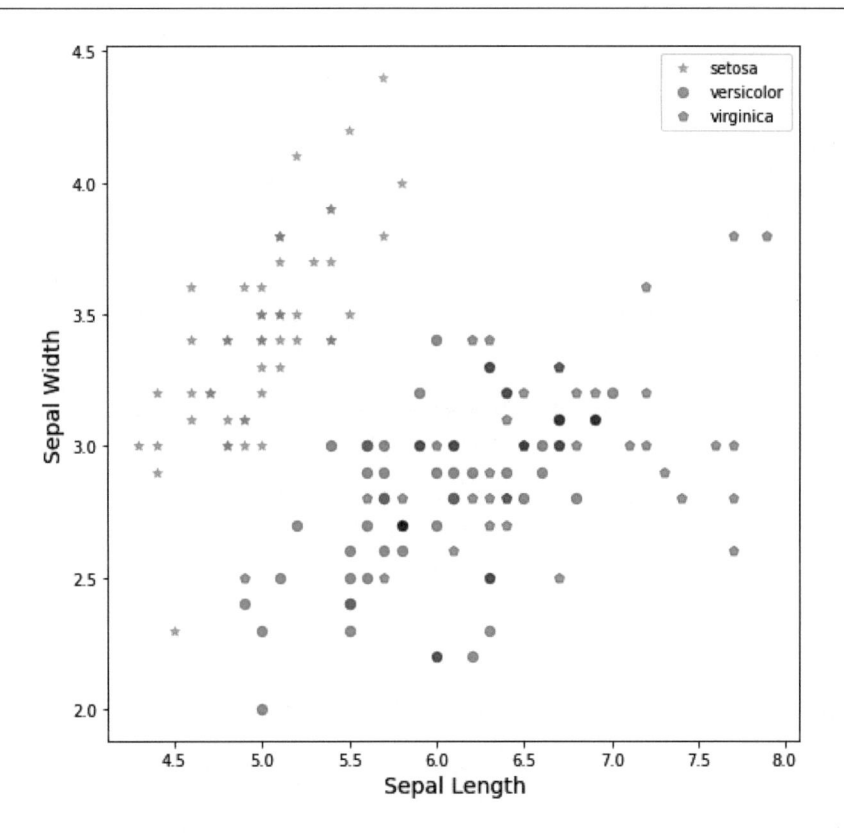

Figure 9.6 – Iris flower species scatterplot

Observing that there is no linear separation between the classes (`setosa`, `versicolor`, and `virginica`), a QDA would be a better approach than an LDA.

The `min` and `max` values are (4.3, 7.9) for `sepal_length` and (2.0, 4.4) for `sepal_width`:

```
df1['sepal_length'].min(), df1['sepal_length'].max()

(4.3, 7.9)

df1['sepal_width'].min(), df1['sepal_width'].max()

(2.0, 4.4)
```

Figure 9.7 – sepal_length and sepal_width min and max values

We train a new QDA model using only these two variables, `sepal_length` and `sepal_width`, on all observations in the Iris dataset, as follows:

```
df1['species'] = df1['target']
df1['species'] = df1['species'].apply(lambda x: 'setosa'
      if x==0 else('versicolor' if x==1 else 'virginica'))
X = df1.drop(['target','species'], axis=1)
y = df1['target']
model_qda = QuadraticDiscriminantAnalysis(store_covariance=True)
model_qda.fit(X,y)
```

Then, we create a generic dataset with 500 observations for `sepal_length` and `sepal_width` within the min and max range of each variable for a prediction:

```
sepal_length = np.linspace(4, 8, 500)
sepal_width = np.linspace(1.5, 4.5, 500)
sepal_length,sepal_width  = np.meshgrid(sepal_length, sepal_width)
prediction = np.array([model_qda.predict( np.array([[x,y]]))
for x,y in zip(np.ravel(sepal_length), np.ravel(sepal_width))
]).reshape(sepal_length.shape)
```

The following Python code is executed for the visualization of boundary separation between classes:

```
fig = sns.FacetGrid(df1, hue="species", size=8, palette =
'colorblind').map(plt.scatter, "sepal_length", "sepal_width").add_
legend()
figax = fig.ax
figax.contourf(sepal_length,sepal_width, prediction, 2, alpha = .1,
colors = ('red','green','blue'))
figax.contour(sepal_length,sepal_width, prediction, 2, alpha = 1,
colors = ('red','green','blue'))
figax.set_xlabel('Sepal Length')
figax.set_ylabel('Sepal Width')
figax.set_title('QDA Visualization')
plt.show()
```

We get the resulting plot as follows:

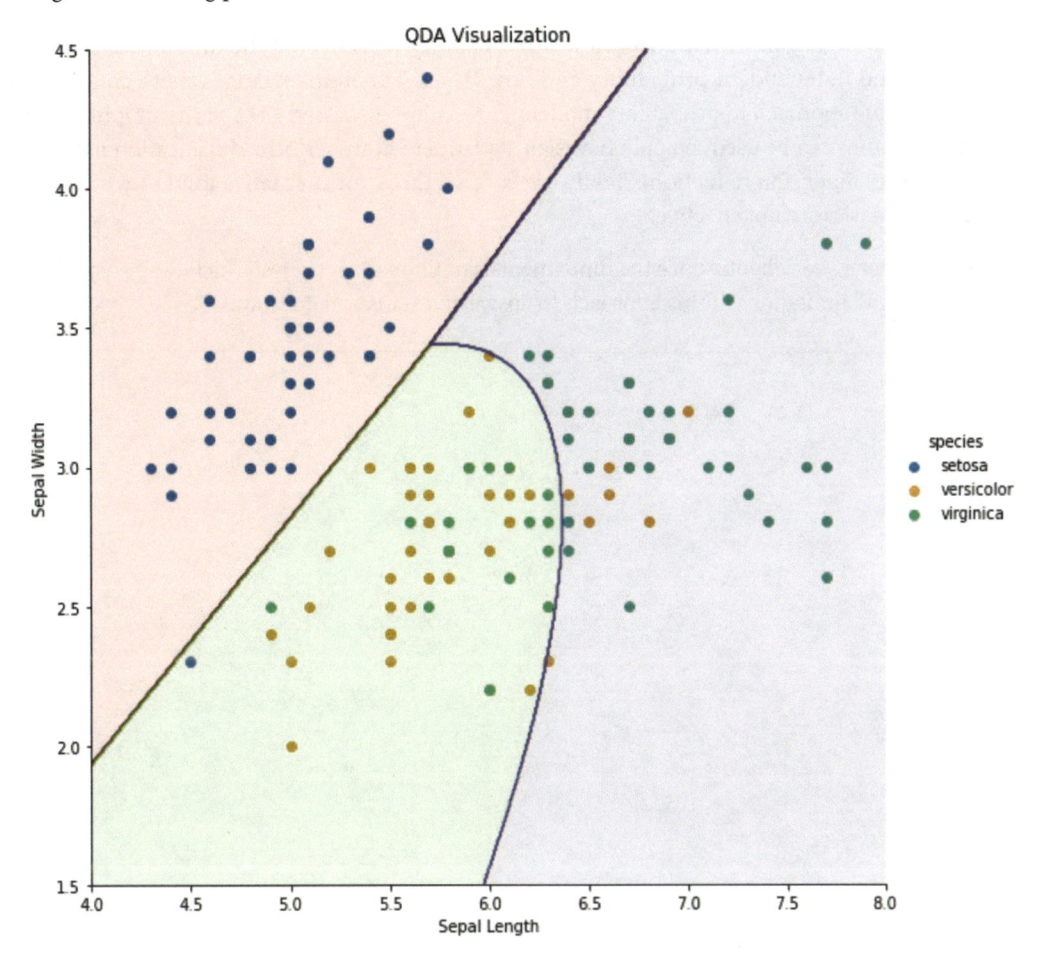

Figure 9.8 – QDA decision boundaries for Iris dataset

There is a good separation between `setosa` and these other two classes but not between the `versicolor` and `virginica` classes. The separation between classes will be better in higher dimensions, which means that in this example, we consider all four independent variables: `sepal_length`, `sepal_width`, `petal_length`, and `petal_width`. In other words, this analysis is better conducted in four dimensions, but it is not possible to have a four-dimensional visualization for human beings.

Summary

In this chapter, we began with an overview of probability. We covered the differences between conditional and independent probability and how Bayes' Theorem leverages these concepts to provide a unique approach to probability modeling. Next, we discussed LDA, its assumptions, and how the algorithm can be used to apply Bayesian statistics to both perform classification modeling and supervised dimension reduction. Finally, we covered QDA, an alternative to LDA when linear decision boundaries are not effective.

In the next chapter, we will introduce the fundamentals of time-series analysis, including an overview of the depths and limitations of this approach to answering statistical questions.

Part 4: Time Series Models

The objective of this part is to learn how to analyze and create forecasts for univariate and multivariate time series data.

It includes the following chapters:

- *Chapter 10, Introduction to Time Series*
- *Chapter 11, ARIMA Models*
- *Chapter 12, Multivariate Time Series*

10
Introduction to Time Series

In *Chapter 9*, *Discriminant Analysis*, we concluded our overview of statistical classification modeling by introducing conditional probability using Bayes' theorem, **Linear Discriminant Analysis** (**LDA**), and **Quadratic Discriminant Analysis** (**QDA**). In this chapter, we will introduce time series, the underlying statistical concepts, and how to apply them in everyday analysis. We will introduce the topic with the distinction between time-series data and what we have discussed up to this point in the book. We then provide an overview of what to expect with time-series modeling and the goals it can be leveraged to achieve. Within the context of time series, we then reintroduce the mean and variance statistical parameters, in addition to correlation. We provide an overview of **linear differencing**, **cross-correlation**, and **autoregressive** (**AR**) and **moving average** (**MA**) properties and how to identify their ordering using **autocorrelation function** (**ACF**) and **partial ACF** (**PACF**) plots. After, we provide an overview of the introductory white-noise model. We conclude the chapter with a detailed, formal overview of the concept of stationarity, arguably the most important precursor to successful time-series forecasting.

In this chapter, we're going to cover the following main topics:

- What is a time series?
- Goals of time-series analysis
- Statistical measurements
- The white-noise model
- Stationarity

What is a time series?

In this chapter and the next few chapters, we will work with a type of data called time-series data. Up until this point, we have worked with independent data—that is, data consisting of samples that are not related. A time series is typically a measurement of the same sample taken over time, which makes the samples in this type of data related. There are many time series present around us every day. A few common examples of time series are daily temperature measurements, stock price ticks, and the heights of ocean tides. While a time series does not need to be measured at fixed intervals, in this book, we will primarily be concerned with measurements taken at fixed intervals, such as daily or every second.

Let's look at some notation. In the following equation, we have a variable x that is repeatedly sampled over time. The subscripts enumerate the sample points (sample 1 through sample t), and the whole series of samples is denoted X. The subscript value is commonly called the **lag** of the variable. For example, x_2 could be referred to as the second lag of the variable x:

$$X = x_1, x_2, \ldots x_{t-1}, x_t$$

In general, the x points can be univariate or multivariate. For example, we could take a temperature reading over time that would result in a **univariate time series**, meaning each x term would correspond to a single temperature value. We could also take a more holistic set of weather readings such as temperature, humidity, rainfall, and sunshine, which would result in a **multivariate time series**, meaning each x term would correspond to a temperature, humidity, rainfall, and sunshine value.

Our discussions on time series start with single variable analysis in this chapter and *Chapter 11, ARIMA Models*. But as with the chapters on regression, we will start with a single-variable approach and proceed to work with multivariate time series. Just as with the chapters on regression, we may find the multiple variables in a time series are related to the outcome of a variable of interest. In *Chapter 12, Multivariate Time Series*, we will deal with the extra complexity of multivariate time series and extend the single-variable models discussed in *Chapter 11* to multiple variables in *Chapter 12*. In this chapter, we will cover the basics of time series.

Time series typically exhibit a property called **serial correlation**, which means that previous knowledge provides some knowledge about the future of the time series. We can measure serial correlation by comparing the current value of a time series to the previous values in the series to determine if present values are correlated with previous values. This type of correlation is also called **autocorrelation**. We will discuss more on autocorrelation later in this chapter, including how to determine if a series exhibits autocorrelation. Later in this chapter, we will look at how to make the calculations to determine whether data exhibits serial correlation and to calculate autocorrelation. However, let's first discuss what we intend to achieve with time-series analysis.

Goals of time series analysis

There are two goals in time-series analysis:

- Identifying any patterns in the time series
- Forecasting future values of the time series

We can use time-series analysis methods to uncover the nature of a time series. At the most basic level, we may want to know if a series appears to be random or if a time series appears to exhibit a pattern. If a time series has a pattern, we can determine if it has seasonal behavior, cyclical patterns, or exhibits trending behavior. *We will investigate the behaviors of time series both by observation and by the results of fitting models.* Models can provide insight into the nature of a series and allow us to forecast the future values of a time series.

The other goal of time-series analysis is **forecasting**. We see examples of forecasting in many common situations, such as weather forecasting and stock price forecasting. It is important to keep in mind that the methods of forecasting we cover in this book are not infallible. Great care should be taken when communicating results from forecasting models. Predictions from models are always uncertain. *Predictions should always be contextualized with an understanding of the model uncertainty.* We will endeavor to reinforce this concept through the use of prediction intervals and model error rates.

Now, with the context of what we hope to achieve with time-series analysis, let's start looking at tools for analyzing univariate time series.

Statistical measurements

When using time-series models to work with serially correlated data sets, we need to understand mean and variance – within the context of time – in addition to autocorrelation and cross-correlation. Understanding these variables helps build an intuition about how time-series models work and when they are more useful than models that do not account for time.

Mean

In time-series analysis, the sample mean of a series is the sum of all values across each point in time in the series divided by the count of values. Where t represents each discrete time step and n is the total number of time steps, we can calculate the sample mean of a time series as follows:

$$\overline{X} = \frac{1}{n} \sum_{t=1}^{n} x_t$$

There are two types of processes generating time series; one is an ergodic process and the other is non-ergodic. An ergodic process has consistent output independent of time, whereas a non-ergodic process does not necessarily have consistent output over time. The sample mean of an ergodic process converges to the true population mean as the sample size increases. However, the sample mean of a non-ergodic process does not converge as the sample size increases; the sample mean of one end-to-end phase of a process's output may not converge toward the process population mean similarly to the sample mean of another end-to-end phase of the same process. One example of a non-ergodic process is a machine that requires frequent recalibration as output quality diminishes due to factors such as moisture or vibration. The tools introduced in this chapter, and expanded in the next, will help an analyst overcome limitations presented by this natural constraint presented in process-driven time-series data.

In time-series analysis, the mean is commonly referred to as the **signal**. With respect to forecasting, the mean must be constant across time. We discussed in *Chapter 6, Simple Linear Regression* – as we will later in this chapter and in the next – the concept of first-order differencing, which is a **low-pass linear filter** used to *remove* high-frequency data from our output and *pass through* low-frequency data. When a signal is not constant, as when it is monotonically increasing or decreasing, for example, a first-order difference can often be applied—and repeated as needed—to produce a constant. This is one requirement of a **stationary** time series. We will discuss all components of stationarity in the *Stationarity* section of this chapter. Once a mean is constant, the variance around it can be assessed for autocorrelation. If autocorrelation exists within the variance around the mean, we can produce models to learn the patterns of the process producing it. We can also forecast the process's future patterns. A model with a constant mean and no autocorrelation can often be forecasted with the mean using a white-noise model. For all times, *t*, in a series, the formulation for a first-order linear difference is this:

$$Y'_t = Y_t - Y_{t-1}.$$

Since the first-order difference is a numerical differentiation, it will result in the removal of one data point from the time series, so must be applied prior to any modeling. Here is an example of a first-order difference in tabular data:

Raw Data	First-Order Difference
1.7	
1.4	-0.3
1.9	0.5
2.3	0.4
2.1	-0.2

Figure 10.1 – First-order difference in tabular data

Here, we have an example of data transformed with a first-order difference. The numpy `diff()` function can be used as follows, where `n=1` prescribes a first-order difference:

```
numpy.diff(array_x, n=1)
```

Variance

Variance is a statistic for measuring the dispersion of data around a mean for a given distribution. Within the context of time-series analysis, variance is distributed around the mean across time. If we have a discrete and stationary process, we can calculate the variance of the sample mean as follows:

$$Var(\overline{X}) = \frac{\sigma^2}{n} \sum_{k=-(n-1)}^{n-1} \left(1 - \frac{|k|}{n}\right) \rho_k$$

Here, n is the length of the time series, k is the number of lags to be included in the series autocorrelation (serial correlation) calculation and ρ_k is the autocorrelation for that lookback horizon. This variance calculation can be obtained through model construction, which we will walk through in *Chapter 11, ARIMA Models*. It is in model construction – where we build the **characteristic equations** for the time series – that the measurement of variance becomes most important. Otherwise, we focus on autocorrelation to build intuition about the variance and the process producing it. Oftentimes, we have what is referred to as white-noise variance, which is a random distribution of variance generated from a stochastic process. White-noise variance has no autocorrelation across the time horizon. In the case of white-noise variance, we have only the following calculation for variance, which is the same calculation for data that is not serially correlated:

$$Var(\overline{X}) = \frac{\sigma^2}{n}$$

Data exhibiting white-noise variance can often be modeled with an average as there are no correlated errors. However, there may be transformations required in order to use an average, such as a first-order difference or a seasonal difference.

One common hypothesis test used in time-series analysis to assess if the variance in a series is white noise is the Ljung-Box test, created by Greta Ljung and George Box. The Ljung-Box test has the following hypotheses:

- H_o: Data points are independently distributed with no serially correlated errors
- H_a: Data points are not independently distributed and thus present serially correlated errors

This test can be applied to the residuals of any time-series model—for example, a linear model that uses time as an input or an **autoregressive integrated moving average** (**ARIMA**) model. If the result of the Ljung-Box test is the validation of the null hypothesis, the model tested is assumed to be valid. If the null hypothesis is rejected, a different model may be required. The Ljung-Box test statistic is shown here:

$$Q = n(n+2)\sum_{k=1}^{h}\frac{\hat{\rho}_k^2}{n-k}$$

Here, n is the sample size, k corresponds to each lag in the test, h is the total time horizon being tested, and $\hat{\rho}_k$ is the sample autocorrelation for each lag. The test follows the Chi-Square (χ^2) distribution and thus places more emphasis on more recent lags than those in the distant past. The Ljung-Box test statistic is compared to χ^2 distribution with h degrees of freedom, as follows:

$$Q > \chi^2_{1-\alpha,h}$$

Here, h is the tested horizon. A Q-statistic greater than the χ^2 critical value results in rejecting the null hypothesis. The Ljung-Box test can also be applied to data without a model to test whether it is constant, zero-mean data with white-noise variance. Otherwise, the test is performed on the residuals of a model. Let's generate a random, normally distributed sample of 1,000 data points having a mean of 0 and a standard deviation of 1 so that we can test using Ljung-Box whether the data is **stationary white noise**:

```
import numpy as np
import matplotlib.pyplot as plt
from statsmodels.graphics.tsaplots import plot_acf
random_white_noise = np.random.normal(loc=0, scale=1, size=1000)
```

We can observe based on the raw data that the mean is constant. We can also see that the autocorrelation structure appears to have no significant lags, which is a strong indication the variance is randomly distributed white noise. In model development, this lack of autocorrelation is what you would want to see from the residuals to help verify a model fits the process's data well:

```
fig, ax = plt.subplots(1,2, figsize=(10, 5))
ax[0].plot(random_white_noise)
ax[0].axhline(0, color='r')
ax[0].set_title('Raw Data')
plot_acf(random_white_noise, ax=ax[1])
```

In *Figure 10.2*, we can see the original white noise data and the ACF plot, which exhibits no statistically significant autocorrelation across lags:

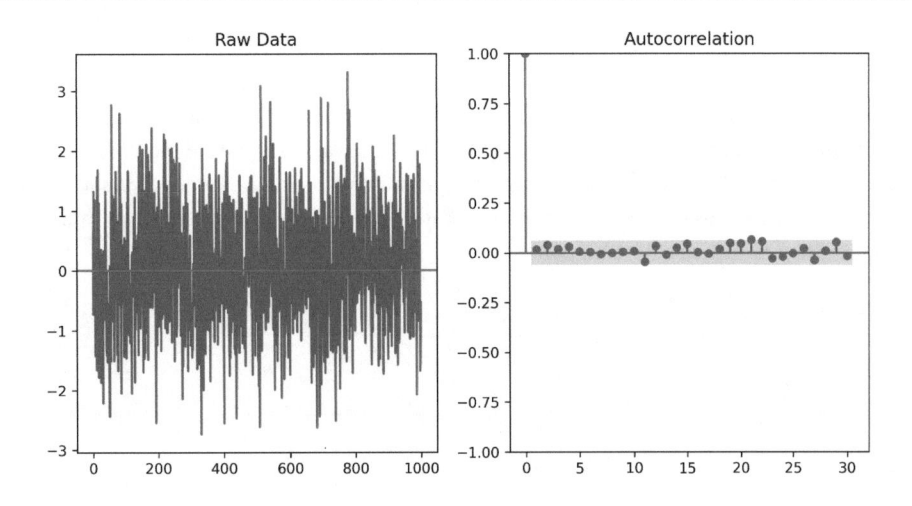

Figure 10.2 – Visual analysis of random white noise

Now, let's use the Ljung-Box test to check our assumptions about autocorrelation. We perform this test with the `acorr_ljungbox()` function from `statsmodels`. We apply the `lags=[50]` argument to test if the autocorrelation is 0 out to 50 lags:

```
from statsmodels.stats.diagnostic import acorr_ljungbox
acorr_ljungbox(random_white_noise, lags=[50], return_df=True)
```

The test returns an insignificant p-value, seen here. Therefore, we can assert that at a 95% level of confidence, the data has no autocorrelation and is thus white noise:

	lb_stat	lb_pvalue
50	51.152656	0.428186

Figure 10.3 – Ljung-Box test results for autocorrelation on white noise data

Autocorrelation

As we've mentioned to this point, autocorrelation, also called serial correlation, is a measure of correlation for a given point corresponding to previous points in the time-constrained sequence of data. It is called "auto" as it refers to a variable's correlation with itself at a previous lag. Autocorrelation across all preceding lags in the specified horizon—as opposed to for a specific lag—is referred to as **autocorrelation structure**. Here, we have, for any given lag k greater than zero, the ACF, r_k:

$$r_k = \frac{\sum_{t=k+1}^{n}(y_t - \bar{y})(y_{t-k} - \bar{y})}{\sum_{t=1}^{n}(y_t - \bar{y})^2}$$

We discussed autocorrelation in *Chapter 6, Simple Linear Regression*, within the context of identifying serial correlation as a violation of the required assumption of sampling independence for linear regression. Here, we note that autocorrelation is a core component of time-series data. We previously visually explored this data in both *Chapter 6* and this chapter using the ACF plot. We also discussed first-order differencing. Let's explore these two concepts in depth using the *United States Macroeconomic* data set from statsmodels, which we used in *Chapter 6*. Here, we select realinv and realdpi, converting both variables to a 32-bit float:

```
import numpy as np
import pandas as pd
import statsmodels.api as sm
import matplotlib.pyplot as plt

df = sm.datasets.macrodata.load().data

df['realinv'] = round(df['realinv'].astype('float32'), 2)
df['realdpi'] = round(df['realdpi'].astype('float32'), 2)

df_mod = df[['realinv','realdpi']]
```

Next, let's plot the data and its ACF using 50 lags (lags=50) and a 5% level of significance (alpha=0.05):

```
from statsmodels.graphics.tsaplots import plot_acf
from statsmodels.graphics.tsaplots import plot_pacf

fig, ax = plt.subplots(2,2, figsize=(15,10))
plot_acf(df_mod['realinv'], alpha=0.05, lags=50, ax=ax[0,1])
ax[0,1].set_title('Original ACF')
ax[0,0].set_title('Original Data')
ax[0,0].plot(df_mod['realinv'])
plot_acf(np.diff(df_mod['realinv'], n=1), alpha=0.05, lags=50,
ax=ax[1,1])
ax[1,1].set_title('Once-Differenced ACF')
ax[1,0].set_title('Once-Differenced Data')
ax[1,0].plot(np.diff(df_mod['realinv'], n=1))
```

We get the resulting plot as follows:

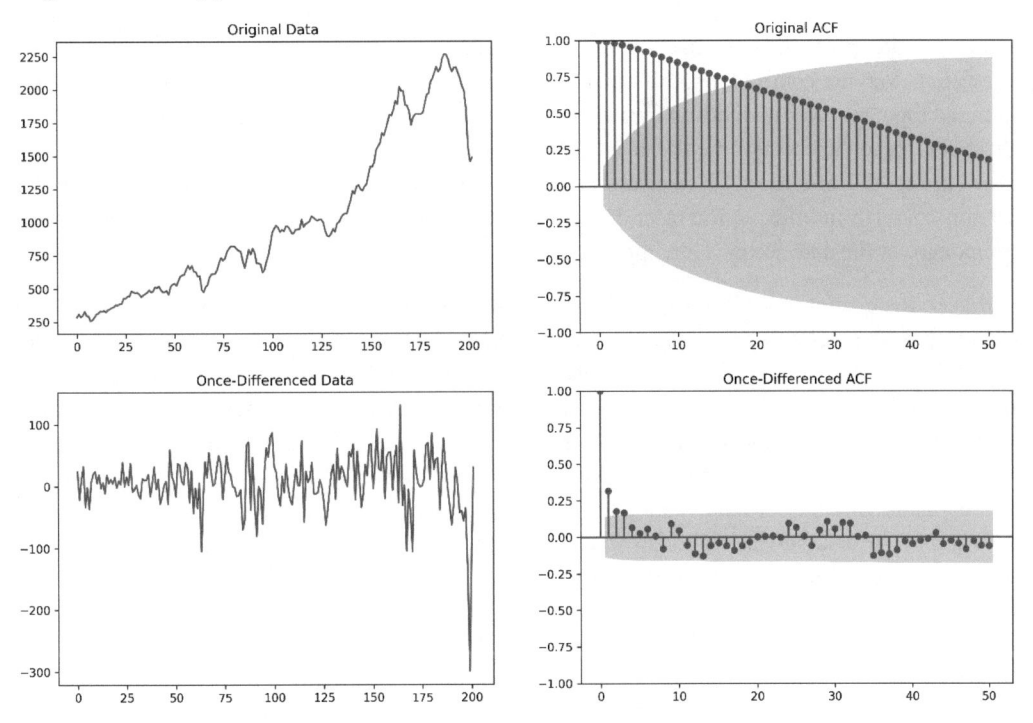

Figure 10.4 – *realinv*: original and first-difference data and ACFs

We can observe in the first plot of *Figure 10.4* a positive, deterministic signal highlighting an overall upward trend in investment. We can also observe that—especially as the time nears lag 0—some variance around the data. These two components of the data are what we would initially attempt to model. In the original data's ACF, we can see a significant, dampening autocorrelation structure, which is characteristic of exponential growth. However, because of the strong trend that dominates the autocorrelation, we aren't able to observe any information about the correlation in the variance, such as potential seasonality, for example. To remove the strong signal, we apply a first difference using the `numpy.diff()` function so that we can observe the variance's autocorrelation structure. Looking at the ACF for the differenced data, we can see some autocorrelation in the variance extending to lag 3 when applying a 95% confidence level.

At this point, it is worth mentioning **autoregressive moving average** (**ARMA**) modeling, which we will use in *Chapter 11, ARIMA Models*. As we can see in the ACF plot, there is a long horizon of autocorrelation present. Going back to the autocorrelation equation we noted earlier, we can see the calculation does not control for specific lags in the time series since the errors across each point are all summed together prior to dividing by the variance. This is why the original ACF shows a dampening effect; it is reasonable to assume that lag 0 is not smoothly serially correlated, individually, with each previous lag. Because the ACF does not control for autocorrelation between specific lags, we can use this correlation information to build a moving average function that can help model the noise component of the data across a long period.

However, if we want to be able to construct a component of a model that allows us to define the relationship from point to point within the framework of ARMA models, we will want to also observe the PACF, which does control for the correlation between lags. Note in *Figure 10.5* the original data and its PACF. It shows a much different level of granularity than the ACF plot does; whereas in the PACF we don't see a continued significant correlation with lag 0 and those beyond lag 4 until we near lag 30, we do see significant correlation in the ACF until almost lag 20. However, the original data is not very helpful because it is not stationary, which is why we see almost as much correlation between lag 0 and lag 45 as we do lag 0 and lag 1, which suggests we would not be able to build a model that converges without transformation (the first-order difference, in this case):

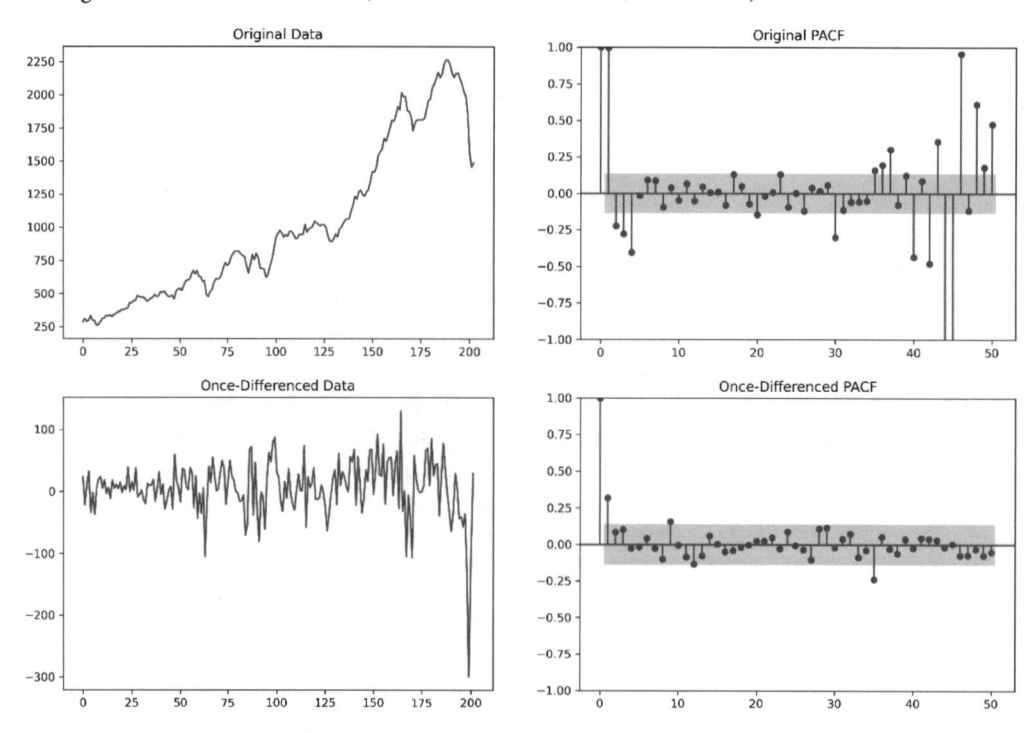

Figure 10.5 – *realinv*: original and first-difference data and PACFs

The result of the ACF in *Figure 10.4* for the *transformed (differenced) data* suggests a moving average component up to an order of 3 (MA(3)) could be useful (although an MA(1) may be better). The result of the PACF in *Figure 10.5* for the *transformed data* suggests an autoregressive component of order 1 (AR(1)) may be useful. Using the PACF, we could also conclude an AR(10) may be useful, but selecting an order this high will often result in overfitting. A rule of thumb is to not use an AR or MA component with an order greater than approximately 5 for process modeling. Model order, however, depends on the analyst and the process details. While we will discuss this concept in depth in *Chapter 11, ARIMA Models*, if we are building an ARIMA model for this data, we would have a model of order (2,1,3) following the *AR(p)* and *MA(q)* construct we selected with an integrated difference, *d*, for the first-order difference we applied illustrated as an order **(p,d,q)** model. The orders are used to build **characteristic polynomial equations** based on the model that we can then use to assess **stationarity** and **invertibility**, which helps identify **model uniqueness** and assess its ability to **converge** toward a solution, using the roots of the equations' factored forms.

Cross-correlation

Moving on, we have, for any given lag *k* greater than zero, the **cross-correlation function** (CCF) that can be used to identify the correlation between two variables across points in time. This analysis can be used to facilitate the identification of leading or lagging indicators. The CCF for two one-dimensional vectors, *i* and *j*, for a given lag, *k*, $\hat{p}_{i,j}(k)$, is shown here:

$$\hat{p}_{i,j}(k) = \frac{\sum_{t=1}^{n-k}(x_{t,i} - \bar{x}_i)(x_{t+k,j} - \bar{x}_j)}{\sqrt{\sum_{t=1}^{n}(x_{t,i} - \bar{x}_i)^2}\sqrt{\sum_{t=1}^{n}(x_{t+k,j} - \bar{x}_j)^2}}$$

In considering a case of linear regression where we have two ordered, sequence-based input variables predicting an ordered, sequenced-based dependent variable, we evaluate the input and output variables at the same time, *t*. However, if using cross-correlation to identify a leading or lagging indicator, we can identify the lag, *k*, at which the leading variable leads and modify its series position by *k* indices. For example, if the time unit of our model is in weeks and we build a regression model to predict sales using advertising as an input, we may find that in any given week, advertising has no impact on sales. However, after performing a cross-correlation analysis, we find a strong lag of 1 exists between advertising and sales such that investment in advertising in 1 week directly impacts sales in the following week. We then use this information to shift forward the advertising expenditure variable by 1 week and rerun the regression model to leverage the strong correlation between advertising expense and sales to predict market behavior. This is what we can refer to as a **lag effect**.

Let's look at an example of this in practice with Python. First, we need to assess the data. Let's load the data to begin. We need to convert the values to a `float` type. We can round to two decimals and select `realinv` and `realdpi`, the variables for real gross private domestic investment and real private disposable income, respectively:

```
import numpy as np
import pandas as pd
import statsmodels.api as sm
import matplotlib.pyplot as plt

df = sm.datasets.macrodata.load().data

df['realinv'] = round(df['realinv'].astype('float32'), 2)
df['realdpi'] = round(df['realdpi'].astype('float32'), 2)

df_mod = df[['realinv','realdpi']]
```

After plotting the data, here, we can see both series have what appears to be a strong, deterministic signal, which is the primary influence on the autocorrelations. It can easily be argued that the two variables are positively correlated, and both are increasing over time, which is mostly true. However, there is more to the data than meets the eye, in that regard; the mean (signal) is deterministic, but we need to assess the variance around the mean to truly understand how the two processes are correlated beyond the trend component. Note that in the following code, we use a lag of 50 when calculating the ACFs. A general rule of thumb in measuring ACFs is to not exceed a lag of 50 points, although this may differ based on context:

```
from statsmodels.graphics.tsaplots import plot_acf

fig, ax = plt.subplots(2,2, figsize=(20,8))
fig.suptitle('Raw Data')
ax[0,0].plot(df_mod['realinv'])
ax[0,0].set_title('Realization')
ax[1,0].set_xlabel('realinv')
ax[0,1].plot(df_mod['realdpi'])
ax[0,1].set_title('Realization')
ax[1,1].set_xlabel('realdpi')

plot_acf(df_mod['realinv'], alpha=0.05, lags=50, ax=ax[1,0])
plot_acf(df_mod['realdpi'], alpha=0.05, lags=50, ax=ax[1,1])
```

We get the following plot:

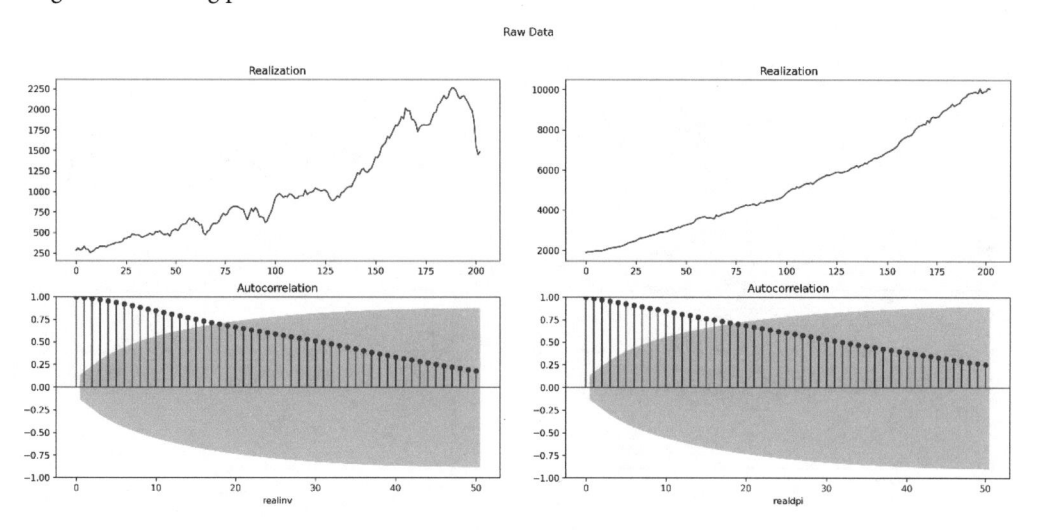

Figure 10.6 – Comparing *realinv* to *realdpi*: raw data and ACF plots

We can see in *Figure 10.6* that in addition to the strong deterministic signal, the ACF plots for both have exponentially dampening autocorrelations that are significant at the 95% confidence (outside of the shaded area) level. Based on this information in the ACFs, we should perform at least a **first-order difference**. We may need to perform two first-order differences if the ACFs exhibit the same behavior after one **first-order linear difference**.

Let's create the differenced data:

```
df_diff = pd.DataFrame()
df_diff['realinv'] = np.diff(df_mod['realinv'], n=1)
df_diff['realdpi'] = np.diff(df_mod['realdpi'], n=1)
```

Now, reusing the previous plotting code, we can see the data in *Figure 10.7*:

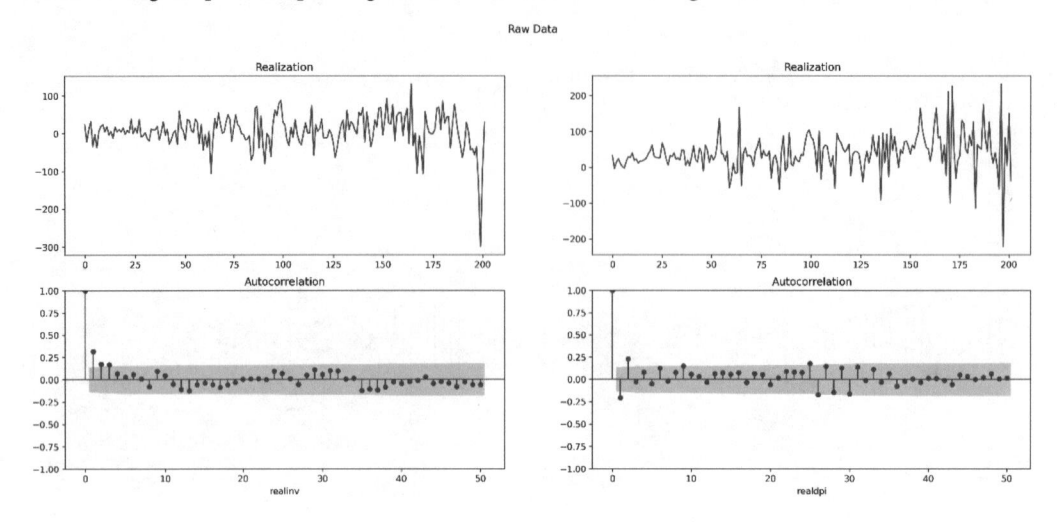

Figure 10.7 – Comparing differenced *realinv* to differenced *realdpi*

Based on *Figure 10.7*, we can see *the deterministic signal that was previously dominating the autocorrelation in Figure 10.6 is now removed*. There are a few points slightly outside the 95% confidence interval, although the level of significance is small (we do not consider lag 0 as lag 0 is 100% correlated with itself). Because some autocorrelation remains, we can consider their behaviors to not be entirely random. Thus, they could be cross-correlated. Since we addressed the issue of the deterministic signal, we can now assess the cross-correlation between the two series. Note that the means of the two series have been differenced to a constant zero. This is one condition of stationarity. Another is constant variance. We can see the differenced data has minimal autocorrelation, but that it is beyond the 95% confidence limit. Thus, we can ascertain additional ARIMA modeling may be useful, but we could also argue for using an average for modeling. However, we can also see that the third requirement of stationarity—constant covariance between periods—does not seem to be met; as the series continues across time steps, the variance fluctuates. We will discuss stationarity in more depth in the following section. For now, let's turn our attention to cross-correlation.

Now that we have removed the deterministic signal and the variance is the dominating behavior in the ACF plots, let's compare the two time series to see if they have a lagging or leading relationship. Here, we construct a CCF for graphically plotting that. We have three options for the confidence interval using a dictionary, `zscore_vals`, to build the z-scores we use to build these confidence intervals—90%, 95%, and 99%:

```
from scipy.signal import correlate
import matplotlib.pyplot as plt

def plot_ccf(data_a, data_b, lag_lookback, percentile):
```

```python
    n = len(data_a)
    
    ccf = correlate(data_a - np.mean(data_a), data_b - np.mean(data_b), method='direct') / (np.std(data_a) * np.std(data_b) * n)
    
    _min = (len(ccf)-1)//2 - lag_lookback
    _max = (len(ccf)-1)//2 + (lag_lookback-1)
    
    zscore_vals={90:1.645,
                 95:1.96,
                 99:2.576}
    
    plt.figure(figsize=(15, 5))
    
    markers, stems, baseline = plt.stem(np.arange(-lag_lookback,(lag_lookback-1)), ccf[_min:_max], markerfmt='o', use_line_collection = True)
    
    plt.setp(baseline, color='r', linewidth=1)
    baseline.set_xdata([0,1])
    baseline.set_transform(plt.gca().get_yaxis_transform())
    
    z_score_95pct = zscore_vals.get(percentile)/np.sqrt(n) #1.645 for 90%, 1.96 for 95%, and 2.576 for 99%
    
    plt.title('Cross-Correlation')
    plt.xlabel('Lag')
    plt.ylabel('Correlation')
    plt.axhline(y=-z_score_95pct, color='b', ls='--')# Z-statistic for 95% CL LL
    plt.axhline(y=z_score_95pct, color='b', ls='--')# Z-statistic for 95% CL UL
    plt.axvline(x=0, color='black', ls='-')
    ;
import numpy as np
import pandas as pd
import statsmodels.api as sm
import matplotlib.pyplot as plt

df = sm.datasets.macrodata.load().data
df['realinv'] = round(df['realinv'].astype('float32'), 2)
df['realdpi'] = round(df['realdpi'].astype('float32'), 2)
df_mod = df[['realinv','realdpi']]
```

```
df_diff = pd.DataFrame()
df_diff['realinv'] = np.diff(df_mod['realinv'], n=1)
df_diff['realdpi'] = np.diff(df_mod['realdpi'], n=1)
plot = plot_ccf(data_a=df_diff['realdpi'], data_b=df_diff['realinv'],
 lag_lookback=50, percentile=95)
```

We can observe in *Figure 10.8* many points of correlation beyond the 95% confidence interval we applied. However, we observe the highest level of correlation is at lag 0. There may be multiple correct answers depending on the data domain or forecast error when a model is applied, but using the statistics, our study indicates the series are most correlated at lag 0 and thus, neither is a leading nor lagging indicator of the other. It is important to note the practical significance of a roughly 0.20 correlation is low; it explains very little variance. Therefore, short-term influence is minimal:

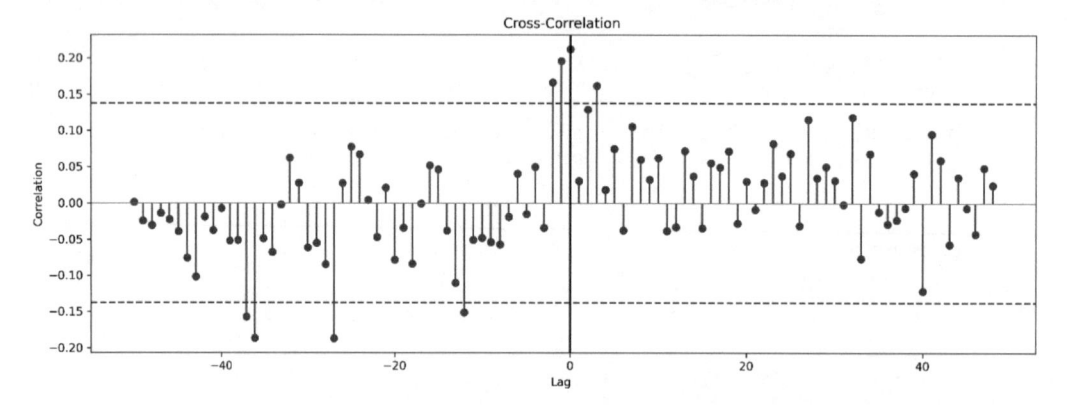

Figure 10.8 – Cross-correlation between *realinv* and *realdpi*

Since we've assessed that short-term cross-correlation between the two series' variances is minimal and we might still be interested in the strength of correlation overall, we could at this point use Pearson's correlation coefficient to compare the two trends, which would measure correlation between the two long-run linear trends. Recall the equation for Pearson's correlation coefficient to observe the relationship to the long-run mean:

$$r = \frac{\sum_{i=1}^{n}(x_i - \bar{x})(y_i - \bar{y})}{\sqrt{\sum_{i=1}^{n}(x_i - \bar{x})^2 \sum_{i=1}^{n}(y_i - \bar{y})^2}}$$

Let's suppose, however, there was a leading indicator present in our data. Let's shift `realdpi` forward by one place. Notice the pandas function `shift()` here, performing this operation:

```
plot = plot_ccf(data_a=df_diff['realdpi'].shift(1).iloc[1:], data_b=df_diff['realinv'].iloc[1:], lag_lookback=50, percentile=95)
```

We get the result as follows:

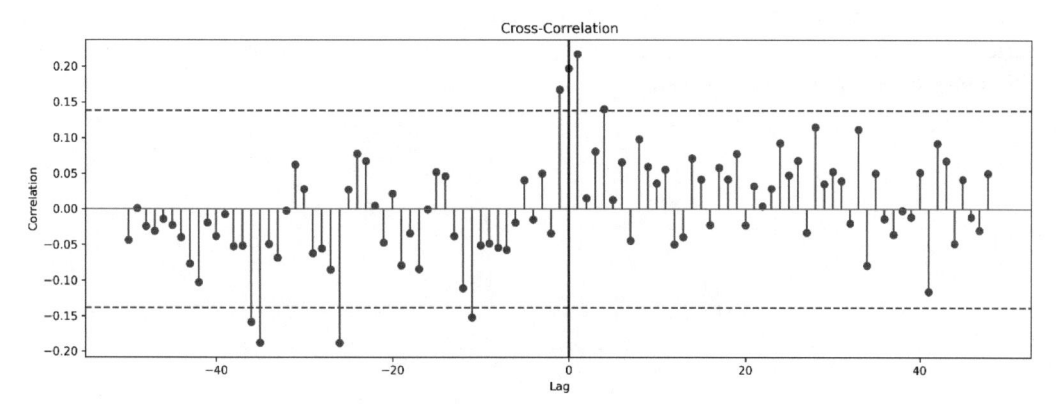

Figure 10.9 – Shifted cross-correlation with leading indicator

After shifting `realdpi` forward one position, we can identify in *Figure 10.9* that `realinv` is now a leading indicator, by a lag of one. If this were our original, differenced data, we might decide to apply a shift to the `realinv` variable—keeping in mind the practicality of the level of correlation—then use `realdpi` and the shifted `realinv` variable when using one variable to forecast the values of the other.

The white-noise model

Any time series can be considered to process two fundamental elements: signal and noise. We can present this mathematically as follows:

$$y(t) = signal(t) + noise(t)$$

The signal is some predictable pattern that we can model with a mathematical function. But the noise element in a time series is unpredictable and so cannot be modeled. Thinking of a time series this way leads to two consequential points:

1. Before attempting to model, we should verify that the time series *is not consistent* with noise.
2. Once we have fit a model to a time series, we should verify that the residuals *are consistent* with noise.

Regarding the first point, if a time series is consistent with noise, there is no predictable pattern to model, and attempting to model the time series could lead to misleading results. About the second point, if the residuals of a time-series model are not consistent with noise, then there are additional patterns we can further model, and the current model is not sufficient to explain the patterns in the signal. Of course, to make these assessments, we first need to understand what noise is. In this section, we will discuss the **white-noise model**.

White noise is a time series where the samples are independent and have a fixed variance with a mean of zero. This means that each sample of the time series is random. So, how do we assess whether a series is random? Let's work through an example. Take a look at the series in *Figure 10.10*. Do you think this series is random or a signal?

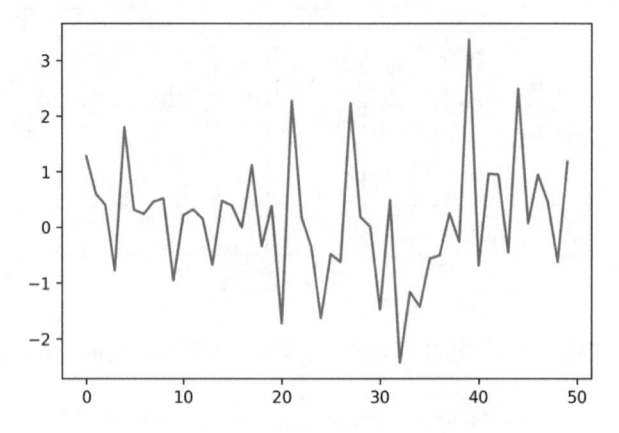

Figure 10.10 – A sample time series

The time series in *Figure 10.10* is a random series generated with numpy. This is not clear from just observing the time series. Let's take a look at how to determine whether a series is noise.

As mentioned previously, the samples in a time series are independent. This means that the sample should not be autocorrelated. We can check the autocorrelation of this series with an ACF plot. The results are shown in *Figure 10.11*:

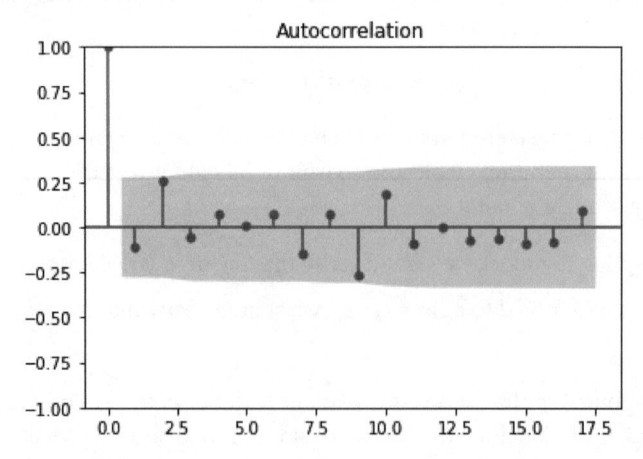

Figure 10.11 – ACF plot of time series in Figure 10.10

The ACF plot shows that the series does not appear to exhibit autocorrelation. This is a strong indication that the time series does not have a pattern to model and may just be noise.

Another assessment we have is the Ljung-Box test for autocorrelation. This is a statistical test for autocorrelation in the lags of a time series. The null hypothesis is there is no autocorrelation, and the alternative hypothesis is there is a correlation. Since autocorrelation is appear at any lag of the time series, the Ljung-Box test provides a p-value for the autocorrelation at each lag. Performing the test on this series, we get the following p-values for the first 10 lags: `[0.41, 0.12, 0.21, 0.31, 0.44, 0.53, 0.5, 0.57, 0.26, 0.2]`. Each of these values is large, which indicates this series is likely noise.

Both methods discussed previously indicate that the series shown is noise, which is the correct result (we know this series is randomly generated). As we progress with modeling time series, we will use these methods to determine whether a series is noise as part of model assessment. We will close this chapter with a discussion on the concept of time series stationarity.

Stationarity

In this section, we provide an overview of stationary and non-stationary time series. Broadly speaking, the main difference between these two types of time series is the statistical properties such as mean, variance, and autocorrelation. They do not vary across time in stationary time series but do change through time in non-stationary time series. Particularly, time series with a trend or seasonality is non-stationary because the trend or seasonality will affect the statistical properties. The following examples illustrate the behaviors of stationary versus non-stationary time series [1]:

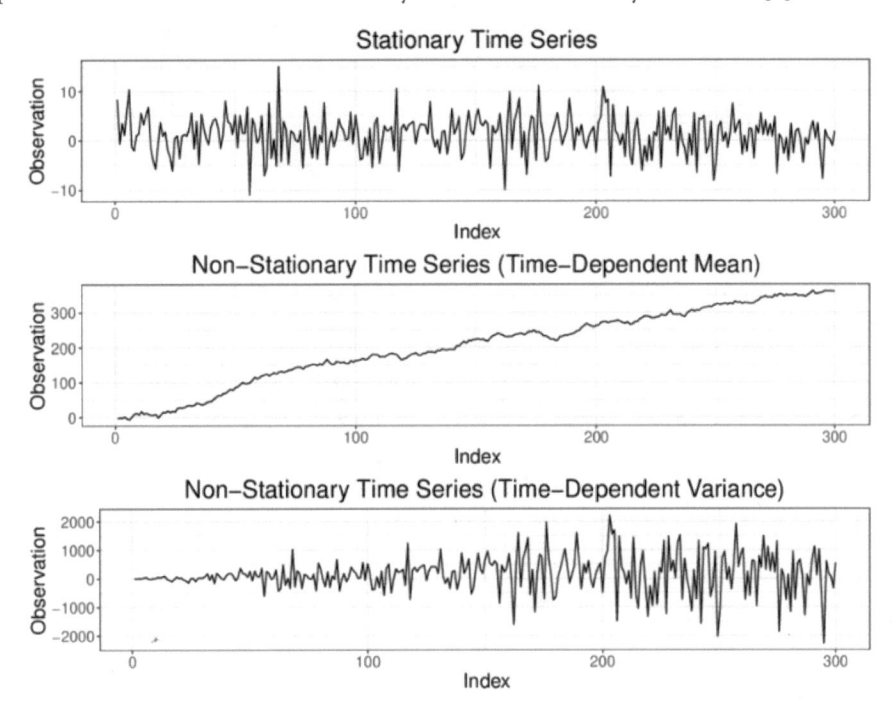

Figure 10.12 – Examples of stationary and non-stationary time series

In order to check the stationary properties, we will check the three following conditions:

- The mean is independent of time:

$$E[X_t] = \mu \text{ for all } t$$

- The variance is independent of time:

$$Var[X_t] = \sigma^2 \text{ for all } t$$

- No autocorrelation with time—the correlation between X_{t_1} and X_{t_2} only depends on how far apart they are on time, $t_2 - t_1$

We will now explore the analysis in Python using the *Air Passengers* dataset (https://www.kaggle.com/datasets/chirag19/air-passengers) providing monthly totals of US airline passengers from 1949 to 1960 that can be downloaded from *Kaggle* or can be found in GitHub repository of the book (https://github.com/PacktPublishing/Building-Statistical-Models-in-Python/blob/main/chapter_10/airline-passengers.csv). First, we import the data to a Python notebook, change the index type to datetime, and plot the dataset:

```
import pandas as pd
import matplotlib.pyplot as plt
data = pd.read_csv('airline-passengers.csv', header=0, index_col =0)
data.index = pd.to_datetime(data.index, format='%Y-%m-%d')
plt.plot(data)
```

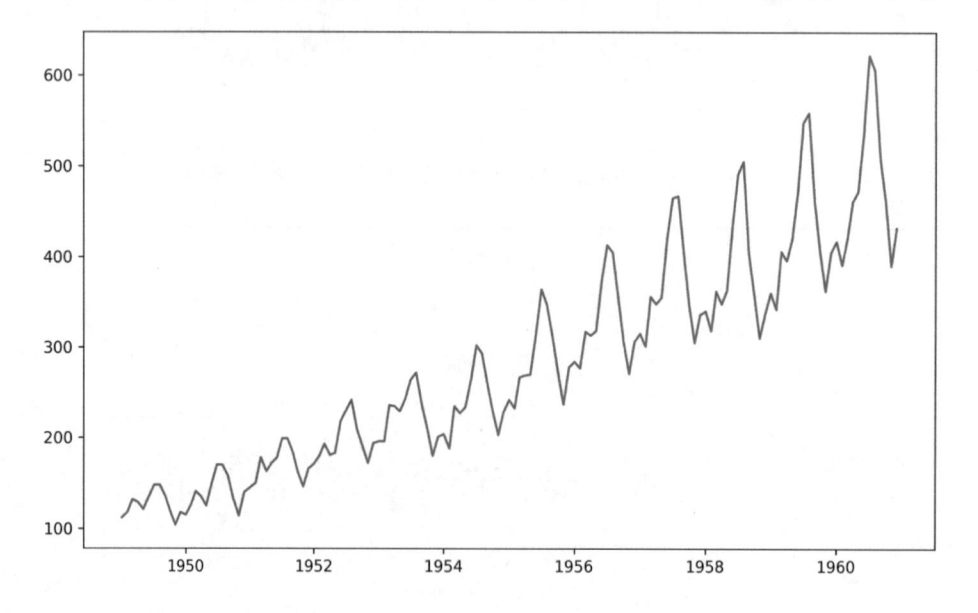

Figure 10.13 – Visualization of passengers using US airlines 1949-1960

From the plot, we can see there are trends and seasonal effects here. Then, it is clearly non-stationary. In the `statsmodels` package, there is a function called `seasonal_decompose` to help us break the original data into different plots for visualization purposes. You can see it in action here:

```
from statsmodels.tsa.seasonal import seasonal_decompose
season_trend = seasonal_decompose(data)
season_trend.plot()
plt.show()
```

We get the resulting plot as follows:

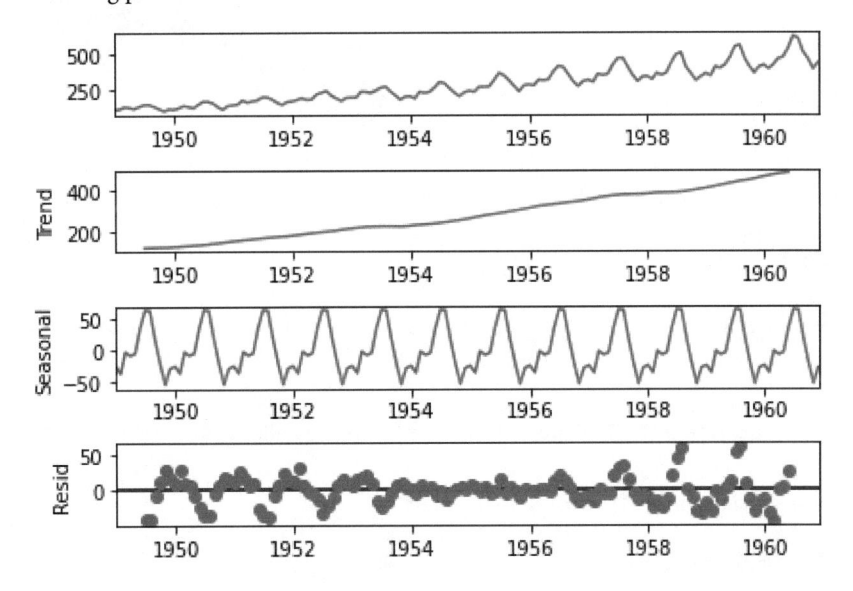

Figure 10.14 – Trend and seasonality visualization of passengers using US airlines 1949-1960

The trend plot shows that the number of passengers using US airlines increases over time. It appears to oscillate seasonally, with summers being the peak. There is a level of dependency between the data points and time, and the variances seem to be smaller earlier (fewer passengers using US airlines in the early 1950s) and greater later (more passengers using the services in the late 1950s). These observations show us that conditions 1 and 2 (constant mean and constant variance with time) are violated. Looking at the non-constant variance differently, we can create boxplots for each year from 1949 to 1960 for visualization, as follows:

```
import seaborn as sns
fig, ax = plt.subplots(figsize=(24,10))
sns.boxplot(x = data.index.year,y = data['Passengers'], ax = ax, color
= "cornflowerblue")
ax.set(xlabel='Year', ylabel='Number of Passengers')
```

For the preceding code, we get the following result:

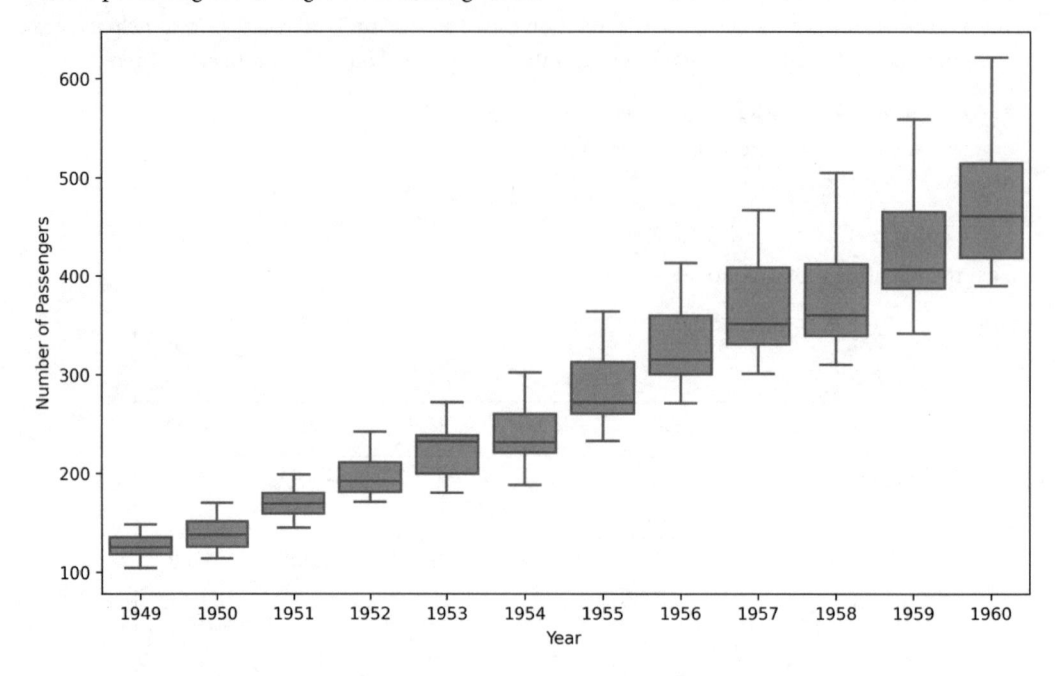

Figure 10.15 – Boxplot: passengers using US airlines 1949-1960

To check autocorrelation, we use the following code:

```
from statsmodels.graphics.tsaplots import plot_acf
plot_acf(data, lags= 20, alpha=0.05)
plt.show()
```

We get the result as shown in *Figure 10.16*:

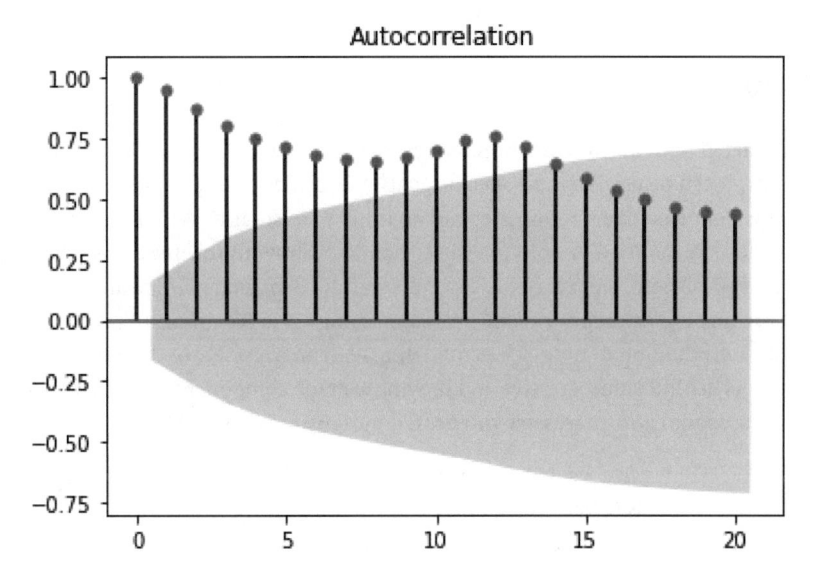

Figure 10.16 – ACF visualization of passengers using US airlines 1949-1960

However, the trend dominates the data. After removing the trend and rerunning the code, we get the *Figure 10.17* ACF plot. There appears to be a strong seasonal component. The autocorrelation cycles are similarly repeated after some lag counts and spaces over time steps:

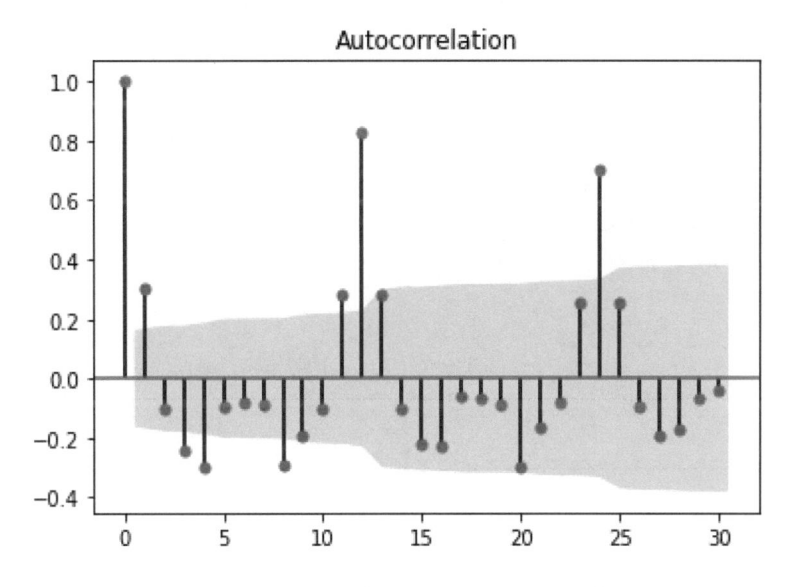

Figure 10.17 – ACF visualization of passengers using US airlines 1949-1960

The Ljung-Box test that we discussed in the *Variance* and *Autocorrelation* sections can also be used to check autocorrelation. Using that test, we get `lb_pvalue = 0`. Therefore, the data has autocorrelation.

Summary

This chapter started with an introduction to time series. We provided an overview of what a time series is and how it can be used to meet specific goals. We also discussed the criteria for differentiating time-series data from data that does not depend on time. We also discussed stationarity, which factors are important for stationarity, how to measure them, and how to resolve cases where stationarity does not exist. From there, we were able to understand the primary functions of ACF and PACF analysis and for making inferences about processes using variance around the mean. Additionally, we provided an introduction to time-series modeling with an overview of the white-noise model and the basic concepts behind autoregressive and moving average components, which help form the basis of ARIMA and **seasonal autoregressive integrated moving average** (**SARIMA**) time-series models.

In *Chapter 11, ARIMA Models*, we will also move deeper into the discussion of autoregressive, moving average, and ARMA models with conceptual overviews and step-by-step examples in Python. We also work on integrated SARIMA models, as well as methods for the evaluation of fit for these models.

References

[1] André Bauer, *Automated Hybrid Time Series Forecasting: Design, Benchmarking, and Use Cases*, University of Chicago, 2021.

11
ARIMA Models

In this chapter, we will discuss univariate time series models. These are models that only consider a single variable and create forecasts based only on the previous samples in the time series. We will start by looking at models for stationary time series data and then progress to models for non-stationary time series data. We will also discuss how to identify appropriate models based on the characteristics of time series. This will provide a powerful set of models for forecasting time series.

In this chapter, we're going to cover the following main topics:

- Models for stationary time series
- Models for non-stationary time series
- More on model evaluation

Technical requirements

In this chapter, we use two additional Python libraries for time series analysis: sktime and pmdarima. Please install the following versions of these libraries to run the provided code. Instructions for installing libraries can be found in *Chapter 1, Sampling and Generalization*.

- sktime==0.15.0
- pmdarima==2.02

More information about sktime can be found at this link: https://www.sktime.org/en/stable/get_started.html

More information about pmdarima can be found at this link: http://alkaline-ml.com/pmdarima/

Models for stationary time series

In this section, we will discuss **Autoregressive (AR)**, **Moving Average (MA)**, and **Autoregressive Moving Average (ARMA)** models that are useful for stationary data. These models are useful when modeling patterns and variance around process means that output over time. *When we have data that does not exhibit autocorrelation, we can use statistical and machine learning models that do not make assumptions about time, such as Logistic Regression or Naïve Bayes, so long as the data supports such use cases.*

Autoregressive (AR) models

The AR(p) model

In *Chapter 10, Introduction to Time Series* we considered how the **Partial Auto-Correlation Function (PACF)** correlates one data point to another lag, controlling for those lags between. We also discussed how inspection of the PACF plot is a frequently used method for assessing the ordering of an autoregressive model. Thereto, the autoregressive model is one that considers specific points in the past to be directly correlated to the value of a given point at lag zero. Suppose we have a process y_t with random, normally distributed white noise, ϵ_t, where $t = \pm 1, \pm 2, \ldots$. If – using real constants of $\phi_1, \phi_2, \ldots, \phi_p$ where $\phi_p \neq 0$ – we can formulate the process in the following way:

$$y_t - \mu - \phi_1(y_{t-1} - \mu) - \phi_2(y_{t-2} - \mu) - \ldots - \phi_p(y_p - \mu) = \epsilon_t$$

Letting μ represent the overall process sample mean (in our examples, we will consider **zero-mean** processes), we can consider this to be an autoregressive process of order p, or AR(p) [1]. We can define the autocorrelation for the AR(p) model as follows:

$$\rho_k = \phi_1^{|k|}$$

There is also this example:

$$\rho_k = \phi_1 \rho_{k-1} + \ldots + \phi_p \rho_{k-p}$$

In the preceding example, where ρ_k is the lag k autocorrelation. ϕ_1 is both the slope and autocorrelation for an AR(1) process.

> **AR(p) model structure and components**
>
> To prevent confusion, note that with the equation $y_t - \mu - \phi_1(y_{t-1} - \mu) - \phi_2(y_{t-2} - \mu) - \ldots - \phi_p(y_p - \mu) = \epsilon_t$, we are attempting to build a mathematical model that represents the process such that if perfectly modeled, all that remains is the random, normally distributed white noise, ϵ_t. This effectively means the model leaves zero residual error (in other words, a perfect fit). Each y_{t-k} term – where k is a lag in time – represents the value at that point in time and each corresponding value of ϕ is the coefficient value required for the y_{t-k} such that when taken in combination with all other values of y, the model statistically approximates zero error.

The AR(1) model

The **backshift operator notation**, or simply **operator notation**, is a simplified, shorthand method of formulating models. It is called "backshift" because it shifts time back one lag from t to $t-1$. The purpose is to avoid the necessitation of writing the subscript (y_{t-k}) following every φ coefficient and instead writing B^{k-1} while including only y_t once, which is handy when writing AR(p) models with high orders of p. In the following equation, the zero-mean form of an AR(1) follows the following structure:

$$y_t - \mu - \phi_1(y_{t-1} - \mu) = \epsilon_t$$

The equation reduces to the following example:

$$y_t - \phi_1(y_{t-1}) = \epsilon_t$$

In backshift operator notation, we can say this:

$$(1 - \phi_1 B)y_t = \epsilon_t$$

> **Note on $|\phi_1|$ in an AR(1)**
>
> It's worth noting at this point that an AR(1) process is stationary if $|\phi_1| < 1$. That is, when the absolute value of the lag-one autocorrelation is < 1, the AR(1) process is stationary. When $|\phi_1| = 1$, an ARIMA model may still be useful, but when $|\phi_1| > 1$, the process is considered to be explosive and should not be modeled. This is because a value $|\phi_1| < 1$ means the root is outside of, and not bounded by, the unit circle. A value of $|\phi_1| = 1$ is on the unit circle but can be differenced to remove the unit root. The root for the case of an AR(1) can be calculated as $z = 1/\phi_1$. A set of data producing $|\phi_1| > 1$ cannot be filtered in a way that puts its root outside the unit circle.
>
> When all roots of an AR(p) are outside the unit circle, the given realization (one time series sampled from a stochastic process) will converge to the mean, have constant variance, and be independent of time. This is an ideal scenario for time series data.

Let's walk through an example of an AR(1) process with $|\phi_1| < 1$ and therefore a stationary root. Assume we have identified the following first-order autoregressive process:

$$y_t - 0.5 y_{t-1} = \epsilon_t$$

This is converted to operator notation:

$$(1 - 0.5B)y_t = \epsilon_t$$

When looking for roots, we can use the following notation:

$$(1 - 0.5z) = 0$$

This gives us a root of z:

$$z = \frac{1}{\phi_1} = \frac{1}{0.5} = 2$$

Therefore, since the root is greater than 1 and thus outside the unit circle, the AR(1) representation of the process is stationary. In Python, we can build this process using the upcoming code. First, we build the AR(1) parameters, which we want to have a 0.5. Because we substitute 0.5 into the model $X_t - \phi_1(y_{t-1}) = \epsilon_t$, we insert *0.5* and not *-0.5* for `arparams`. Also, note based on $\rho_k = \phi_1^{|k|}$ that *0.5* is the lag-1 autocorrelation. The process we build will have an arbitrary sample size of `nsample=200`. We build the $(1 - 0.5B)$ component of $(1 - 0.5B)y_t = \epsilon_t$ using the `np.r_[1, -arparams]` step:

```
from statsmodels.graphics.tsaplots import plot_acf, plot_pacf
import matplotlib.pyplot as plt
import statsmodels.api as sm
import numpy as np
arparams = np.array([0.5])
ar = np.r_[1, -arparams]
ar_process = sm.tsa.ArmaProcess(ar)
y = ar_process.generate_sample(nsample=200)
```

Now that we have the code to create the AR(1) we looked at the equation for, let's see the roots and compare it to our manually calculated *z*:

```
ar_process.arroots
```

```
array([2.])
```

We can see the Python output, `2.` is the same as the calculation we performed. We know that since the absolute value of the root is greater than 1, the AR(1) process is stationary, but let's confirm with Python:

```
ar_process.isstationary
```

```
True
```

We can observe by looking at the PACF that this is an autoregressive of order $p = 5$. We can also observe by looking at the ACF that the value of ϕ_1 is approximately 0.5. For an autoregressive model, the PACF is used to identify the number of significant lags to include as the order of the AR, and the ACF is used to determine the values of the coefficients, ϕ_k, included in that order. It is simple to observe the values for an AR(1) using the ACF, but less obvious when p > 1 since ACF does not control for individual lags when compared to the most recent point (lag zero) as the PACF does. Let's generate the plots with Python:

```
fig, ax = plt.subplots(1,3, figsize=(20,5))
ax[0].set_title('Realization')
ax[0].plot(y)
plot_acf(y, alpha=0.05, lags=50, ax=ax[1])
ax[1].set_title('ACF')
plot_pacf(y, alpha=0.05, lags=50, ax=ax[2])
ax[2].set_title('PACF')
```

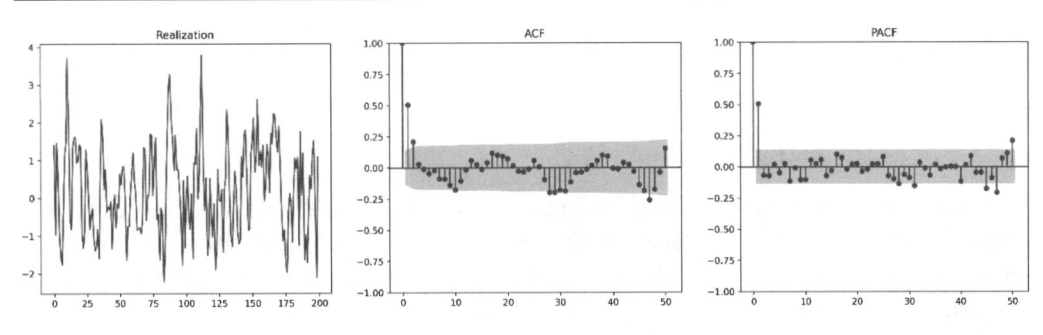

Figure 11.1 – The AR(1) process

Common for AR(1) processes, we see in *Figure 11.1* a single significant partial autocorrelation in the PACF plot, excluding lag zero. Note there is some significance as we near lag 45, but because of the insignificance between lag 1 and those points, including those lags and constructing something such as an AR(50) would result in extreme overfitting; the coefficients from lag 2 through roughly lag 45 would fall between roughly 0 and ±0.15. As referenced in *Chapter 4*, *Parametric Tests*, the correlation between about ±0.1 and ±0.3 is generally considered a weak correlation.

The AR(2) model

Let's look at the following stationary AR(2) process:

$$y_t - 0.8 y_{t-1} - 0.48 y_{t-2} = \epsilon_t$$

Converted to backshift operator notation, we have the following:

$$(1 - 0.8B - 0.48 B^2) y_t = \epsilon_t$$

We also have this:

$$(1 - 0.8z - 0.48 z^2) = 0$$

Since we're focusing on Python in this book, we won't walk through the steps, but it may be useful to know second-order polynomials – such as AR(2) – follow the quadratic equation, $ax^2 + bx + c$ ($-0.48 z^2 - 0.8z + 1$ for our process). Therefore, we can use the quadratic formula:

$$\frac{-b \pm \sqrt{b^2 - 4ac}}{2a}$$

This is what we'll use to find the roots. In Python, we can find the roots of this model using the following:

```
arparams = np.array([-0.8, -0.48])
ar = np.r_[1, -arparams]
ar_process = sm.tsa.ArmaProcess(ar)
print('AR(2) Roots: ', ar_process.arroots)
print('AR(2) Stationarity: ', ar_process.isstationary)
```

Here we can see the unit roots identified using the `statmodels` ArmaProcess function:

```
AR(2) Roots:    [-0.83333333-1.1785113j  -0.83333333+1.1785113j]
AR(2) Stationarity:   True
```

We can observe the roots are in complex conjugate form $a \pm bi$. When a process has roots in complex conjugate form, it is expected the autocorrelations will exhibit an oscillatory pattern, which we can see in the ACF plot in *Figure 11.2*. We can also observe the two significant lags in the PACF, which support the case of order p=2:

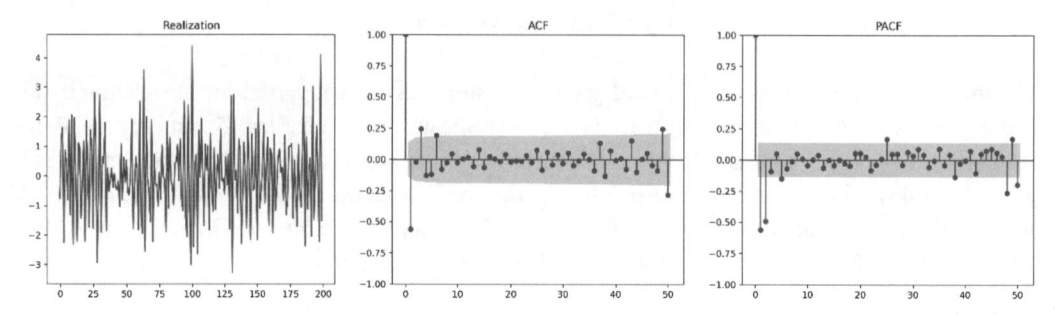

Figure 11.2 – AR(2) with complex conjugate roots

To mathematically test that complex conjugate roots are stationary (outside the unit circle), we take the magnitude of the vector of each root's real and imaginary parts and check if it is greater than 1. The magnitude for complex conjugate roots, z, following form $a \pm bi$ is the following equation:

$$\|z\| = \sqrt{a^2 + b^2}$$

The magnitude of our roots is this:

$$\sqrt{-0.8333^2 \pm 1.1785^2} = 1.4433$$

Since 1.4433 > 1, we know our AR(2) model is stationary.

> **Identifying order p for AR models using the PACF**
>
> When identifying a lag order p for an autoregressive process, AR(p), based on the PACF plot, we take the maximum lag where significant partial autocorrelation exists as the order for p. In observing *Figure 11.3*, because the PACF dampens after lag 4 and through about lag 30, we will cut off order consideration after lag 4 because using more lags (consider them as features for a time series model) will likely result in overfitting. The order selected using PACF is based on the last significant lag before the partial autocorrelations dampen. While lags 2 and 3 seem to be small and may not be significant, lag 4 is. Therefore, we may get the best model using an AR of order 4. Typically, we test our assumptions using errors with information criteria, such as AIC or BIC.

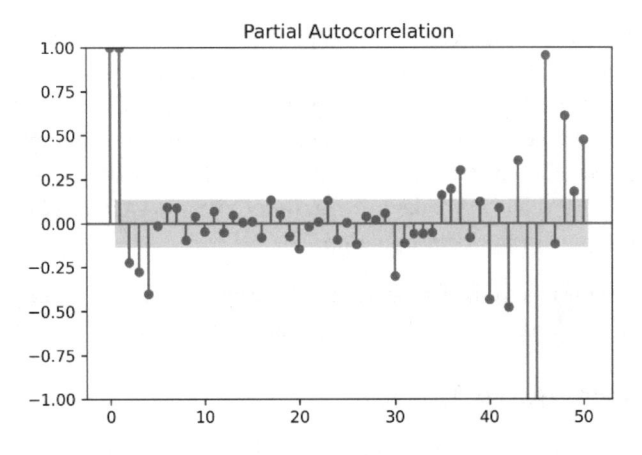

Figure 11.3 – AR(p) order identification

AR(p) end-to-end example

Let us walk through an end-to-end example of AR(p) modeling in Python. First, we need to generate a dataset produced by an AR(4) process. We will use this data as the process we will attempt to model:

```
arparams = np.array([1.59, -0.544, -0.511, 0.222])
ar = np.r_[1, -arparams]
ar_process = sm.tsa.ArmaProcess(ar)
y = ar_process.generate_sample(nsample=200)
```

For the following steps, let us assume data y is the output of a machine about which we know nothing.

Step 1 - visual inspection

We first visualize the original data and its ACF and PACF plots using the code we provided earlier in the chapter:

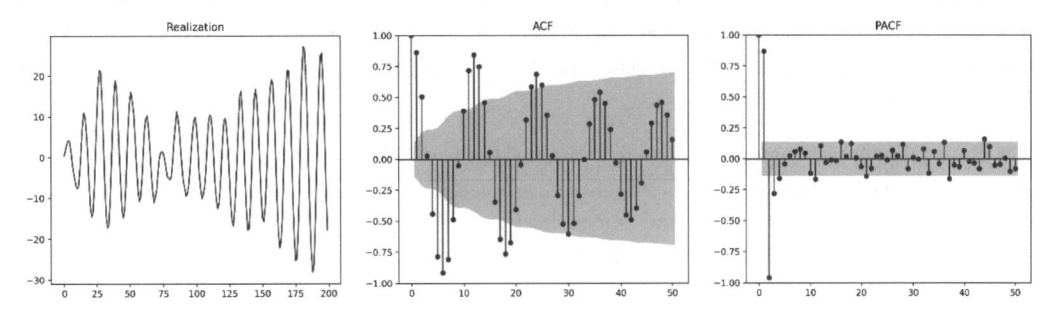

Figure 11.4 – Step 1 in model development: visual inspection

We can see based on the PACF plot we have what appears to be an AR(2), but possibly an AR(4). After lag 4, the partial autocorrelations lose statistical significance at the 5% level of significance. We can see, however, when considering the statistical significance of lag 4 in the PACF, however slight, lag 4 in the ACF is significant. While the value at lag 4 is not the value of the coefficient, its significance is useful in helping determine order p. Nonetheless, an AR(4) may overfit and fail to generalize as well as an AR(2). Next, we will use **Aikake Information Criterion (AIC)** and **Bayesian Information Criterion (BIC)** to help make our determination.

Based on the constant mean and the fact we do not have an exponentially dampening (which would also need to be significant) ACF, there does not appear to be a trend.

Step 2 - selecting the order of AR(p)

Since we are uncertain based on the visual inspection of the order we should use for the AR(p) model, we will use AIC and BIC to help our decision. The AIC and BIC process will fit models using order zero up to the `max_ar` value provided in the upcoming code. The models will fit the entire dataset. The order with the lowest error is generally the best. Their error calculations are as follows:

$$AIC = 2k - 2ln(\hat{L})$$

$$BIC = kln(n) - 2ln(\hat{L})$$

This is where k is the number of lags – up to the maximum order tested – for the data, $\hat{L}$ is the maximum likelihood estimate, and n is the sample size (or length of the dataset being tested). For both tests, the lower error is better.

We will import `arma_order_select_ic` from `Statsmodels` and test it using up to a maximum of 4 lags based on our observation in the PACF plot in *Figure 11.4*. As noted, based on our visual inspection, we do not appear to have a trend. However, we can verify this statistically with an OLS-based unit root test called the **Dickey-Fuller test**. The **null hypothesis** of the Dickey-Fuller test is that a unit root (and therefore, trend) is present at some point in the maximum number of lags tested (maxlag). The alternative hypothesis is that there is no unit root (no trend) in the data. For reference, the alternative hypothesis states that the data is an order zero - **I(0)** - integrated process while the null hypothesis states that the data is an order one - **I(1)** - integrated process. If the absolute value of the test statistic is greater than the critical value or the p-value is significant, we can conclude there is no trend present (no unit root).

The Dickey-Fuller test considers each data point out to the number of lags included in the regression test. We will need to analyze the ACF plot for this; because we want to consider as far out as a trend could be possible, we must choose the longest lag that has significance. The idea is that if we have a strong trend in our data, such as growth, for the most part, each sequential value will lead to another increasing subsequent value for as long as the trend exists. In our case, the maximum significant lag in the ACF plot is approximately 25. The Dickey-Fuller test has relatively low statistical power (prone to Type II error or failing to reject the null when the null should be rejected) so a high order of lags is not concerning so long as it is practical; the risk is failing to include enough lags.

> **Dickey-Fuller unit roots**
>
> The Dickey-Fuller tests only if there is a trend unit root, but not if there is a seasonal unit root. We discuss the difference between trend and seasonal unit roots in the ARIMA section of this chapter.

In the upcoming code block, we add `maxlag=25` for our 25 lags from the ACF plot in *Figure 11.4*. We will also include `regression='c'`, which adds a constant (or intercept) into the OLS it performs; we will not need to manually add the constant in that case:

```
from statsmodels.tsa.stattools import adfuller
dicky_fuller = adfuller(y, maxlag=25, regression='c')
print('Dickey-Fuller p-value: ', dicky_fuller[1])
print('Dickey-Fuller test statistic: ', dicky_fuller[0])
print('Dickey-Fuller critical value: ', dicky_fuller[4].get('5%'))
```

We can see based on the Dickey-Fuller test that we should reject the null hypothesis and conclude the process is order-zero integrated and therefore does not have trend:

```
Dickey-Fuller p-value:    1.6668842047161513e-06
Dickey-Fuller test statistic:    -5.545206445371327
Dickey-Fuller critical value:    -2.8765564361715534
```

We can therefore insert into our `arma_order_select_ic` function that `trend='n'` (otherwise, we may want to difference the data, which we will show in the ARIMA section of the chapter):

```
from statsmodels.tsa.stattools import arma_order_select_ic
model_ar = arma_order_select_ic(y=y, max_ar=4, max_ma=0,
                                ic=['aic','bic'], trend='n')
print('AIC Order Selection: ', model_ar.aic_min_order)
print('AIC Error: ', round(model_ar.aic.min()[0], 3))
print('BIC Order Selection: ', model_ar.bic_min_order)
print('BIC Error: ', round(model_ar.bic.min()[0], 3))
```

Here we can see the AR and MA orders identified to produce a fit with the lowest overall error according to our AIC and BIC tests:

```
AIC Order Selection:    (4, 0)
AIC Error:    586.341
BIC Order Selection:    (2, 0)
BIC Error:    597.642
```

We can see AIC selected an AR(4) and BIC selected an AR(2). It is preferable that both tests select the same term orders. However, as we noted already, the AR(2) may be less likely to overfit. Since the best order isn't completely clear, we will proceed to test both models (using AR(2) and AR(4)) by comparing their errors and log-likelihood estimates.

Step 3 - building the AR(p) model

In this step, we can add our arguments to the `statsmodels`' ARIMA function and fit it to the data with our prescribed AR(4). To be clear, an AR(4) is the same as an ARIMA(4,0,0). We want to include `enforce_stationarity=True` to ensure our model will produce useful results. If not, we will receive a warning and need to address the issue by either differencing, using a different model – such as SARIMA, changing our sampling method, changing our time binning (from days to weeks, for example), or abandoning time series modeling for the data altogether:

```
from statsmodels.tsa.arima.model import ARIMA
ar_aic = ARIMA(y, order=(4,0,0),
               enforce_stationarity=True).fit()
print(ar_aic.summary())
```

In our model output, we can see the *SARIMAX Results* title and *Model: ARIMA(4,0,0)*. This can be disregarded. A SARIMAX with no seasonal component and no exogenous variables (in our case) is simply an ARIMA. Further, an ARIMA of order (4,0,0) is an *AR(4)*:

```
                              SARIMAX Results
==============================================================================
Dep. Variable:                      y   No. Observations:              200
Model:                   ARIMA(4,0,0)   Log Likelihood              -287.919
Date:                Fri, 13 Jan 2023   AIC                          587.837
Time:                        19:12:18   BIC                          607.627
Sample:                             0   HQIC                         595.846
                                - 200
Covariance Type:                  opg
==============================================================================
                 coef    std err          z      P>|z|      [0.025      0.975]
------------------------------------------------------------------------------
const         -0.197      0.278     -0.709      0.478      -0.742       0.348
ar.L1          1.6217     0.068     23.92       0.000       1.489       1.755
ar.L2         -0.6877     0.131     -5.258      0.000      -0.944      -0.431
ar.L3         -0.3066     0.145     -2.114      0.034      -0.591      -0.022
ar.L4          0.1158     0.081      1.425      0.154      -0.043       0.275
sigma2         0.9981     0.113      8.843      0.000       0.777       1.219
===================================================================================
Ljung-Box (L1) (Q):                   0.02   Jarque-Bera (JB):              0.96
Prob(Q):                              0.88   Prob(JB):                      0.62
Heteroskedasticity (H):               1.23   Skew:                          0.04
Prob(H) (two-sided):                  0.39   Kurtosis:                      2.67
```

Figure 11.5 – AR(4) model results

The AR(4) process we modeled (the simulated process we built prior to step 1 is

$$y_t - 1.59 y_{t-1} + 0.544 y_{t-2} + 0.511 y_{t-3} - 0.222 y_{t-4} = \epsilon_t$$

and the AR(4) model we produced using the data from the input process is the following:

$$y_t - 1.6217 y_{t-1} + 0.6877 y_{t-2} + 0.3066 y_{t-3} - 0.1158 y_{t-4} = \epsilon_t$$

In backshift operator notation, we have the following equation:

$$(1 - 1.6217 B + 0.6877 B^2 + 0.3066 B^3 - 0.1158 B^4) y_t = \epsilon_t$$

Notably, the term for lag 4 is not significant and the confidence interval contains 0. Therefore, including this term is a known risk for overfitting and something worth weighing if considering alternative models. It would be prudent to compare an AR(2) and even an AR(3) to our AR(4) based on AIC and BIC and choose a different model if the results are much improved, but we will skip this process for the sake of time.

Regarding the model summary metrics, we discussed the **Ljung-Box test** in the last chapter so will not cover the details here, but the high p-value *(Prob(Q))* for that test indicates there is not correlated error at lag 1. Typically, if there is serial correlation in the residuals of a model fit, the residuals will have lag 1 autocorrelation. The **Jarque-Bera test** assumes the errors are normally distributed under the null hypothesis and not normally distributed under the alternative hypothesis. The high p-value *(Prob(JB))* for that test suggests the error is normally distributed. The test for **heteroskedasticity** tests the null that the residuals are constant (homoscedastic) with the alternative hypothesis being that they are non-constant, which is an issue for a time series regression fit. Here, our Heteroskedasticity p-value *(Prob(H))* is high so we can assume our model's residuals have constant variance. A **skew** score between [-0.5, 0.5] is considered not skewed, whereas between [-1, -0.5] or [0.5, 1] is moderately skewed and > ±2 is high. A perfect score for **kurtosis** is 3. Kurtosis > ±7 is high. Because our skew is 0.04 and our kurtosis score is 2.58, we can assume our residuals are normally distributed.

Step 4 - test forecasting

Another method for validating a model is to forecast existing points using data leading up to those points. Here, we use the model to forecast the last 5 points using the full dataset excluding those last 5 points. We then compare to get an idea of model performance. Note that we generated 200 samples and the index for those samples starts at 0. Therefore, our 200th sample is at positional index 199:

```
df_pred = ar_aic.get_prediction(start=195, end=199).summary_
frame(alpha=0.05)
df_pred.index=[195,196,197,198,199] # reindexing for 0 index
```

ARIMA Models

In the following table, the *mean* column is the forecast. We manually appended the *actuals* column with the last 5 values in our data to compare to the forecast. *mean_se* is our mean squared error for our estimates compared to actuals. *y* is our index and the *ci* columns are for our 95% forecast confidence interval since we used `alpha=0.05` in the previous code.

y	mean	mean_se	mean_ci_lower	mean_ci_upper	actuals
195	24.70391	0.99906	22.74579	26.662035	25.5264
196	19.36453	0.99906	17.4064	21.322652	18.8797
197	7.525904	0.99906	5.567779	9.484028	7.4586
198	-5.8744	0.99906	-7.83252	-3.916274	-7.1316
199	-19.5785	0.99906	-21.5366	-17.620356	-17.9268

Figure 11.6 – AR(4) model outputs versus actuals

We can see based on our mean squared error (0.999062) that our model provides a reasonable fit across a forecast horizon of 5 points on test data. Using the following code, we plot our test forecast against the corresponding actuals:

```
fig, ax = plt.subplots(1,1,figsize=(20,5))
ax.plot(y, marker='o', markersize=5)
ax.plot(df_pred['mean'], marker='o', markersize=4)
ax.plot(df_pred['mean_ci_lower'], color='g')
ax.plot(df_pred['mean_ci_upper'], color='g')
ax.fill_between(df_pred.index, df_pred['mean_ci_lower'], df_pred['mean_ci_upper'], color='g', alpha=0.1)
ax.set_title('Test Forecast for AR(4)')
```

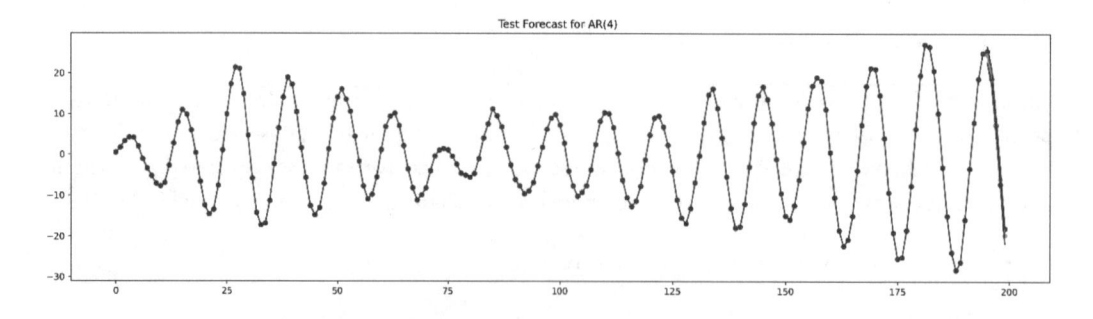

Figure 11.7 – The AR(4) test forecast

Step 5 - building a forecast

Determining a reasonable forecast horizon is highly dependent on at least the data and the process it represents, and the lag used for modeling, in addition to model error. The time series practitioner should weigh all factors of model performance and business needs versus risks before providing forecasting to stakeholders. Adding the following code, we re-run the plot to see our true forecast with a 5-point horizon:

```
df_forecast = ar_aic.get_prediction(start=200, end=204).summary_
frame(alpha=0.05)
df_forecast.index=[200, 201, 202, 203, 204]
forecast = np.hstack([np.repeat(np.nan, len(y)), df_pred['mean']])
```

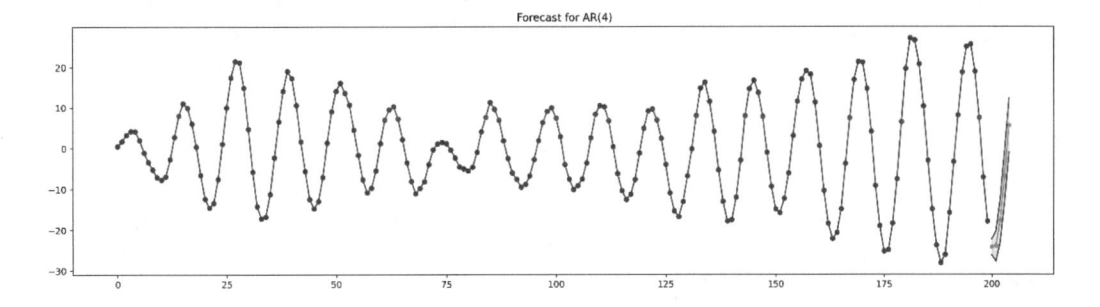

Figure 11.8 – The AR(4) forecast horizon = 5

We will cover additional steps in the model evaluation section of this chapter.

Moving average (MA) models

The MA(q) model

Whereas the AR(p) model is a direct function of the correlation between lag zero and specific individual lags of order p over time, the moving average model of order q, MA(q), is a function of autocorrelation between lag zero and all previous lags included in order q. It acts as a low-pass filter that models errors to provide a useful fit to data.

Let us take a process, y_t, that has a mean of zero and a random, normally distributed white noise component, ϵ_t, where $t = \pm 1, \pm 2, \ldots$. If we can write this process as

$$y_t - \mu = \epsilon_t - \Theta_1 \epsilon_{t-1} - \ldots - \Theta_q \epsilon_{t-q}$$

and $\Theta_1, \Theta_2, \ldots, \Theta_q$ are real constants and $\Theta_q \neq 0$, then we can call this a moving average process having order q, or MA(q). In backshift operator notation, we have the following:

$$y_t - \mu = (1 - \Theta_1 B - \ldots - \Theta_q B^q) \epsilon_t$$

We can define the autocorrelations (ρ_k) for the MA(q) model thusly:

$$\rho_k = \frac{-\Theta_k + \sum_{j=1}^{q-k} \Theta_j \Theta_{j+k}}{1 + \sum_{j=1}^{q} \Theta_j^2}$$

For all lags k in $1, 2, \ldots, q$. Where $k > q$, we have $\rho_k = 0$.

When discussing the AR(p) models, we explained how the roots of the AR model must be outside the unit circle. When considering MA(q) models, we have the concept of invertibility. **Invertibility** essentially ensures a *logical and stable correlation with the past*. Typically, this means the current point in time is more closely related to nearby points in the past than those more distant. The inability to model a process using invertible roots means we cannot ensure our model provides a unique solution for the set of model autocorrelations. If points in the distant past are more relevant to the current point than those nearby, we have a randomness in the process that cannot be reliably modeled or forecasted.

> **Identifying MA(q) model invertibility**
>
> For a moving average model to be invertible, all roots must be outside the unit circle and non-imaginary; *all* roots must be greater than 1. For an MA(1) process, it is invertible when $|\Theta_1| < 1$. An invertible MA(q) process is equivalent to an infinite-order, converging AR(p). An AR(p) model converges if its coefficients converge to zero as lags k approach p. If an MA(q) is invertible, we can say $y_t = \Theta(B)\epsilon_t$ and $\Theta^{-1}(B)y_t = \epsilon_t$ [1].

The MA(1) model

For an MA(q) model of order 1, we have the following process:

$$\rho_0 = 1$$

$$\rho_1 = \frac{\Theta_1}{1 + \Theta_1^2}$$

$$\rho_{k>1} = 0$$

For a MA(q) model with zero autocorrelation, the pattern of whose process we are attempting to model is random, normally distributed white noise variance, which can – at best – only be modeled by its mean. It is important to note that as $\Theta_1 \to 0$, $\rho_1 \to 0$ and for an MA(1) process, this means the process can be approximated by white noise.

Let us consider the following MA(1) zero-mean model following the form $y_t - \mu = \epsilon_t - \Theta_1 \epsilon_{t-1}$:

$$y_t - 0 = a_t - 0.8\epsilon_{t-1}$$

In backshift notation we have the following:

$$y_t = (1 - 0.8B)\epsilon_t$$

We know this process is invertible because $|\Theta_1| < 1$. Let us confirm this with Python using the `ArmaProcess` function from the `statsmodels tsa` module:

```
from statsmodels.graphics.tsaplots import plot_acf, plot_pacf
import matplotlib.pyplot as plt
import statsmodels.api as sm
import numpy as np
maparams = np.array([0.8])
ma = np.r_[1, -maparams]
ma_process = sm.tsa.ArmaProcess(ma=ma)
print('MA(1) Roots: ', ma_process.maroots)
print('MA(1) Invertibility: ', ma_process.isinvertible)
```

```
MA(1) Roots:    [1.25]

MA(1) Invertibility:    True
```

Contrary to the AR(p) model, the MA(q) model's order is identified using the ACF plot. Because the ACF does not control for lags and is a composite autocorrelation measure across all lags up to the lag whose autocorrelation measure is considered, the function is used to identify the relevant lag order for the moving average component. In *Figure 11.9*, we can see the significant correlation at lag 1 for our MA(1) process. There are two additional significant correlations at lags 6 and 7, but using lags this far out typically results in overfitting, especially when taken alongside the fact that lags 2 through 5 are not significant at the 5% level of significance:

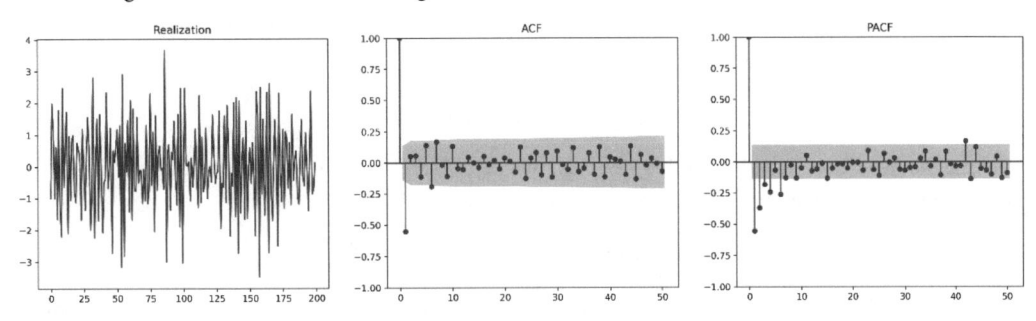

Figure 11.9 – MA(1) plots

> **Dampening ACFs and PACFs**
>
> For an invertible moving average model, we can observe that the ACF will cut off at the order of significance, but the PACF will typically continue and dampen *overall* to statistical zero over time. It is not necessarily expected to happen smoothly and all at once as all data sets are different, but it is expected that over time, more and more lags will dampen to zero. We explain the reason for this behavior from the ACFs and PACFs in the ARMA section of this chapter, but it is worth noting this is to be expected for invertible processes. Conversely, stationary autoregressive processes are expected to cut off at the order of significance in the PACF plots while the ACFs dampen to zero over time.

The MA(2) model

For an MA(q) model of order 2, we have this:

$$\rho_0 = 1$$

$$\rho_1 = \frac{-\Theta_1 + \Theta_1\Theta_2}{1 + \Theta_1^2 + \Theta_2^2}$$

$$\rho_2 = \frac{-\Theta_2}{1 + \Theta_1^2 + \Theta_2^2}$$

$$\rho_{k>2} = 0$$

The following is an MA(2) example we will look at:

$$y_t = (1 - 1.6B + 0.9B^2)\epsilon_t$$

Using the model's polynomial in the quadratic equation form, we can find the approximate roots using the quadratic formula:

$$0.888 \pm 0.567i$$

Because we have two complex conjugate roots, we can take the same L^2 *norm* we did for the AR(p) process, using the form of $a \pm bi$:

$$\sqrt{0.888^2 + 0.567^2} \approx 1.054$$

Because 1.054 is greater than 1, we can confirm the MA(2) has invertible roots and is thus capable of producing a unique solution and a model whose values are logically serially correlated to past values. Let us perform the same analysis in Python:

```
maparams = np.array([1.6, -0.9])
ma = np.r_[1, -maparams]
ma_process = sm.tsa.ArmaProcess(ma=ma)
```

```
print('MA(2) Roots: ', ma_process.maroots)
print('MA(2) Invertibility: ', ma_process.isinvertible)
```

The output we see highlighted in green confirms our calculated findings and the fact that since the magnitude of the complex conjugate roots is greater than 0, we have an invertible MA(2) process:

```
MA(2) Roots:    [0.88888889-0.56655772j  0.88888889+0.56655772j]

MA(2) Invertibility:    True
```

We can see in the ACF in *Figure 11.10* that this is a second-order moving average process:

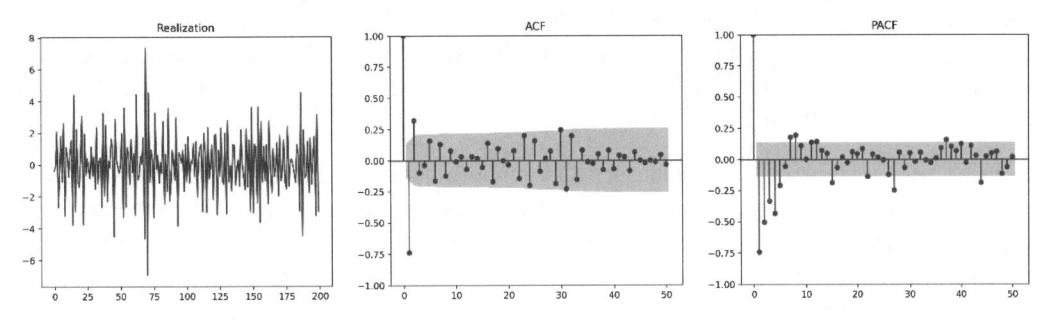

Figure 11.10 – MA(2) plots

The process and code for identifying model ordering, building the model, and generating a forecast are the same for the MA(q) model as it is for the AR(p) model. We have discussed that the order selection based on visual inspection for the MA(q) is performed using the ACF, whereas for the AR(p) this is done using the PACF, and that it is the only major difference in the process between the two models. Aside from that, `enforce_invertibility` should be set to equal `True` for MA(q) models in place of `enforce_stationarity=True`. Providing a `max_ar` or `max_ma` order higher or lower than useful in the `arma_order_select_ic` function may result in a *convergence warning or an invertibility warning*. One reason for these warnings is there was a higher order provided than possible to fit (such as when there is no order possible). Another reason is the presence of a **unit root**. If there is an apparent trend in the data, it *must be removed* before modeling. If there is no trend, it is possible to receive this error due to seasonality in the data, which presents a different order of unit root. We will discuss modeling in the case of unit roots associated with trend and seasonality in the ARIMA and seasonal ARIMA section of this chapter. It is also worth specifying that the Dickey-Fuller test can be used for moving average data since moving average processes can be influenced by trends.

Autoregressive moving average (ARMA) models

In the autoregressive model section, we discussed how an AR(p) model is used to model process output values using autocorrelation controlling for individual lags. The goal of the AR(p) model is to estimate exact values for points corresponding to lags in the future using the values for the same specific lags in the context of a past horizon. For example, the value at two points in the future is strongly correlated with the value at two points in the past. In the moving average model section, we discussed how MA(q) models act as low-pass filters that help a model explain noise in a process. Rather than seeking to model exact points, we use the MA(q) to model variance around the process.

Consider an example of a four-cylinder car engine that produces constant output. Let us assume we have a worn-down motor mount near the fourth cylinder. We can expect consistent output vibration related to each cylinder firing, but the vibration will increase slightly for each stroke that is closer to the worn motor mount. Using an AR-only model would assume each cylinder vibrates a certain amount and be able to account for that, but we would be missing information. Adding a MA component would be able to model the fact that starting at cylinder one, each subsequent stroke up through cylinder four will have additional vibration related to the worn motor mount and thus explain much more of the overall process. This would reasonably be an ARMA(4,4) model. Suppose we replace the worn-down motor mount with a mount of equal wear compared to the other mounts; we would then have an ARMA(4,0) (or AR(4)) process.

In many cases, we find we have significant peaks in both autocorrelation and partial autocorrelation. Rather than using only MA(q) or AR(p) modeling, respectively, we can combine the two. This combination, represented as an ARMA(p,q), enables us to model the process as well as any noise component around the process that may correlate to specific lags. Because an ARMA(p,q) typically has fewer parameters (lower order) for each AR and MA component than AR or MA models do, the ARMA is considered a **parsimonious model**, which is a model that uses as few explanatory variables (in this case, time lags) as possible to achieve the desired level of performance. When y_t is an invertible and stationary process, we can define it as an ARMA(p,q):

$$y_t - \mu = \Phi_1(y_{t-1} - \mu) - \ldots - \Phi_p(y_{t-p} - \mu) = \epsilon_t - \Theta_1 \epsilon_{t-1} - \ldots - \Theta_q \epsilon_{t-q}$$

Where $\Phi_p \neq 0$ and $\Theta_q \neq 0$, we can re-write the equation for ARMA(p,q) in backshift operator notation:

$$\Phi B(y_t - \mu) = \Theta(B)\epsilon_t$$

Practically speaking, we can expect that for an invertible moving average process, we see significant lags in the ACF up to the magnitude of the order q, but then the PACF will taper off thereafter, typically in lags beyond the order of the moving average process identified in the ACF. This is because a finite moving average process can be represented as an infinite-order autoregressive process. Conversely, because the moving average process that has this behavior is invertible, the inverse must also be true; that a finite autoregressive process can be represented as an infinite-order moving average process. Therefore, the PACF will dampen to zero for an invertible moving average process, and for a stationary autoregressive process, the ACF will dampen to zero. Because invertibility is a requirement of an ARMA process, it allows us to re-write the equation as an infinite-order MA process in general linear form:

$$y_t = \Phi^{-1}(B)\Theta(B)\epsilon_t$$

It also allows us to do so as an infinite-order AR process:

$$\Theta^{-1}(B)\Phi(B)y_t = \epsilon_t$$

Let us walk through an example in Python. First, using the same imports we did before in this chapter, let us generate an invertible and stationary ARMA(2,1) process dummy dataset that satisfies the following equation:

$$(1 - 1.28B + 0.682 B^2) y_t = (1 - 0.58B) \epsilon_t$$

```
arparams = np.array([1.2, -0.6])
ar = np.r_[1, -arparams]
maparams = np.array([0.5])
ma = np.r_[1, -maparams]
arma_process = sm.tsa.ArmaProcess(ar=ar, ma=ma)
```

Let us confirm stationary and invertibility:

```
print('AR(2) Roots: ', arma_process.arroots)
print('AR(2) Invertibility: ', arma_process.isstationary)
print('MA(1) Roots: ', arma_process.maroots)
print('MA(1) Invertibility: ', arma_process.isinvertible)
```

We can use the quadratic formula to test, but we can trust the code to confirm:

```
AR(2) Roots:    [1.-0.81649658j 1.+0.81649658j]

AR(2) Stationarity:    True

MA(1) Roots:    [2.]

MA(1) Invertibility:    True
```

Now that we have a stationary and invertible process, let us generate 200 samples from it:

```
y = arma_process.generate_sample(nsample=200)
```

Step 1 – Visual inspection

Let us take a look at the plots we have been using to build intuition about the process that generated the data:

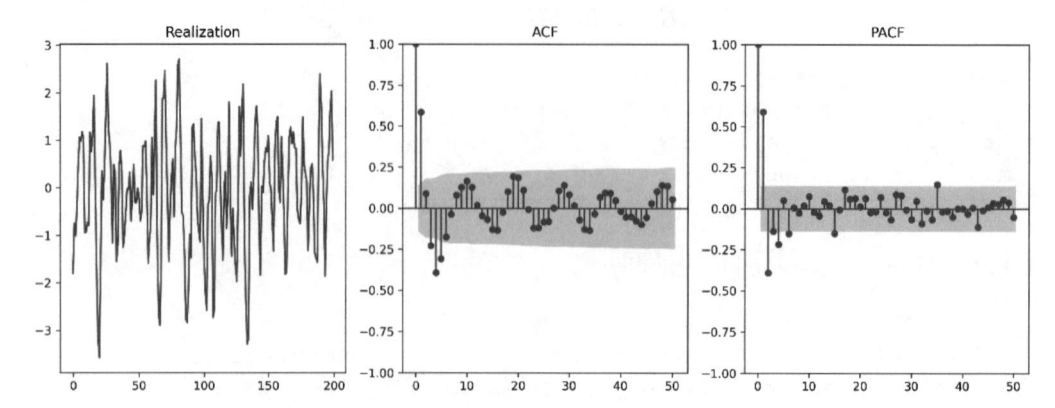

Figure 11.11 – ARMA(p,q) process sample data

We can see using the ACF that we have what appears to be an MA(1) component. Based on the PACF, it looks as if we could have either an AR(2) or an AR(4). The realization appears to be a process that satisfies stationarity.

Step 2 – Select order of ARMA(p,q)

Before we decide on an order for our ARMA model, let us use the Dickey-Fuller test to check if our data has any trend:

```
from statsmodels.tsa.stattools import adfuller
dicky_fuller = adfuller(y, maxlag=25, regression='c')
print('Dickey-Fuller p-value: ', dicky_fuller[1])
print('Dickey-Fuller test statistic: ', dicky_fuller[0])
print('Dickey-Fuller critical value: ', dicky_fuller[4].get('5%'))
```

We can see the statistical significance that confirms we do not have a unit root in the lags provided with `maxlag` (remember, H_o : *the data has a unit root* and H_a : *the data does not have a unit root*). Therefore, we can use the ARMA model without any first-order differencing, which would require at least an ARIMA:

```
Dickey-Fuller p-value:    6.090665062133195e-16

Dickey-Fuller test statistic:    -9.40370671340928

Dickey-Fuller critical value:    -2.876401960790147
```

Now let us use `statmodels arma_order_select_ic` to see what AIC and BIC select for the ARMA(p,q) order. We know the maximum order for an MA(q) is one, but since we are not sure if this is an AR(2) or an AR(4), we can use `max_ar=4`:

```
from statsmodels.tsa.stattools import arma_order_select_ic
model_arma = arma_order_select_ic(y=y, max_ar=4, max_ma=1,
ic=['aic','bic'], trend='n')
print('AIC Order Selection: ', model_arma.aic_min_order)
print('AIC Error: ', round(model_arma.aic.min()[0], 3))
print('BIC Order Selection: ', model_arma.bic_min_order)
print('BIC Error: ', round(model_arma.bic.min()[0], 3))
```

We can see that AIC selected an ARMA(4,1) and BIC selected an ARMA(2,1):

```
AIC Order Selection:    (4, 1)

AIC Error:   548.527

BIC Order Selection:    (2, 1)

BIC Error:   565.019
```

The ARMA(4,1) has a lower error, but we know from this and previous chapters in the book that models with lower error on training data may be more likely to have more variance and thus be more likely to overfit. However, let us use ARMA(4,1).

Step 3 – Building the AR(p) model

Now let us build our ARMA(4,1) model. Note the 0 is for the integrated first-order difference for an ARIMA(p,d,q) model. Since we do not have a trend-based unit root, we do not need to difference to remove any trend. Thus, d=0:

```
from statsmodels.tsa.arima.model import ARIMA
arma_aic = ARIMA(y, order=(4,0,1),
                 enforce_stationarity=True, enforce_
invertibility=True).fit()
print(arma_aic.summary())
```

Here, we can see the model provides a reasonable fit based on the model metrics. However, there is one issue; our first three AR coefficients have no statistical significance (high p-values and confidence intervals containing zero). This is a big problem and confirms our model is overfitted. Our model includes terms it is not getting benefit from. Therefore, our model would most certainly fail to generalize well on unseen data and should be reconstructed:

```
                               SARIMAX Results
==============================================================================
Dep. Variable:                      y   No. Observations:                  200
Model:                   ARIMA(4,0,1)   Log Likelihood                -267.418
Date:                Fri, 20 Jan 2023   AIC                            548.836
Time:                        20:32:18   BIC                            571.924
Sample:                             0   HQIC                           558.180
                                - 200
Covariance Type:                  opg
==============================================================================
                 coef    std err          z      P>|z|      [0.025      0.975]
------------------------------------------------------------------------------
const         -0.0952      0.091     -1.041      0.298      -0.274       0.084
ar.L1          0.2234      0.209      1.067      0.286      -0.187       0.634
ar.L2          0.0506      0.172      0.294      0.769      -0.287       0.388
ar.L3         -0.1207      0.110     -1.096      0.273      -0.337       0.095
ar.L4         -0.2744      0.068     -4.046      0.000      -0.407      -0.141
ma.L1          0.5493      0.213      2.581      0.010       0.132       0.966
sigma2         0.8444      0.099      8.528      0.000       0.650       1.038
===================================================================================
Ljung-Box (L1) (Q):                   0.01   Jarque-Bera (JB):                 2.40
Prob(Q):                              0.90   Prob(JB):                         0.30
Heteroskedasticity (H):               0.80   Skew:                             0.04
Prob(H) (two-sided):                  0.37   Kurtosis:                         2.47
===================================================================================
```

Figure 11.12 – ARIMA(4,0,1) model results

Let us re-run this as an ARMA(2,1), which is the same as an ARIMA(2,0,1) since there is no differencing to be integrated. This is what we visually identified, and BIC selected the following code:

```
from statsmodels.tsa.arima.model import ARIMA
arma_aic = ARIMA(y, order=(2,0,1),
                 enforce_stationarity=True, enforce_
invertibility=True).fit()
print(arma_aic.summary())
```

We can see now a much better fit for our variables. All coefficients are significant, and the model metrics remain sufficient. The model we have identified corresponds to the following equation:

$$(1 - 1.2765B + 0.6526\,B^2)\,y_t \;=\; (1 + 0.58B)\,\epsilon_t$$

We can compare that to our dummy process:

$$(1 - 1.28B + 0.682\,B^2)\,y_t \;=\; (1 - 0.58B)\,\epsilon_t$$

```
                               SARIMAX Results
==============================================================================
Dep. Variable:                      y   No. Observations:                  200
Model:                     ARIMA(2,0,1)   Log Likelihood                -270.382
Date:                Fri, 20 Jan 2023   AIC                            550.765
Time:                        21:08:18   BIC                            567.256
Sample:                             0   HQIC                           557.439
                                -200
Covariance Type:                  opg
==============================================================================
                 coef    std err          z      P>|z|      [0.025      0.975]
------------------------------------------------------------------------------
const         -0.0960      0.075     -1.287      0.198      -0.242       0.050
ar.L1          1.2765      0.094     13.616      0.000       1.093       1.460
ar.L2         -0.6526      0.058    -11.241      0.000      -0.766      -0.539
ma.L1         -0.5807      0.120     -4.849      0.000      -0.815      -0.346
sigma2         0.8704      0.099      8.797      0.000       0.676       1.064
==============================================================================
Ljung-Box (L1) (Q):                   0.09   Jarque-Bera (JB):              1.03
Prob(Q):                              0.77   Prob(JB):                      0.60
Heteroskedasticity (H):               0.83   Skew:                         -0.06
Prob(H) (two-sided):                  0.46   Kurtosis:                      2.67
```

Figure 11.13 – ARIMA(2,0,1) model results

Step 4 – Test forecasting

Now, let us cross-validate our model by training the model on data up through the last five points, then forecasting the last five points so that we can compare them to the actuals. Recall that our indexing starts at 0 so our dataset ends at index 199:

```
df_pred = arma_aic.get_prediction(start=195, end=199).summary_
frame(alpha=0.05)
df_pred.index=[195,196,197,198,199]
```

We can see our predicted values in the *mean* column in the following table:

y	mean	mean_se	mean_ci_lower	mean_ci_upper	actuals
195	-0.01911	0.932933	-1.84762	1.80940631	0.559875
196	0.58446	0.932933	-1.24406	2.412975242	0.778127
197	0.479364	0.932933	-1.34915	2.307879057	1.695218
198	0.914009	0.932933	-0.91451	2.74252465	2.041826
199	0.80913	0.932933	-1.01939	2.637645206	0.578695

Figure 11.14 – AR(4) model outputs versus actuals

ARIMA Models

Let's print out our model's **Average Squared Error** (ASE):

```
print('Average Squared Error: ', np.mean((df_pred['mean'] -
y[195:])**2))
```

Here we see the ASE:

```
Average Squared Error:   0.6352208223437921
```

Our test forecast plot is shown in *Figure 11.15*. Note that our estimate appears conservative. Using ARMA(4,1) may have produced a closer, but less generalizable fit. One method for improving forecasting would be to build the model using only recent points (relative to subject matter knowledge of the process). Including a larger set of data will produce a fit that generalizes more to the overall process rather than to a possibly more relevant timeframe:

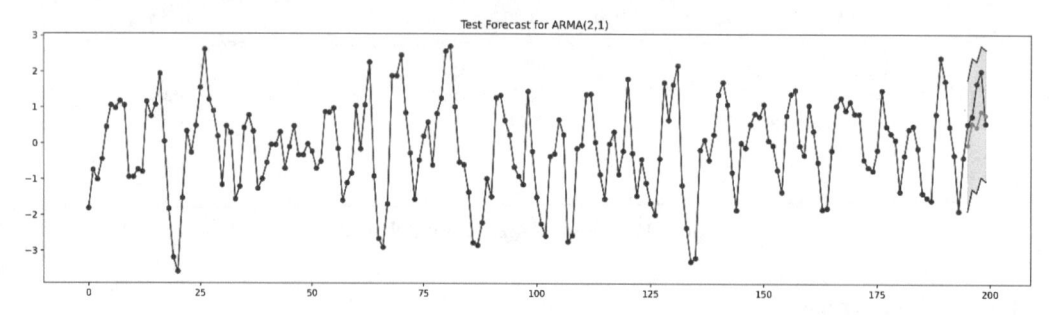

Figure 11.15 – ARMA(2,1) test forecast

Step 5 – Building a forecast

Now, let us forecast five ahead:

```
df_forecast = arma_aic.get_prediction(start=200, end=204).summary_
frame(alpha=0.05)
df_forecast.index=[200, 201, 202, 203, 204]
forecast = np.hstack([np.repeat(np.nan, len(y)), df_pred['mean']])
```

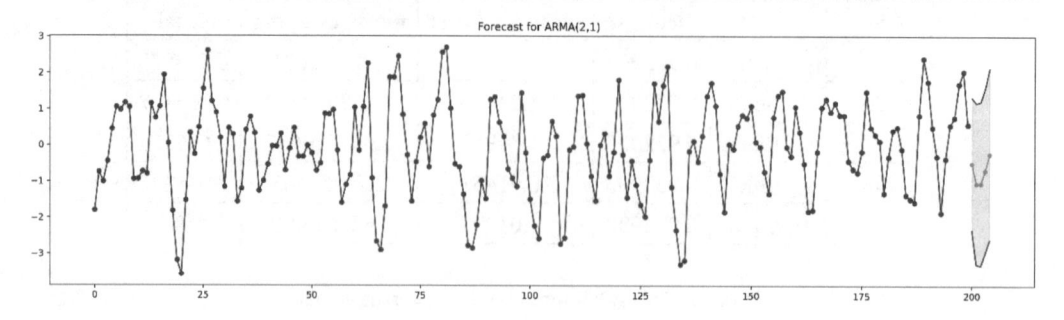

Figure 11.16 – ARMA(2,1) forecast horizon = 5

On a final note, regarding ARMA models, *we always assume process stationarity*. If stationarity cannot be assumed, neither autoregressive nor moving average models can be used. In the next section of this chapter, we will discuss integrating into ARMA models first-order differencing as a method for conditionally stationarizing a process to overcome limitations of non-stationarity.

Models for non-stationary time series

In the previous section, we discussed ARMA models for stationary time series data. In this section, we will look at non-stationary time series data and extend our model to work with non-stationary data. Let us start by taking a look at some sample data (shown in *Figure 11.17*). There are two series: US GDP (left) and airline passenger volume (right).

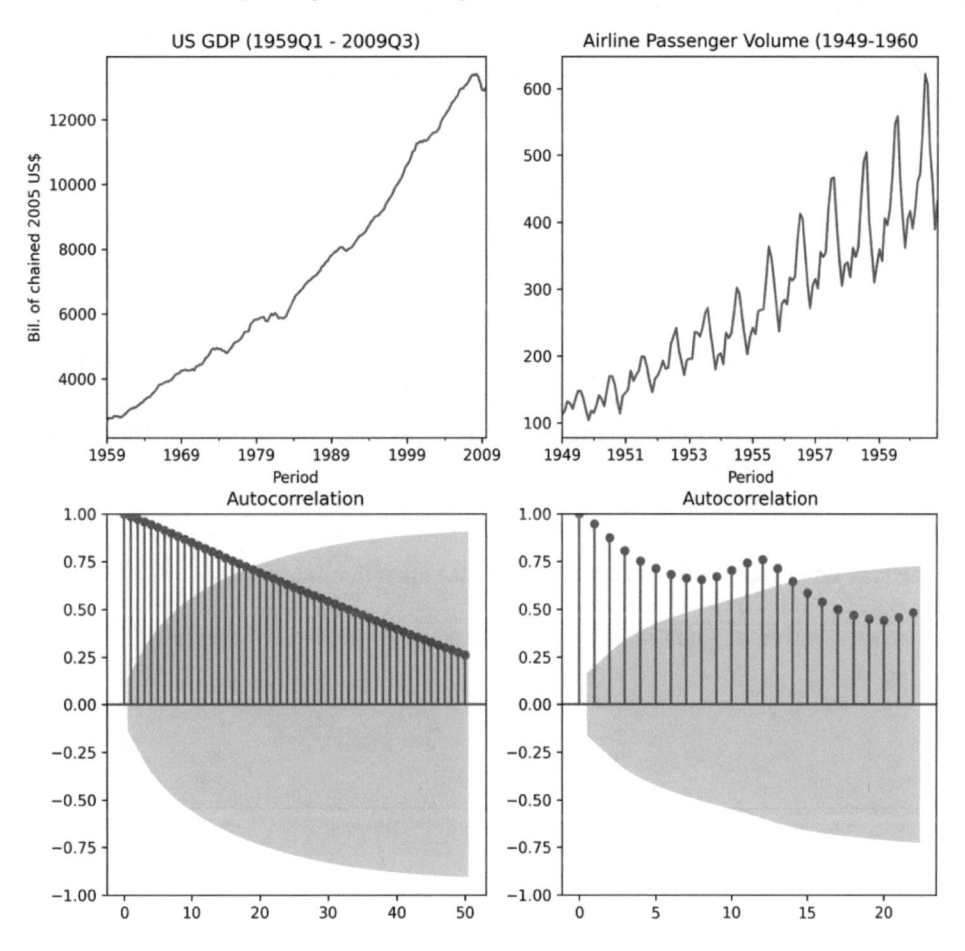

Figure 11.17 – US GDP (left) and airline passenger (right) time series

The US GDP series appears to exhibit an upward trend with some variations in the series. The airline passenger volume series also exhibits an upward trend, but there also appears to be a repeated pattern in the series. The repeated pattern in the airline series is called **seasonality**. Both series are non-stationary because of the apparent trend. Additionally, the airline passenger volume series appears to exhibit non-constant variance. We will model the GDP series with ARIMA, and we will model the seasonal ARIMA. Let's take a look at these models.

ARIMA models

ARIMA is an acronym for **AutoRegressive Integrated Moving Average**. This model is a generalization of the ARMA model that can be applied to non-stationary time series data. The new part added to this model is "integrated," which is a **differencing** operation applied to the time series to **stationarize** (to make stationary) the time series. After the time series is stationarized, we can fit an ARMA model to the differenced data. Let's take a look at the mathematics of this model. We will start with understanding how differencing works, and then put the whole model ARIMA model together.

Differencing

Differencing data is computing the difference between consecutive data points. The resulting data from differencing represents the *change* between each data point. We can write the difference as such:

$$y'_t = y_t - y_{t-1}$$

This equation is the first-order difference, meaning it is the first difference between the data points. It may be necessary to make additional differences between the data points to stationarize the series. The second difference represents the *change of changes* between the data points. The second-order difference can be written thusly:

$$y''_t = y'_t - y'_{t-1} = (y_t - y_{t-1}) - (y_{t-1} - y_{t-2})$$

The "order" is simply the number of times a difference operation is applied.

The ARIMA model

As mentioned earlier, the ARIMA model is ARMA with the addition of differencing to make the time series stationary (stationarize the time series). Then we can express an ARIMA model mathematically as follows, where y'_t is the differenced series, differenced d times until it is stationary:

$$y'_t = c + \phi_1 y'_{t-1} + \ldots + \phi_p y'_{t-p} + \epsilon_t + \theta_1 \epsilon_{t-1} + \ldots + \theta_q \epsilon_{t-q}$$

The ARIMA model has three orders, which are denoted ARIMA(p,d,q):

- p – the autoregressive order
- d – the differencing order
- q – the moving average order

With more complicated models such as ARIMA, we will tend to describe them with backshift notation since it is easier to express these models with backshift notation. An ARIMA model will take the following form using backshift notation:

$$\left(1 - \phi_1 B - \ldots - \phi_p B^p\right)(1 - B)^d y_t = c + \left(1 + \theta_1 B + \ldots + \theta_q B^q\right) \epsilon_t$$

$$\underset{AR(p)}{\uparrow} \quad \underset{d \text{ differences}}{\uparrow} \quad \underset{MA(q)}{\uparrow}$$

Notice the term for the differences in the equation: $(1 - B)^d$. In the previous section, we discussed roots as related to stationary models. In that context, the roots were always outside of the unit circle. With an ARIMA model, we add unit roots to the model. To understand the impact of a unit root, let's simulate an AR(1) model and see what happens as the root of the model is moved toward one. These simulations are shown in *Figure 11.18*.

ARIMA Models

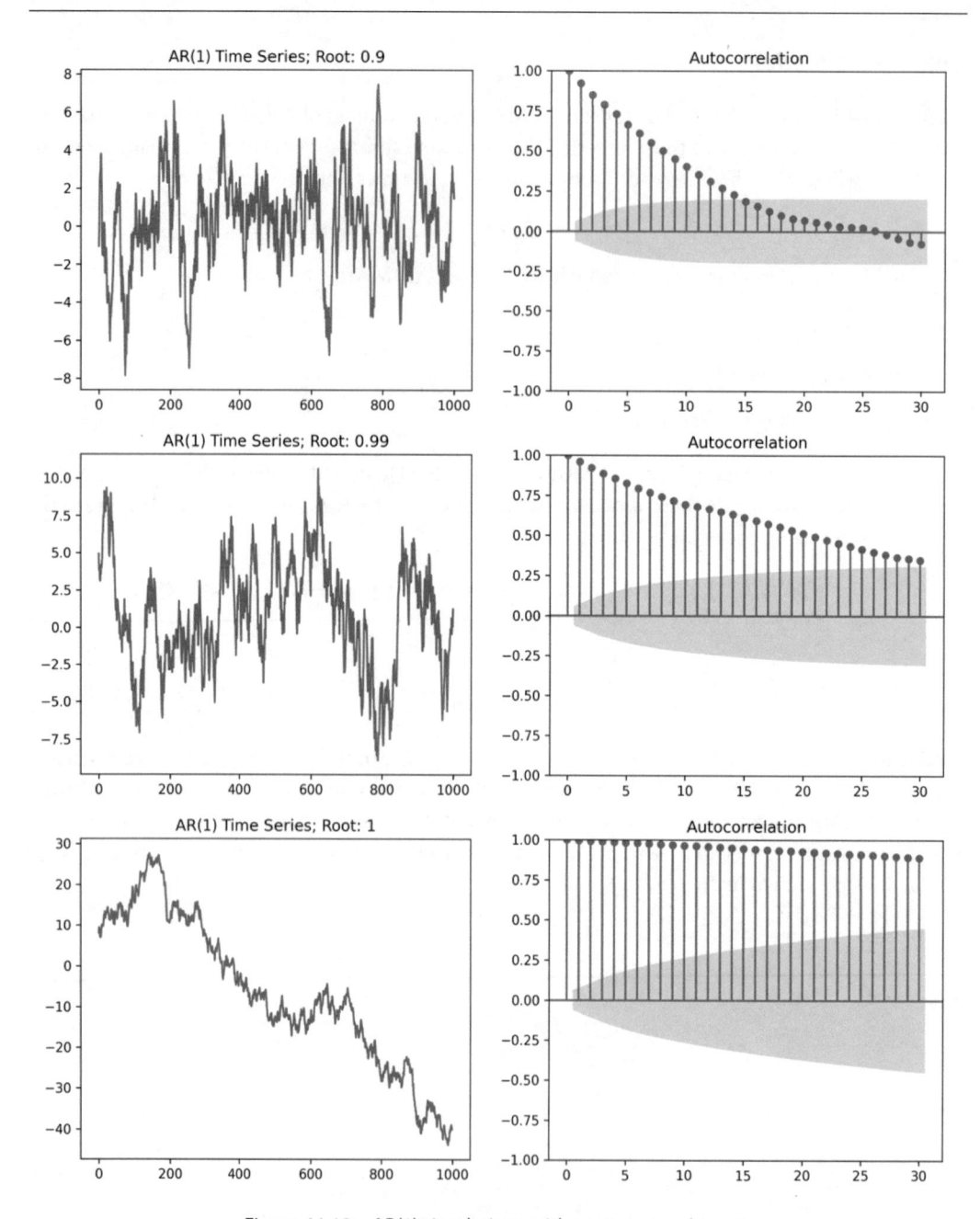

Figure 11.18 – AR(1) simulations with root approaching 1

We can make two observations from the simulations shown in *Figure 11.18*. The first observation is that the time series appear to exhibit more wandering behavior as the root increases toward one. For instance, the middle-time series shows more wandering from the mean than the top-time series. The bottom time series (with a root of one), does not appear to regress toward a mean such as the other two simulations. The second observation is about the autocorrelations. As the root of the AR(1) approaches 1, the autocorrelations get stronger and decrease slower over the lags. These two observations are characteristic of a series with a root near or at one. Additionally, the presence of unit roots will dominate the time series behavior, making it easy to recognize from the autocorrelation plot.

Fitting an ARIMA model

There are two steps to fit an ARIMA model: (1) stationarize the series from differencing to determine the difference order and (2) fit an ARMA model to the resulting series. In the previous section, we discussed how to fit an ARMA model so in this section, we will focus on the first step.

At the beginning of this section, we showed a time series of US GDP values. We will use that time series as a case study for fitting an ARIMA model. First, let's take a look at the series and its autocorrelations again. The series and autocorrelations are shown in *Figure 11.19*.

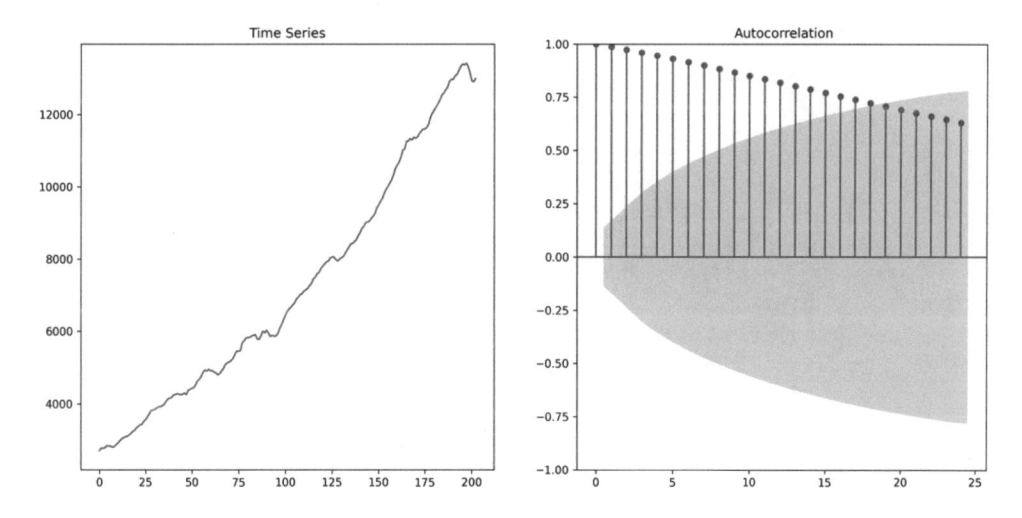

Figure 11.19 – US GDP time series and autocorrelations

From the plots shown in *Figure 11.19*, it appears that the time series of US GPD data is non-stationary time series. The time series exhibits wandering behavior, and the autocorrelations are strong and decrease slowly. As we discussed, this is characteristic behavior of unit roots. For secondary evidence, we can use the Dickey-Fuller test for unit root. The null hypothesis of the Dickey-Fuller test is that a unit root is present in the time series. The following code shows how to use the Dickey-Fuller test from `pmdarima`. The test returns a p-value of 0.74 indicating we cannot reject the null hypothesis, meaning that the time series should be differenced:

```
Import pmdarima as pm
from sktime import datasets
y_macro_economic = datasets.load_macroeconomic()
adf_test = pm.arima.ADFTest()
adf_test.should_diff(y_macro_economic.realgdp.values)
# (0.7423236714537164, True)
```

We can take the first difference of time series using the `diff` function from numpy:

```
first_diff = np.diff(y_macro_economic.realgdp.values, n=1)
```

Taking the first difference, we arrive at a new time series as shown in *Figure 11.20*:

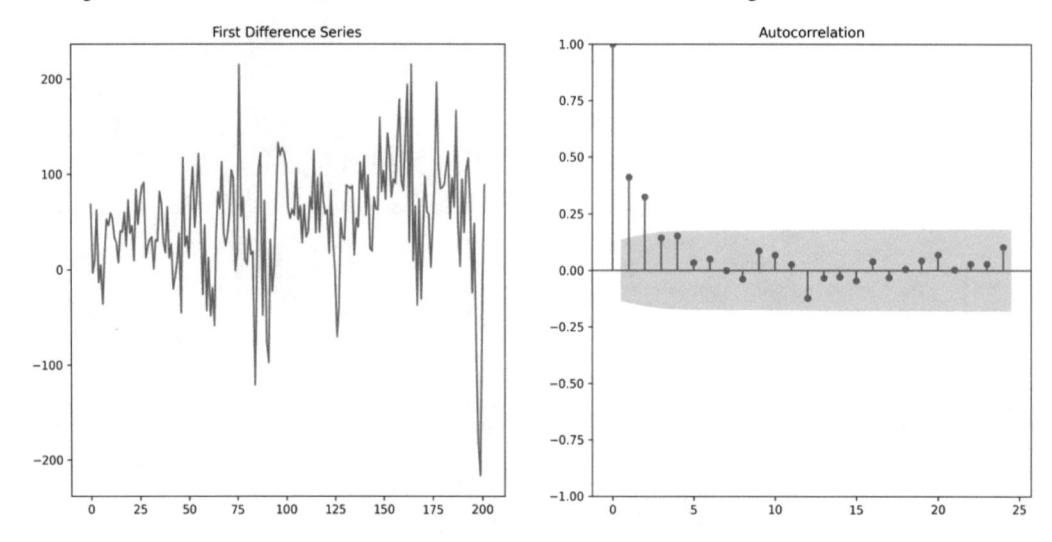

Figure 11.20 – The first difference of US GDP time series

The first difference of the US GDP time series shown in *Figure 11.20* appears to be stationary. In fact, it appears to be consistent with an AR(2) model. We double-check whether we need to take an additional difference using the Dickey-Fuller test on the first differenced data:

```
first_diff = np.diff(y_macro_economic.realgdp.values, n=1)
adf_test.should_diff(first_diff)
# (0.01, False)
```

The Dickey-Fuller test returns a p-value of 0.01 for the first differenced data, which means we can reject the null hypothesis and we can stop differencing the data. That means that our ARIMA model for this data will have a difference order of 1 ($d = 1$).

After finding the difference order, we can fit an ARMA model to the differenced data. Since we have already discussed fitting ARMA models, we will use an automated fitting method provided by pmdarima. pm.auto_arima is a function for automatically fitting an ARIMA model to data, however, in this case, we will use it to fit the ARMA portion from the differenced series. The output of pm.auto_arima for the first difference data is shown in the following code block:

```
pm.auto_arima(
    first_diff, error_action='ignore', trace=True,
    suppress_warnings=True, maxiter=5, seasonal=False,
    test='adf'
)
Performing stepwise search to minimize aic
  ARIMA(2,0,2)(0,0,0)[0]             : AIC=2207.388, Time=0.03 sec
  ARIMA(0,0,0)(0,0,0)[0]             : AIC=2338.346, Time=0.01 sec
  ARIMA(1,0,0)(0,0,0)[0]             : AIC=2226.760, Time=0.02 sec
  ARIMA(0,0,1)(0,0,0)[0]             : AIC=2284.220, Time=0.01 sec
  ARIMA(1,0,2)(0,0,0)[0]             : AIC=2206.365, Time=0.02 sec
  ARIMA(0,0,2)(0,0,0)[0]             : AIC=2253.267, Time=0.02 sec
  ARIMA(1,0,1)(0,0,0)[0]             : AIC=2203.917, Time=0.01 sec
  ARIMA(2,0,1)(0,0,0)[0]             : AIC=2208.521, Time=0.02 sec
  ARIMA(2,0,0)(0,0,0)[0]             : AIC=2208.726, Time=0.02 sec
  ARIMA(1,0,1)(0,0,0)[0] intercept   : AIC=2193.482, Time=0.04 sec
  ARIMA(0,0,1)(0,0,0)[0] intercept   : AIC=2208.669, Time=0.03 sec
  ARIMA(1,0,0)(0,0,0)[0] intercept   : AIC=2195.212, Time=0.02 sec
  ARIMA(2,0,1)(0,0,0)[0] intercept   : AIC=2191.810, Time=0.03 sec
  ARIMA(2,0,0)(0,0,0)[0] intercept   : AIC=2190.196, Time=0.02 sec
  ARIMA(3,0,0)(0,0,0)[0] intercept   : AIC=2191.589, Time=0.03 sec
  ARIMA(3,0,1)(0,0,0)[0] intercept   : AIC=2193.567, Time=0.03 sec
Best model:  ARIMA(2,0,0)(0,0,0)[0] intercept
Total fit time: 0.349 seconds
```

Since the ARMA fit for the differenced data is ARMA(2,0), the ARIMA orders for the original time series would be ARIMA(2,1,0). Next, we will look at forecasting from an ARIMA model.

Forecasting with ARIMA

Once we have a fit model, we can forecast with that model. As mentioned in previous chapters, when making predictions we should create a train-test split, so we have data to compare with the predictions. The model should only fit the training data to avoid data leakage. We can use the `train_test_split` function from `pmdarima` to split the data. Then we proceed with the usual steps: split, train, and predict. The code for this is shown in the following code block:

```
from pmdarima.model_selection import train_test_split
train, test = 
    train_test_split(y_macro_economic.realgdp.values,
    train_size=0.9
)
arima = pm.auto_arima(
    train, out_of_sample_size=10,
    suppress_warnings=True, error_action='ignore',
    test='adf'
)
preds, conf_int = arima.predict(
    n_periods=test.shape[0], return_conf_int=True
)
```

The preceding code fits an ARIMA model with `auto_arima` and then forecasts the size of the test set using the `predict` method of the ARIMA object. The forecasts for the series generated by the code are shown in *Figure 11.21*:

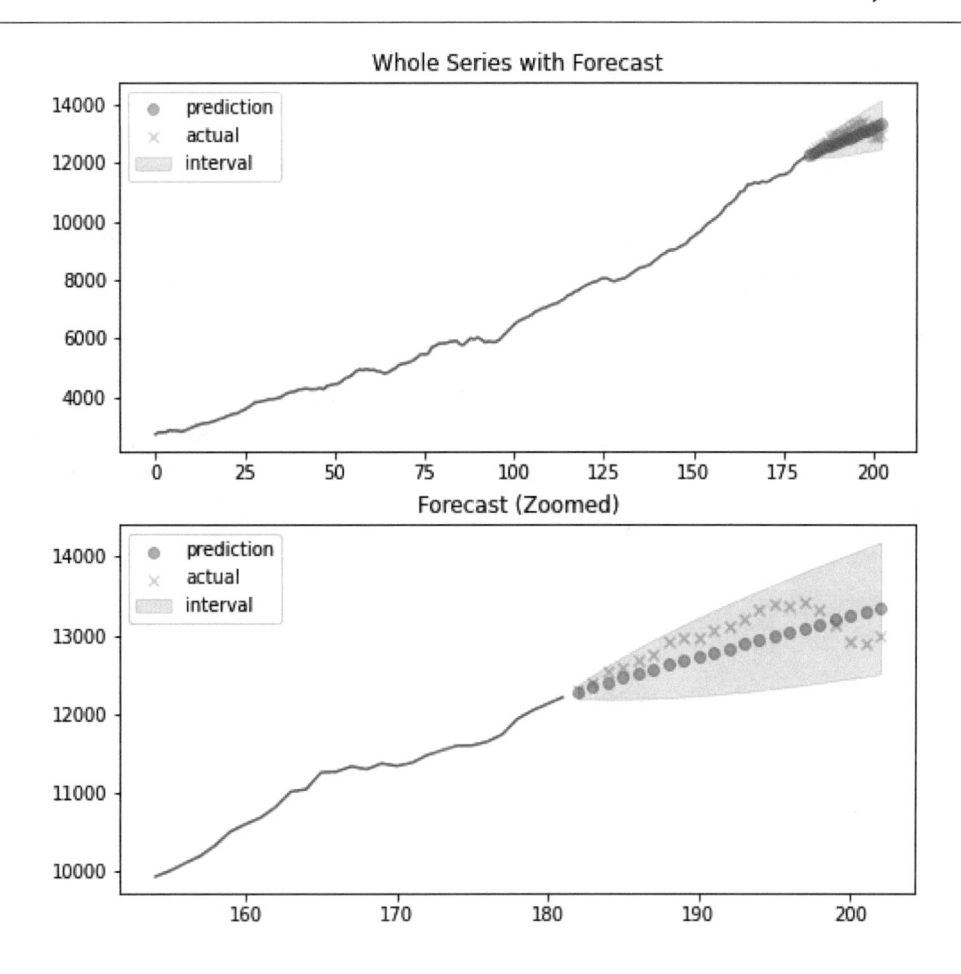

Figure 11.21 – US GDP ARIMA forecast over test split

The forecast of the US GDP in *Figure 11.21* appears to follow the trend of the data but does not capture the small variations in the series. However, the variation is captured in the prediction interval (labeled as "interval"). This model appears to provide a reasonably good prediction of the test data. Note that the interval increases over time. This is because predictions become more uncertain farther in the future. Generally, shorter forecasts are more likely to be accurate.

In this section, we built on the ARMA model and extended it to non-stationary data using differencing, which formed the ARIMA model. In the next section, we will look at non-stationary time series that include seasonal effects and make a further extension to the ARIMA model.

Seasonal ARIMA models

Let's look at another characteristic of time series called **seasonality**. Seasonality is the presence of a pattern in a time series that repeats at regular intervals. Seasonal time series are common in nature. For example, yearly weather patterns and daily sunshine patterns are seasonal patterns. Back at the start of the non-stationary section, we showed an example of a non-stationary time series with seasonality. This time series is shown again in *Figure 11.22* along with its ACF plot.

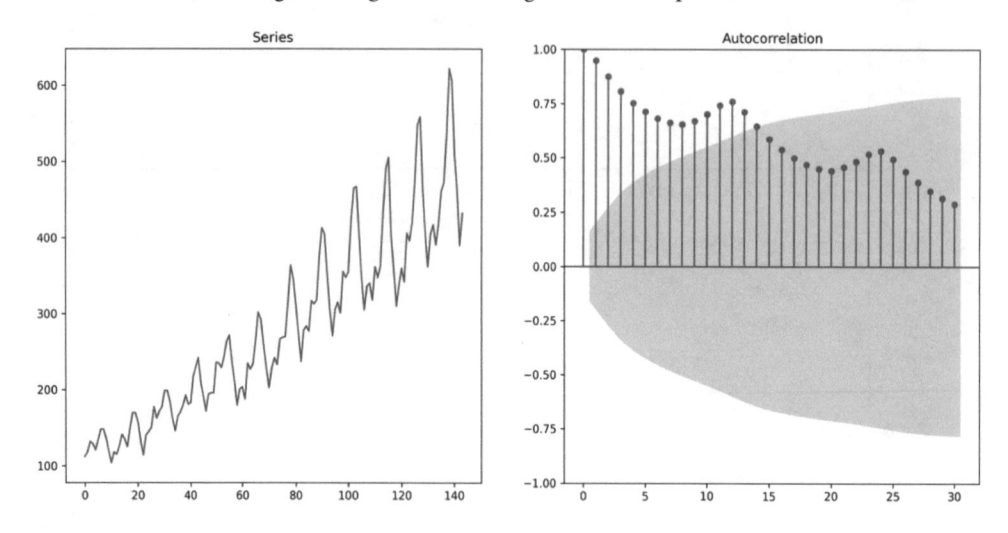

Figure 11.22 – Airline volume data and ACF plot

The time series shown in *Figure 11.22* is the monthly total of international airline passengers from 1949 to 1960 [3]. There is a definite repeated pattern in this time series. To model this type of data, we will need to an additional term to the ARIMA model to account for seasonality.

Seasonal differencing

We will use a similar approach for modeling this type of time series as we did with ARIMA. We will start by using differencing to stationarize the data, then fit an ARMA model to the differenced data. With seasonal time series, we will need to use seasonal differencing to remove the seasonal effects, which we can show mathematically:

$$y'_t = y_t - y_{t-T}$$

Where T is the period of the season. For example, the time series in *Figure 11.22* exhibits monthly seasonality and each data point represents one month; therefore, the $T = 12$ for the airline volume data. Then, for the airline data, we would use the following difference equation to remove seasonality:

$$y'_t = y_t - y_{t-12}$$

Seasonal ARIMA models

We can also identify the seasonality by observing where peaks occur in the ACF plot. The ACF plot in *Figure 11.22* shows a peak at 12, indicating a seasonal period of 12, which is consistent with our knowledge of the time series.

We will see how to apply seasonal differences later in this section using pmdarima. Let's take a look at how seasonality is included in the model.

Seasonal ARIMA

As mentioned in the ARIMA section, we will be differencing the original series, then fitting a stationary model to the differenced data. Then our time series would be described by the following equation where y'_t is the differenced series (including seasonal and sequential differences):

$$y'_t = c + \phi_1 y'_{t-1} + \ldots + \phi_p y'_{t-p} + \epsilon_t + \theta_1 \epsilon_{t-1} + \ldots + \phi_q \epsilon_{t-q}$$

We can express the whole model with backshift notation:

$$\underbrace{(1 - \phi_1 B - \ldots - \phi_p B^p)}_{AR(p)} \underbrace{(1 - B)^d}_{d\ diff} \underbrace{(1 - B^s)}_{seasonal\ diff} y_t = c + \underbrace{(1 + \theta_1 B + \ldots + \theta_q B^q)}_{MA(q)} \epsilon_t$$

We have a new term in the equation that accounts for seasonality: $(1 - B^s)$. We are adding a new order parameter to the model: s. This model is typically denoted ARUMA(p,d,q,s).

> **SARIMA models**
>
> In this section, we are only covering seasonal differencing. There are more complex models that allow for moving average seasonality and autoregressive seasonality called SARIMA and denoted SARIMA(p,d,q)(P,D,Q)[m]. These models are beyond the scope of this chapter. However, we would encourage the reader to explore these models further after mastering the topics found in this chapter and the next chapter. The ARIMA model covered in this chapter is a subset of the SARIMA model, which accounts for seasonal differencing, which is the "D" order of SARIMA(p,d,q)(P,D,Q)[m].

Just as the $(1 - B)^d$ term we added for ARIMA, the $(1 - B^s)$ term adds roots to the unit circle. However, unlike the roots from $(1 - B)^d$, the roots from $(1 - B^s)$ are distributed uniformly around the unit circle. These roots can be calculated and plotted programmatically with numpy and matplotlib or automatically with computational intelligence tools such as Wolfram Alpha (https://www.wolframalpha.com/).

Fitting an ARIMA model with seasonality

We will take the following steps to fit an ARIMA model with seasonality:

- Remove seasonality with differencing.
- Remove additional non-stationarity with differencing.
- Fit a stationary model to the resulting series.

This is essentially the same process we used to fit an ARIMA model, but there is an additional step to handle the seasonal component. Let us walk through an example with the airline data.

We will start with using differencing to remove the seasonal component of the time series. Recall that the seasonal period of the airline time series is 12, meaning that we need to perform differencing at lag 12 as shown with this equation:

$$y'_t = y_t - y_{t-12}$$

We can perform this difference using the `diff` function from `pmdarima`. The following code shows how to perform the 12th lagged difference on the airline data:

```
import pmdarima as pm
from sktime import datasets
y_airline = datasets.load_airline()
series = pm.utils.diff( y_airline.values, lag=12)
```

After performing the seasonal difference, we get the differenced series shown in *Figure 11.23* along with the ACF plot. The seasonal portion of the time series appears to be completely removed. The differenced series does not appear to exhibit any repeating patterns. Additionally, the ACF plot does not show the seasonal peak that was present in the ACF plot of the original data:

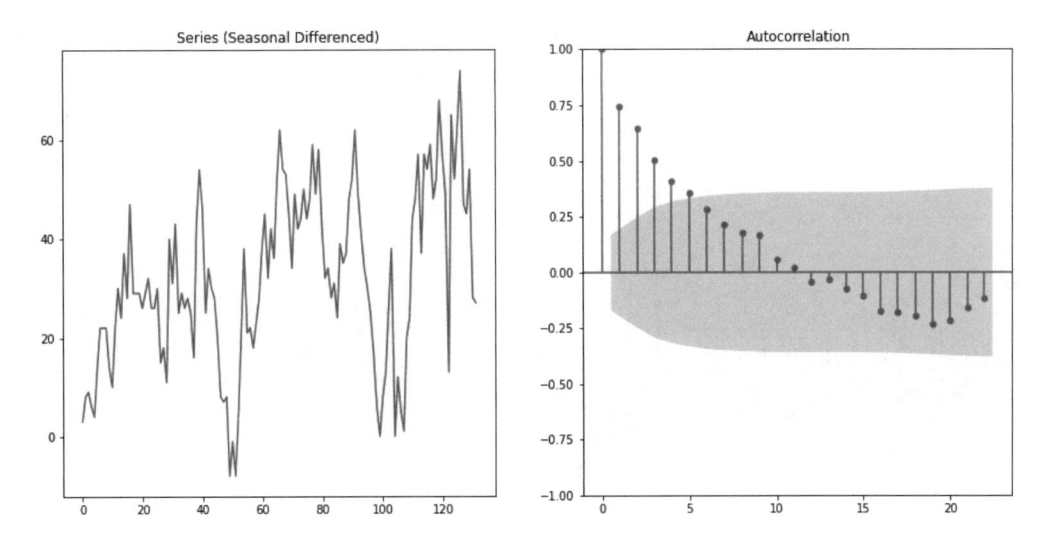

Figure 11.23 – Airline data after seasonal difference

With the seasonal portion of the time series removed, we need to determine whether we need to take any additional differences to stationarize the new time series. The differenced series in *Figure 11.23* appears to exhibit a trend. The original data also exhibited a trend. As before, we can use the Dickey-Fuller test to get additional evidence on whether we should apply additional differences. Running the Dickey-Fuller test on this series will result in a p-value of 0.099, which suggests that we should take a difference in the series to account for a unit root:

```
adf_test = pm.arima.ADFTest()
adf_test.should_diff(series)
# (0.09898694171553156, True)
```

ARIMA Models

Taking the first difference of the series will result in the series shown in *Figure 11.24*. After taking these two differences the series appears to be sufficiently stationarized.

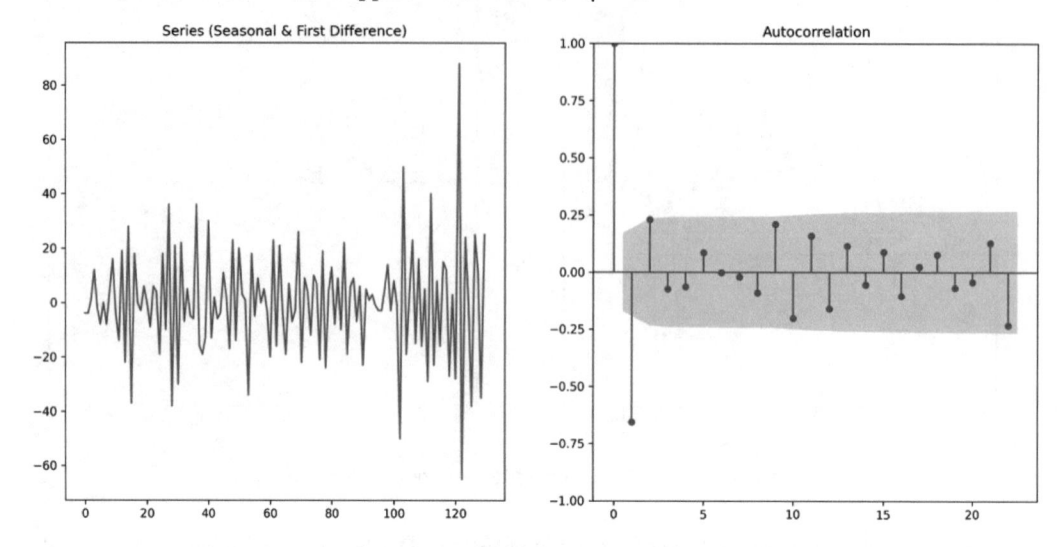

Figure 11.24 – Airline data after seasonal and first difference

The series in *Figure 11.24* shows the stationarized version of the airline data. Based on the ACF plot, we should be able to fit a relatively simple ARMA model to the stationarized series. We will use `auto_arima` function to make an automatic fit as we did in the ARIMA section:

```
pm.auto_arima(
    series, error_action='ignore', trace=False,
    suppress_warnings=True, maxiter=5, seasonal=False,
    test='adf'
)
# ARIMA(1,0,0)(0,0,0)[0]
```

Fitting the differenced data with `auto_arima` returns an AR(1) model. A simple model as we expected.

Putting this all together our resulting model is an ARUMA(1,1,0,12). As with the previous ARIMA example, we could have fit this model with `auto_arima`, but we walked through the differencing steps here to help build intuition for what each difference element does to the series. Let's take a look at the direct fit from `auto_arima` now:

```
pm.auto_arima(
    y_airline.values, error_action='ignore', trace=True,
    suppress_warnings=True, maxiter=5, seasonal=True, m=12,
```

```
    test='adf'
)
# Best model:  ARIMA(1,1,0)(0,1,0)[12]
```

We see that `auto_arima` found the same model that we did using manual differencing. Note the model is denoted in SARIMA format (see earlier callout about SARIMA). The (0,1,0)[12] means seasonality of 12 when one difference for the seasonality. Now that we have a fit model, let's look at forecasting for our seasonal model.

Forecasting ARIMA with seasonality

Once we have a fit model, we can forecast with that model. As mentioned in the section on forecasting with ARIMA, when should we make a train-test split so we have data to compare with the predictions? We will use the same procedure: split the data, train the model, and forecast over the test set size:

```
train, test = train_test_split(y_airline.values, train_size=0.9)
sarima = pm.auto_arima(
    train, error_action='ignore', trace=True,
    suppress_warnings=True, maxiter=5, seasonal=True, m=12,
    test='adf'
)
preds, conf_int = sarima.predict(n_periods=test.shape[0], return_conf_int=True)
```

The preceding code fits a full SARIMA model with `auto_arima` and then forecasts the size of the test set using the `predict` method. The forecasts for the series generated by the code are shown in *Figure 11.25*.

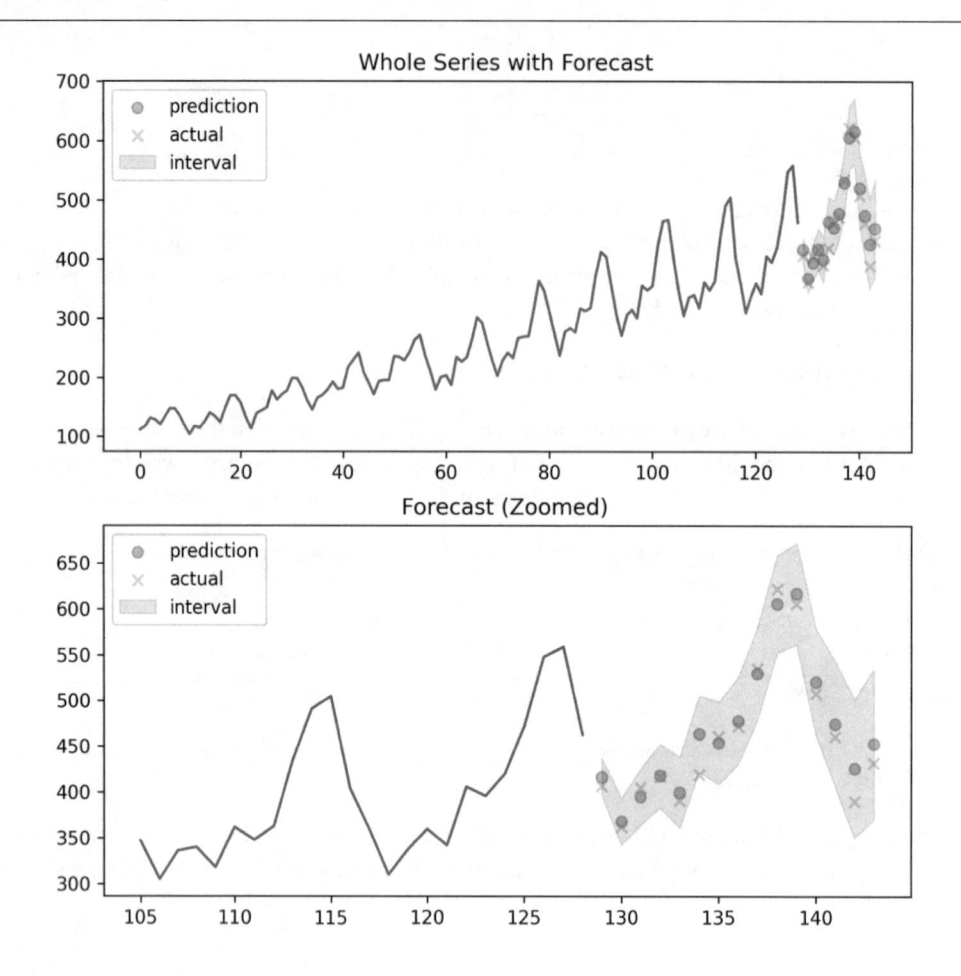

Figure 11.25 – SARIMA forecast of the airline data

The forecast of the airline data in *Figure 11.25* appears to capture the variation of the data very well. This is likely due to the strength of the seasonality component in the time series. Note that the prediction intervals increase over time just as with the ARIMA prediction intervals, but the intervals follow the general pattern of the series. This is an impact of the additional knowledge of seasonality in the series.

In this section, we discussed ARIMA models with seasonality and showed how to remove seasonal components. We also looked at forecasting a model with seasonality. In the next section, we will take a closer look at validating time series models.

More on model evaluation

In the previous sections, we discussed other methods to prepare data, test and validate models. In this section, we will discuss how to validate time series models and introduce several methods for validating time series models. We will cover the following methods for model evaluation: **resampling, shifting, optimized persistence forecasting,** and **rolling window forecasting**.

The real-world dataset considered in this section is Coca Cola stock data collected from Yahoo Finance databases from 01/19/1962 to 12/19/2021 for stock price prediction. This is a time series analysis to forecast the future stock value of a given stock. The reader can download the dataset from the Kaggle platform for this analysis. To motivate the study, we first go to explore the Coco Cola stock dataset:

```
data = pd.read_csv("COCO COLA.csv", parse_dates=["Date"], index_col="Date")
```

Date	Open	High	Low	Close	Adj Close	Volume
1962-01-02	0.263021	0.270182	0.263021	0.263021	0.051133	806400
1962-01-03	0.259115	0.259115	0.253255	0.257161	0.049994	1574400
1962-01-04	0.257813	0.261068	0.257813	0.259115	0.050374	844800
1962-01-05	0.259115	0.262370	0.252604	0.253255	0.049234	1420800
1962-01-08	0.251302	0.251302	0.245768	0.250651	0.048728	2035200

Figure 11.26 – Coco Cola dataset

The Date index is related to 15096 trading days from 01/19/1962 to 12/19/2021. The **High** and **Low** columns here refer to the maximum and minimum prices on each trading day. **Open** and **Close** refer to the stock prices when the market was open and closed on the same trading day. The total amount of trading stocks in each day refers to the **Volume** column and the last column (**Adj Close**) refers to adjusted values (combining with stock splits, dividends, etc.). To illustrate how resampling, shifting, rolling windows, and expanding windows perform, we narrow down to use only the **Open** column from the year 2016:

```
data= data[data.index>='2016-01-01'][['Open']]
```

The data was collected by trading dates. However, we will perform the study monthly. The **resampling** technique is used to aggregate data from days to months. This idea motivates us to introduce this technique.

Resampling

Resampling is a method to deal with time frequency. We use this method to change time series data frequency by a new time frequency. There are **upsampling** and **downsampling**. In upsampling, the *frequency of the samples is increased* using interpolation, for instance, from day frequency to minute or second frequency. On the other hand, downsampling is used to *decrease the sample time frequency* using summary statistics to aggregate the current time frequency in a dataset to a new time frequency. Resampling can be used in feature engineering and also used for problem framing when time frequency in the dataset is not available for our analysis purposes. In Python, we can use the `resample()` function to change time frequencies. The following code illustrates the resampling technique in Python:

```
fig, ax = plt.subplots(4,1, figsize=(12,8))
ax[0].set_title('Original stock price from 2016')
ax[0].plot(data)
ax[1].plot(data.resample('7D').mean())
ax[1].set_title('7 days - Downsampling stock price from 2016')
ax[2].plot(data.resample('M').mean())
ax[2].set_title('Monthly Downsampling  stock price from 2016')
ax[3].plot(data.resample('Y').mean())
ax[3].set_title('Yearly Downsampling stock price from 2016')
fig.tight_layout(pad=5.0)
plt.show()
```

Here is the output of the previous code:

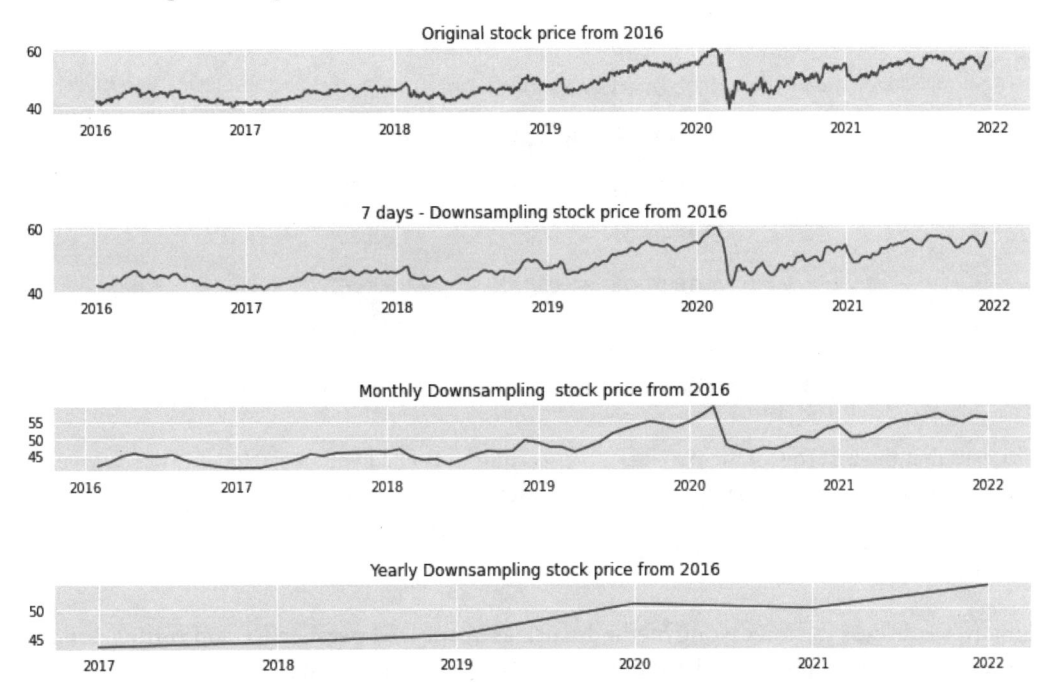

Figure 11.27 – Resampling for Coco Cola dataset

We observe that the plots become smoother when time frequencies decrease. Next, we discuss the shifting method used in time series.

Shifting

In time series analysis, it is not uncommon to **shift** data points to compare lagging and leading futures. We demonstrated using the cross-correlation function in *Chapter 10* as a method to help identify lagging and leading. We use the Coco Cola stock prices from 2016 to the month with the `shift()` function to create new features:

```
data["price_lag_1"] = data["Open"].shift(1)
data.head()
```

	Open	price_lag_1
Date		
2016-01-04	42.340000	NaN
2016-01-05	42.310001	42.340000
2016-01-06	42.200001	42.310001
2016-01-07	41.650002	42.200001
2016-01-08	41.650002	41.650002

Figure 11.28 – First five rows of Coco Cola stock data with price shifted once

Observe that the first row of the `price_lag_1` column is filled with a NaN value. We can replace the missing value with the `fill_value` parameter:

```
data["price_lag_1"] = data["Open"].shift(1, fill_value = data['Open'].mean())
```

Finally, we discuss the forecasting methods such as **optimized persistence** and **rolling window forecasting**. Another resource related to these methods can be found in [3].

Optimized persistence forecasting

We will convert the Coco Cola stock price time frequency to monthly frequency from 2016 using resampling and then we apply an optimized persistence forecasting technique to predict the future value using the previous observation. RMSE scores are considered to evaluate persistence models:

```
from sklearn.metrics import mean_squared_error
import math
train, test = data.resample('M').mean()['Open'][0:-24], data.resample('M').mean()['Open'][-24:]
persistence = range(1, 25)
RMSE_scores = []
for p in persistence:
    history = [x for x in train]
    pred = []
    for i in range(len(test)):
    # Prediction on test set
        yhat = history[-p]
        pred.append(yhat)
    history.append(test[i])
```

```
# RMSE score performance
rmse = math.sqrt(mean_squared_error(test, pred))
RMSE_scores.append(rmse)
print(f'p={p} RMSE={rmse}')
```

The output is as follows:

```
p=1   RMSE=4.491395928336876
p=2   RMSE=3.9349279400485933
p=3   RMSE=4.279904207057928
p=4   RMSE=4.596603380615331
p=5   RMSE=4.15310552002015
p=6   RMSE=3.8368556199724897
p=7   RMSE=3.908843668637612
p=8   RMSE=5.034401761211634
p=9   RMSE=6.156221311108847
p=10  RMSE=7.311228641126667
p=11  RMSE=6.127112381878048
p=12  RMSE=6.063846181348495
p=13  RMSE=5.1342578079826
p=14  RMSE=4.730061082708834
p=15  RMSE=7.119435656445713
p=16  RMSE=7.219009724382379
p=17  RMSE=7.04573891427961
p=18  RMSE=7.928961466851558
p=19  RMSE=9.312774509996125
p=20  RMSE=10.46189505956539
p=21  RMSE=9.01822011343673
p=22  RMSE=9.259350169110826
p=23  RMSE=8.338668174317808
p=24  RMSE=6.526660895698311
```

Figure 11.29 – RMSE scores for Optimized persistence forecasting

ARIMA Models

We observe that when p=6, the RMSE score is the smallest:

Optimized Persistence Forecasting

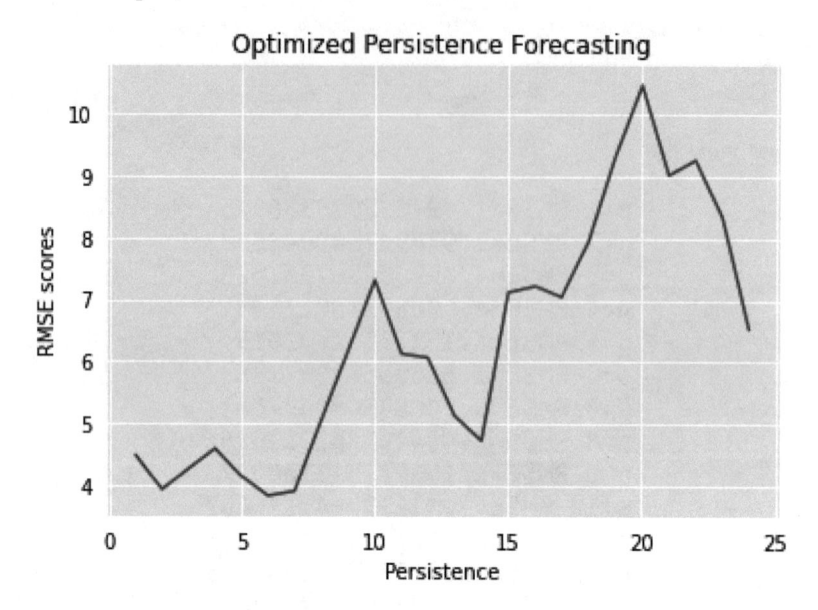

Figure 11.30 – Optimized Persistence Forecasting, test versus prediction

Running the persistence test again with p=6, we can see the following:

```
history = [x for x in train]
pred = []
for i in range(len(test)):
    # Prediction
    yhat = history[-6]
    pred.append(yhat)
    history.append(test[i])
# Plots
plt.plot(list(test))
plt.plot(pred)
plt.show()
```

Then, we can produce a visualization:

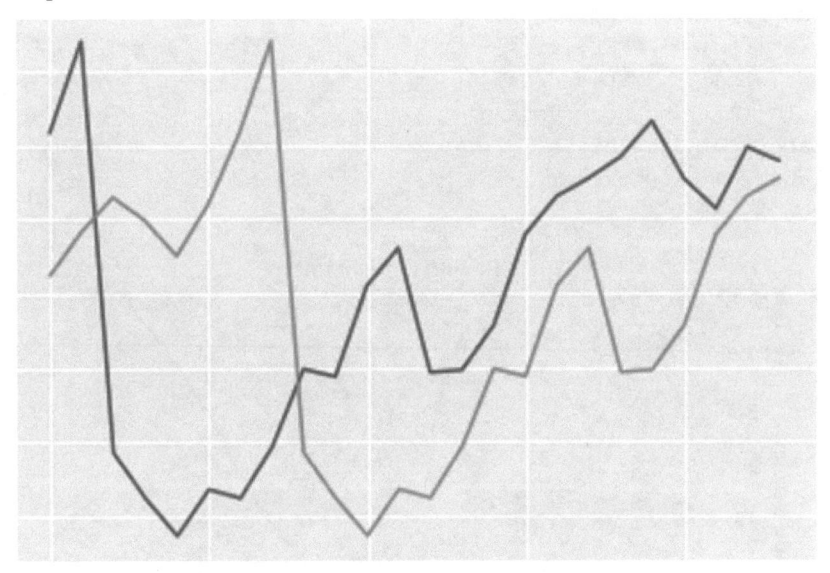

Figure 11.31 – Optimized Persistence Forecasting

The blue curve is the test value, and the orange curve is for the prediction.

Rolling window forecast

This technique creates a **rolling window** with a specified window size and then performs a statistic calculation in this window, using it for forecasting which rolls through the data used in a study. We conduct a similar study as in the last part on the Coco Cola stock price dataset from 2016 using a monthly resampling dataset:

```
from numpy import mean
train, test = data.resample('M').mean()['Open'][0:-24], data.
resample('M').mean()['Open'][-24:]
window = range(1, 25)
RMSE_scores = []
for w in window:
    history = [x for x in train]
    pred = []
    for i in range(len(test)):
    # Prediction on test set
        yhat = mean(history[-w:])
        pred.append(yhat)
    history.append(test[i])
    # RMSE score performance
```

```
        rmse = math.sqrt(mean_squared_error(test, pred))
        RMSE_scores.append(rmse)
        print(f'w={w} RMSE={rmse}')
plt.plot(window, RMSE_scores)
plt.title('Rolling Forecasting')
plt.xlabel('Windows sizes')
plt.ylabel('RMSE scores')
plt.show()
```

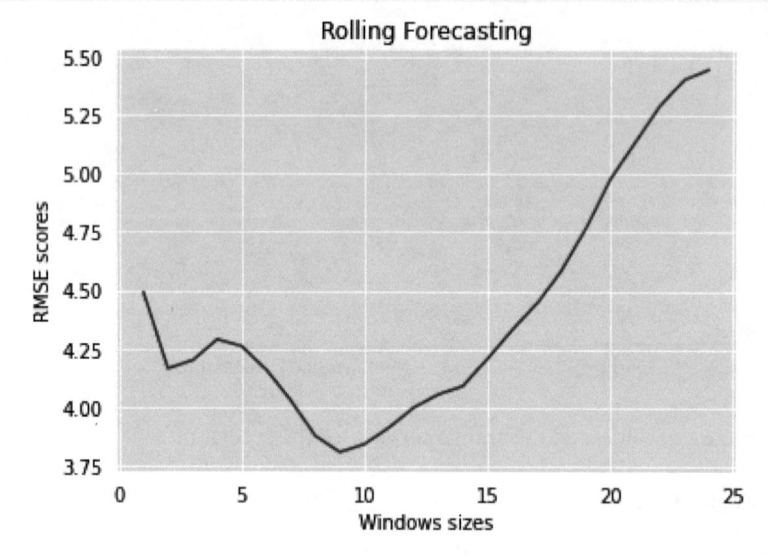

Figure 11.32 – Rolling window forecasting

With window size = 9, the RMSE of 3.808 is the smallest. Run the Python code again with window size = 9 we produce a similar visualization as with the Optimized Persistence Forecast.

Summary

In this chapter, we discussed various methods for modeling univariate time series data from stationary time series models such as ARMA to non-stationary models such as ARIMA. We started with stationary models and discussed how to identify modeling approaches based on the characteristics of time series. Then we built on the stationary models by adding a term in the model to stationarize time series. Finally, we talked about seasonality and how to account for seasonality in an ARIMA model. While these methods are powerful for forecasting, they do not incorporate potential information from other external variables. As in the previous chapter, we will see that external variables can help improve forecasts. In the next chapter, we will look at multivariate methods for time series data to take advantage of other explanatory variables.

References

Please refer to the final word file for how the references should look.

1. *APPLIED TIME SERIES ANALYSIS WITH R*, W. Woodward, H. Gray, A. Elliott. Taylor & Francis Group, LLC. 2017.
2. Box, G. E. P., Jenkins, G. M. and Reinsel, G. C. (1976) *Time Series Analysis, Forecasting and Control.* Third Edition. Holden-Day. Series G.
3. Brownlee, J, (2017) *Simple Time Series Forecasting Models to Test So That You Don't Fool Yourself* (`https://machinelearningmastery.com/simple-time-series-forecasting-models/`).

12
Multivariate Time Series

The models we discussed in the previous chapter only depended on the previous values of the single variable of interest. Those models are appropriate when we only have a single variable in our time series. However, it is common to have multiple variables in time-series data. Often, these other variables in the series can improve forecasting of the variable of interest. We will discuss models for time series with multiple variables in this chapter. We will first discuss the correlation relationship between time-series variables, then discuss how we can model multivariate time series. While there are many models for multivariate time-series data, we will discuss two models that are both powerful and widely used: **autoregressive integrated moving average with exogenous variables** (**ARIMAX**) and **vector autoregressive** (**VAR**). Understanding these two models will extend the reader's model toolbox and provide building blocks for the reader to learn more about multivariate time-series models.

In this chapter, we're going to cover the following main topics:

- Multivariate time series
- ARIMAX
- VAR modeling

Multivariate time series

In the previous chapter, we discussed models for **univariate time series** or a time series of one variable. However, in many modeling situations, it is common to have multiple time-varying variables that are measured together. A time series consisting of multiple time-varying variables is called a **multivariate time series**. Each variable in the time series is called a **covariate**. For example, a time series of weather data might include temperature, rain amount, wind speed, and relative humidity. Each of these variables, in the weather dataset, is a univariate time series, and together, a multivariate time series and each pair of variables are covariates.

Mathematically, we typically represent a multivariate time-series as a vector-valued series, as follows:

$$X = \begin{matrix} x_{0,0} & x_{1,0} & & x_{t,0} \\ x_{0,1}, & x_{1,1}, & \ldots, & x_{t,1} \\ \vdots & \vdots & & \vdots \end{matrix}$$

Here, each X instance consists of multiple values at each time step, and there are t steps in the series. The first index in the preceding equation represents the time step (0 through t), and the second index represents the individual series (0 through N variables). For example, if X were a time series of stock prices, at each time step of X, there would be a value of each stock price. In that case, X might look like the following equation if the first two stock symbols in the series were AAPL and GOOGL:

$$X = \begin{matrix} x_{0,\,AAPL} & x_{1,\,AAPL} & & x_{t,\,APPL} \\ x_{0,\,GOOGL}, & x_{1,\,GOOGL}, & \ldots, & x_{t,\,GOOGL} \\ \vdots & \vdots & & \vdots \end{matrix}$$

When we have a multivariate time series, a question arises: *What can we do with this additional information?* To answer that question, we need to understand whether these time-series variables are related. We can determine how time series are related by looking at their cross-correlation. Let's discuss cross-correlation next.

Time-series cross-correlation

We need to understand how time series can be related in order to determine whether covariate time series may be useful for forecasting a time series of interest. In a previous chapter, we discussed how to understand the relationship between two variables using plots and correlation coefficients; we will take a similar approach for the time-series data. Recall the following **cross-correlation function** (**CCF**) discussed in *Chapter 10, Introduction to Time Series*:

$$\hat{p}_k(X, Y) = \frac{\sum_{t=1}^{n-k}(X_t - \overline{X})(Y_{t-k} - \overline{Y})}{\sqrt{\sum_{t=1}^{n}(X_t - \overline{X})^2} \sqrt{\sum_{t=1}^{n}(Y_t - \overline{Y})^2}}$$

Here, X is a univariate time series and Y is a univariate time series. This function tells us how X and Y are related at k lags. In this equation, Y is k time steps behind X. In this situation, we say that X is leading Y by k steps or Y is lagging X by k steps. We will use this CCF to help us determine whether two time series are related and at which time steps they are related. Notice that if $X = Y$, then the CCF reduces to the **autocorrelation function** (**ACF**).

> **Cross-correlation significance**
>
> Just as with the ACF, cross-correlations have a threshold for statistical significance. We consider any cross-correlation value with an absolute value greater than $1.96/\sqrt{N}$ (where N is the sample size of the time series) to be significantly different from zero.

Let's take a look at cross-correlations with some data. This data [1] comes from the *UCI Machine Learning Repository* [2]. The data is weather data that was collected as part of a pollution study conducted in China. This data was sampled hourly over several years. We will subset the data to the first 1,000 data points. The weather data contained in this set consists of temperature (TEMP), pressure (PRES), dew point temperature (DEWP), rain precipitation (RAIN), and wind speed (WSPM). We will look at cross-correlations from the perspective of trying to forecast wind speed, meaning we will look at the cross-correlations of wind speed and the other variables. This may be an important problem for someone who wants to forecast power generation from windmills. The wind speed data and the ACF of the wind speed data are plotted in *Figure 12.1*:

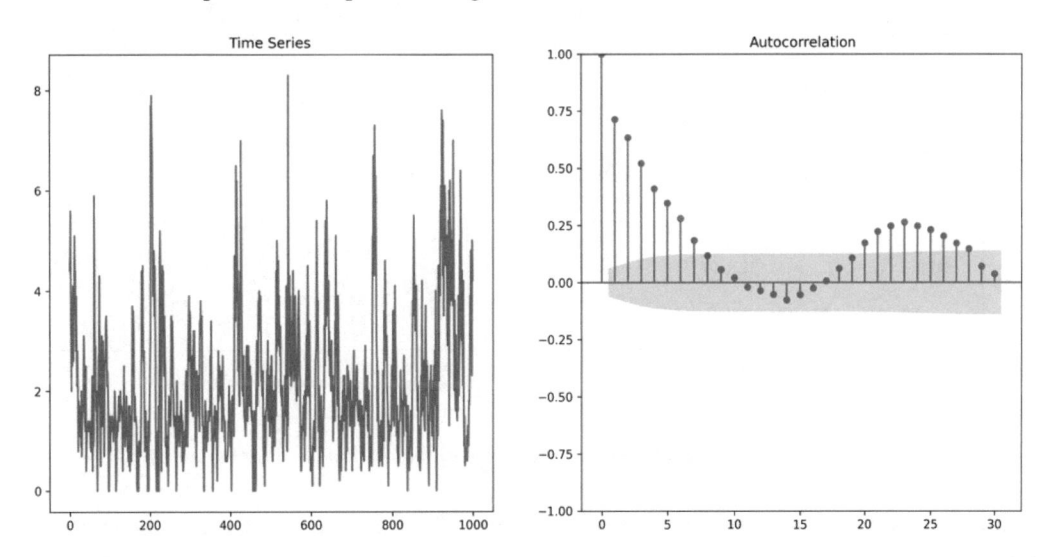

Figure 12.1 – Plot of wind speed time series and its ACF

We can see from the plot in *Figure 12.1* that the wind speed could potentially be modeled by a stationary univariate model, which we discussed in the previous chapter. While there does not appear to be evidence of **autoregressive integrated moving average** (**ARIMA**) behavior, there does appear to be evidence of seasonality at lag 24. Since the data is sampled hourly, the seasonality is likely a daily pattern. We have plotted the time series of the other variables in *Figure 12.2*. These are the other variables we may want to use to help predict the value of wind speed:

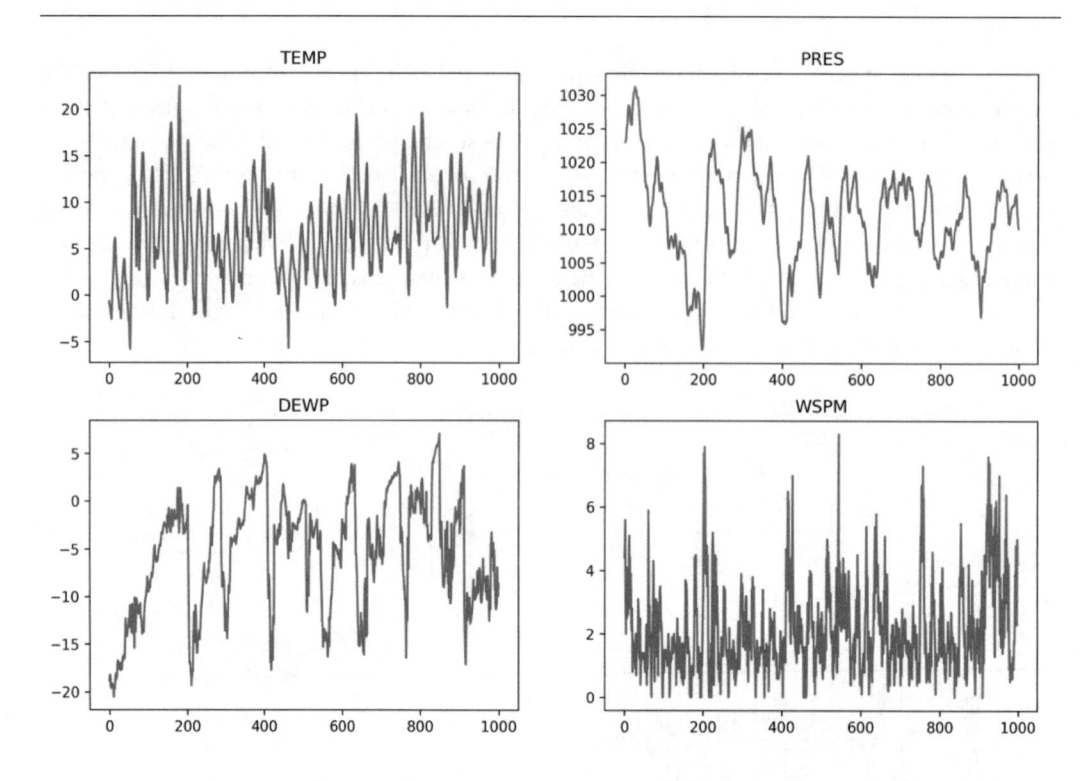

Figure 12.2 – Plot of wind speed time series and three other weather variables

While we have not plotted the ACF for all the time series shown in *Figure 12.2*, these time series have characteristics that suggest they may be stationary. However, it is not clear from the time series alone whether these time series are related. This is where the cross-correlation of time series comes in. We can use a cross-correlation plot to determine whether any of the other time series are related to wind speed. We have plotted the cross-correlation of wind speed and the other variables in *Figure 12.3*. The CCFs are plotted up to 48 lags. Since the data was sampled hourly, this gives us 2 days' worth of lags to see the cross-correlations:

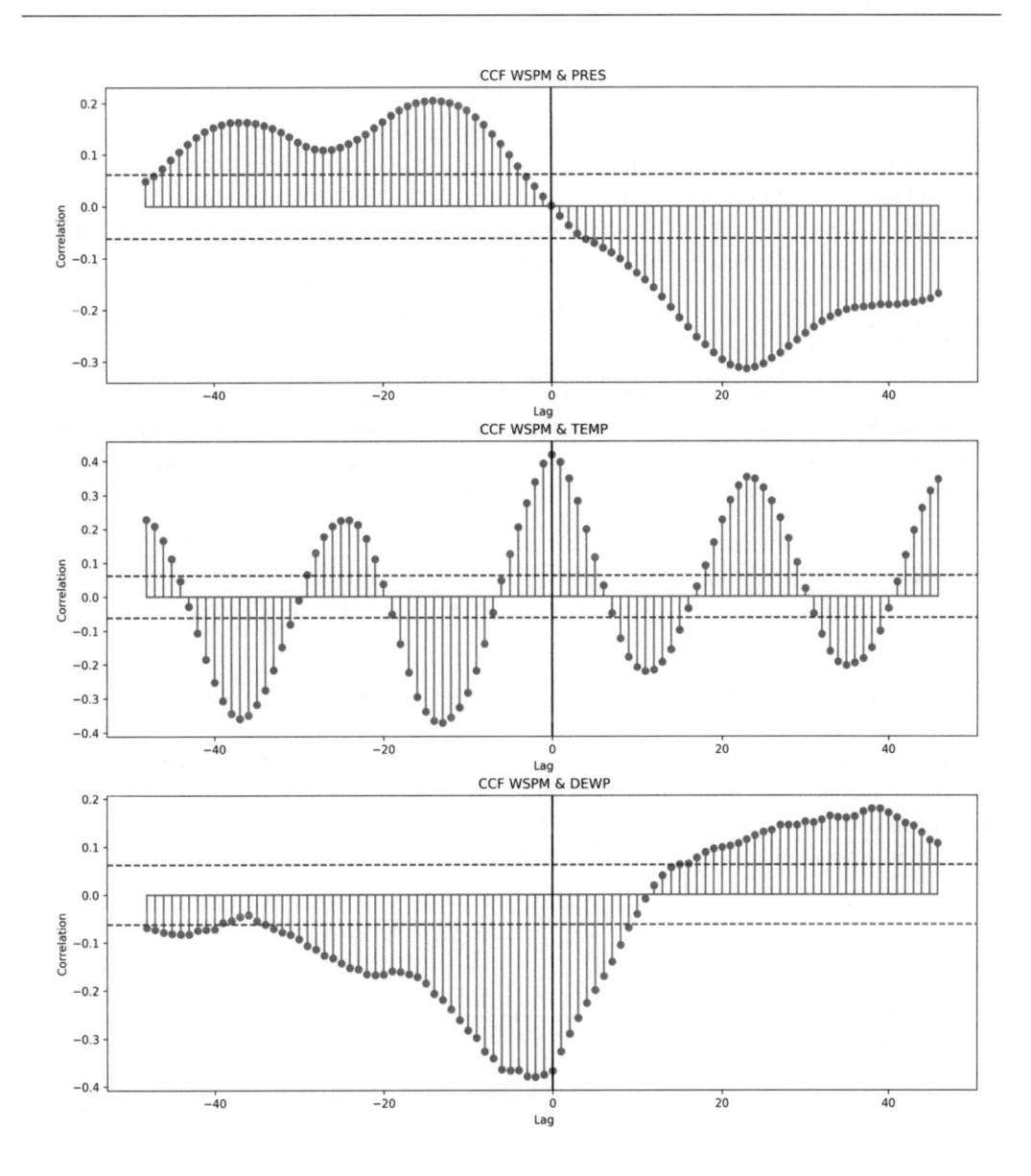

Figure 12.3 – CCF plots of wind speed and three other variables

The plots in *Figure 12.3* show that there are cross-correlations between wind speed and the other variables. While we are not too interested in the shapes of the cross-correlations, the patterns may provide some depth to how the variables are related. For example, the cross-correlation of wind speed and temperature shows an oscillating pattern with a period of about 12 hours; this may correspond to the effect of sunlight on wind speed. The primary finding is that these other variables share a relationship with wind speed and may help us forecast the values of wind speed. Let's take a look at how we can use the additional variables for modeling.

ARIMAX

In the previous chapter, we discussed the ARIMA family of models and demonstrated how to model univariate time-series data. However, as we mentioned in the previous section, many time series are multivariate, such as stock data, weather data, or economic data. In this section, we will discuss how we can incorporate information from covariate variables when modeling time-series data.

When we model a multivariate time series, we typically have a variable we are interested in forecasting. This variable is commonly called the **endogenous** variable. The other covariates in the multivariate time series are called **exogenous** variables. Recall from *Chapter 11, ARIMA Models*, the equation representing the ARIMA model:

$$y'_t = c + \phi_1 y'_{t-1} + \ldots + \phi_p y'_{t-p} + \epsilon_t + \theta_1 \epsilon_{t-1} + \ldots + \phi_q \epsilon_{t-q}$$

Here, y'_t is the differenced series, differenced d times until it is stationary. In this model, y_t is the variable we are interested in forecasting; it is the endogenous variable. Let us rewrite this equation in a more compact form. We collect the terms related to y'_t into one summation and the terms related to ϵ_t into another summation. Let us start with the y'_t terms:

$$\phi_1 y'_{t-1} + \ldots + \phi_p y'_{t-p} = \sum_{i=1}^{p} \phi_i y'_{t-i}$$

Then, we will also collect the ϵ_t terms into different summations:

$$\epsilon_t + \theta_1 \epsilon_{t-1} + \ldots + \phi_q \epsilon_{t-q} = \epsilon_t + \sum_{j=1}^{q} \phi_j \epsilon_{t-j}$$

Putting these summations together, we get the following more compact version of our ARIMA equation:

$$y'_t = c + \sum_{i=1}^{p} \phi_i y'_{t-i} + \epsilon_t + \sum_{j=1}^{q} \phi_j \epsilon_{t-j}$$

Now, let's add the exogenous variables to the model. We will add the exogenous variables by adding a summation term to the model:

$$y'_t = c + \sum_{i=1}^{p} \phi_i y'_{t-i} + \epsilon_t + \sum_{j=1}^{q} \phi_j \epsilon_{t-j} + \sum_{k=1}^{r} \beta_k X_{tk}$$

Here, there are r exogenous variables contained in the vector-valued variable X. Recall from earlier in the chapter, we can represent the vector-valued variable, as shown next, where the first index represents the time step, and the second index represents an individual series within the vector-valued variable:

$$X = \begin{matrix} x_{0,0} \\ \vdots \\ x_{0,k} \end{matrix}, \begin{matrix} x_{1,0} \\ \vdots \\ x_{1,k} \end{matrix}, \ldots, \begin{matrix} x_{t,0} \\ \vdots \\ x_{t,k} \end{matrix}$$

This is how we will incorporate the exogenous variables into the ARIMA model. Now, let's talk about how we would pick the exogenous variables. In *Chapter 11, ARIMA Models*, using an ACF plot, we saw that certain lags of a time series are more influential than others for modeling a univariate time series. We will make similar judgments about exogenous variables using CCF plots. Let's take a look at example CCF plots and discuss how to make these judgments.

The CCF plot for wind speed and temperature is shown in *Figure 12.4*. We have marked where the most significant cross-correlations occur within a lag period of 48 hours. Recall this data is sampled hourly; thus, each lagged time step corresponds to a lag of 1 hour. The plot shows that the most significant cross-correlations occur at approximately lag 13, lag 24, and lag 37:

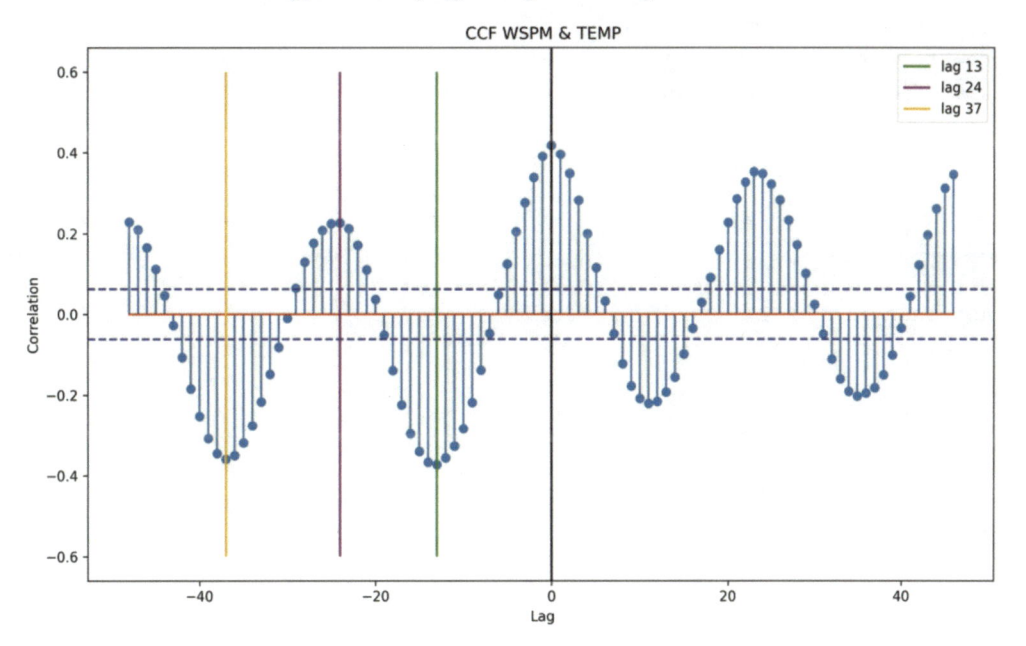

Figure 12.4 – CCF plot of wind speed and temperature variables

Based on the observation of the cross-correlation plot in *Figure 12.4*, we would expect that temperature lags at 13, 24, and 37 could provide useful information for the current predicting wind speed. That is, the variables $x_{t-13,temperature}$, $x_{t-24,temperature}$, and $x_{t-37,temperature}$ may be useful for predicting y_t. We can perform the same CCF analysis with the other two exogenous variables to determine which lags to use in the model. We found that pressure (PRES) had significant cross-correlations at lags 14 and 37, and dew point (DEWP) had significant cross-correlations at lags 2 and 20. Now that we have assessed which lags of the exogenous variables we want to use in the model, let us go ahead and preprocess the data and fit the model.

Preprocessing the exogenous variables

Before we fit the model, we need to preprocess the data to make the exogenous variables we mentioned previously. We need to create lagged versions of the time-series variables we described previously. We can lag a time series using the shift() method of a pandas DataFrame. Let's take a look at a code example.

The following code sample shows how to use the shift() method to lag the pressure variable by 14. The first several lines of code import the library and load the data. The last line in the code sample shifts the PRES variable backward in time by 14 time steps:

```
import pandas as pd
# code to download the data
url = 'https://archive.ics.uci.edu/ml/machine-learning-databases/00501/PRSA2017_Data_20130301-20170228.zip'
file_name = url.split('/')[-1]
r = requests.get(url)
with open(file_name, 'wb') as fout:
    fout.write(r.content)

with zipfile.ZipFile(file_name, 'r') as zip_file:
    zip_file.extractall('./')
df = pd.read_csv(
    './PRSA_Data_20130301-20170228/PRSA_Data_Aotizhongxin_20130301-20170228.csv'
    , sep=','
)
pres_lag_14 = df.PRES.shift(-14)
```

The `shift()` method can be used to shift a series forward or backward in time. In this case, we want our exogenous variables to lag the variable we want to predict. So, we need to provide a negative shift value to move the series backward in time. We will use a similar preprocessing step for the other exogenous variables to create the other lagged variables. This is shown in the following code sample. All of these lagged variables created with the `shift()` method are collected into a new `DataFrame` called X, which we will use in the fit step:

```
temp_lag_13 = df.TEMP.shift(-13)
temp_lag_24 = df.TEMP.shift(-24)
temp_lag_37 = df.TEMP.shift(-37)

pres_lag_14 = df.PRES.shift(-14)
pres_lag_24 = df.PRES.shift(-37)

dewp_lag_13 = df.DEWP.shift(-2)
dewp_lag_24 = df.DEWP.shift(-20)


X = pd.DataFrame({
    "temp_lag_13": temp_lag_13[:1000],
    "temp_lag_24": temp_lag_24[:1000],
    "temp_lag_37": temp_lag_37[:1000],
    "pres_lag_14": pres_lag_14[:1000],
    "pres_lag_24": pres_lag_24[:1000],
    "dewp_lag_2": dewp_lag_13[:1000],
    "dewp_lag_20": dewp_lag_24[:1000],
})
```

In the preceding code sample, we are taking a subset of the lagged variables (the first 1,000 data points) to limit the computation time of the model. With the variables collected in the new X `DataFrame`, the preprocessing is complete, and we can move on to fitting and assessing the model.

Fitting the model

As in the previous chapter, we will use `auto_arima` to fit our ARIMA model. The main difference, in this case, is that we will need to provide the `auto_arima` function with the exogenous variables we created. The code to fit an ARIMA model with exogenous variables is shown in the following code sample:

```
y = df.WSPM[:1000]
arima = pm.auto_arima(
    y, X,
    error_action='ignore',
    trace=True,
```

```
            suppress_warnings=True,
            maxiter=5,
            seasonal=False,
            test='adf'
)
```

This code should look very similar to the code shown in *Chapter 11, ARIMA Models*. In fact, the only difference is the addition is the X variable in the `auto_arima` function. Running this code will produce the following output:

```
Performing stepwise search to minimize aic
 ARIMA(2,0,2)(0,0,0)[0]             : AIC=2689.456, Time=0.24 sec
 ARIMA(0,0,0)(0,0,0)[0]             : AIC=8635.103, Time=0.16 sec
 ARIMA(1,0,0)(0,0,0)[0]             : AIC=2742.728, Time=0.29 sec
 ARIMA(0,0,1)(0,0,0)[0]             : AIC=2923.324, Time=0.32 sec
 ARIMA(1,0,2)(0,0,0)[0]             : AIC=2685.581, Time=0.23 sec
 ARIMA(0,0,2)(0,0,0)[0]             : AIC=2810.239, Time=0.37 sec
 ARIMA(1,0,1)(0,0,0)[0]             : AIC=2683.888, Time=0.23 sec
 ARIMA(2,0,1)(0,0,0)[0]             : AIC=2682.988, Time=0.28 sec
 ARIMA(2,0,0)(0,0,0)[0]             : AIC=2681.309, Time=0.26 sec
 ARIMA(3,0,0)(0,0,0)[0]             : AIC=2682.416, Time=0.32 sec
 ARIMA(3,0,1)(0,0,0)[0]             : AIC=2684.608, Time=0.32 sec
 ARIMA(2,0,0)(0,0,0)[0] intercept   : AIC=2683.328, Time=0.30 sec

Best model:  ARIMA(2,0,0)(0,0,0)[0]
Total fit time: 3.347 seconds
```

Again, this output should look very similar to the output we saw from `auto_arima` in *Chapter 11, ARIMA Models*. It indicates we should use an AR(2) model based on the fit using the **Akaike Information Criteria** (**AIC**). Interestingly, there doesn't appear to be any information regarding the exogenous variables. This is because `auto_arima` only needs to fit a single coefficient for each exogenous variable in the model. It does not need to fit an **autoregressive moving average** (**ARMA**) model for each exogenous variable. We can use the `summary()` method on the object returned from `auto_arima` to see the coefficients and significance tests for the exogenous variables. Running the `summary()` method will output the following **seasonal autoregressive integrated moving average with eXogenous factors** (**SARIMAX**) results:

```
                                    SARIMAX Results
==============================================================================
Dep. Variable:                      y   No. Observations:                 1000
Model:                 SARIMAX(2,0,0)   Log Likelihood               -1330.654
Date:                Wed, 01 Feb 2023   AIC                           2681.309
Time:                        19:17:32   BIC                           2730.386
Sample:                             0   HQIC                          2699.962
                                -1000
Covariance Type:                  opg
==============================================================================
                 coef    std err      z      P>|z|      [0.025      0.975]
------------------------------------------------------------------------------
temp_lag_13    -0.0279    0.021    -1.355    0.175     -0.068       0.012
temp_lag_24     0.0560    0.014     3.887    0.000      0.028       0.084
temp_lag_37    -0.0587    0.020    -2.961    0.003     -0.097      -0.020
pres_lag_14     0.0009    0.012     0.077    0.938     -0.022       0.024
pres_lag_24     0.0010    0.012     0.086    0.931     -0.022       0.024
dewp_lag_2     -0.0636    0.013    -4.802    0.000     -0.090      -0.038
dewp_lag_20    -0.0135    0.015    -0.919    0.358     -0.042       0.015
ar.L1           0.4480    0.027    16.647    0.000      0.395       0.501
ar.L2           0.2400    0.024    10.050    0.000      0.193       0.287
sigma2          0.8400    0.028    29.793    0.000      0.785       0.895
==============================================================================
Ljung-Box (L1) (Q):                   0.01   Jarque-Bera (JB):         486.47
Prob(Q):                              0.91   Prob(JB):                   0.00
Heteroskedasticity (H):               1.16   Skew:                       0.80
Prob(H) (two-sided):                  0.18   Kurtosis:                   6.02
==============================================================================
```

Figure 12.5 – SARIMAX summary for AR(2) model

The output from the `summary()` method shown in the preceding snippet shows a lot of information. Let's focus on the middle section of the output. The middle section shows the coefficients and significance tests for each exogenous variable and each ARIMA coefficient. The coefficients and test p-values are labeled `coef` and `P>|z|`, respectively. The significance test can help us determine whether variables should be included in the model.

Recall back in *Chapter 7, Multiple Linear Regression*, we discussed the idea of multicollinearity, which essentially means two or more predictor variables are highly correlated. We could encounter a similar situation here. For example, consider the `temp_lag_13`, `temp_lag_24`, and `temp_lag_37` variables. Since the temperature time series exhibits autocorrelation, it is possible, even likely, that these lagged versions of the temperature time series could be highly correlated. When we observe multicollinearity, it is a sign that we can remove variables from the model. We would recommend using one of the feature selection methods described in *Chapter 7*, such as recursive feature selection, performance-based selection, or selection by statistical significance.

The preceding output shows only three of the exogenous variables in the current model, which are highlighted. This means we should consider reducing the variable in this model. We can do this with multiple methods; for this example, we will choose to eliminate features based on p-values. We will iteratively remove the feature with the highest p-value until all features have a p-value below 0.05. The code for this selection process is shown here:

```
# initial model
arima = pm.auto_arima(
    y, X,
    error_action='ignore',
    trace=False,
    suppress_warnings=True,
    maxiter=5,
    seasonal=False,
    test='adf'
)

pvalues = arima.pvalues()
iterations = 0

while (pvalues > 0.05).any():
    # get the variable name with the largest p-value
    variable_with_max_pval = pvalues.idxmax()
    # drop that variable from the exogenous variables
    X = X.drop(variable_with_max_pval, axis=1)
    arima = pm.auto_arima(
        y, X,
        error_action='ignore',
        trace=False,
        suppress_warnings=True,
        maxiter=5,
        seasonal=False,
        test='adf'
    )
    pvalues = arima.pvalues()
    print(f"fit iteration {iterations}")
    iterations += 1
```

Performing this selection process will result in a model only containing the `temp_lag_24`, `temp_lag_37`, `pres_lag_24`, and `dewp_lag_2` exogenous variables. The ARIMA portion of the model is still an AR(2) model. With a model selection taken, let's assess the model's performance.

Assessing model performance

We will assess the performance of the model by plotting the forecast versus the actual data points and calculating an error estimate. Let us start by looking at the forecasts from the ARIMAX model for the next 200 observations. The forecasts and test data points are plotted in *Figure 12.6*:

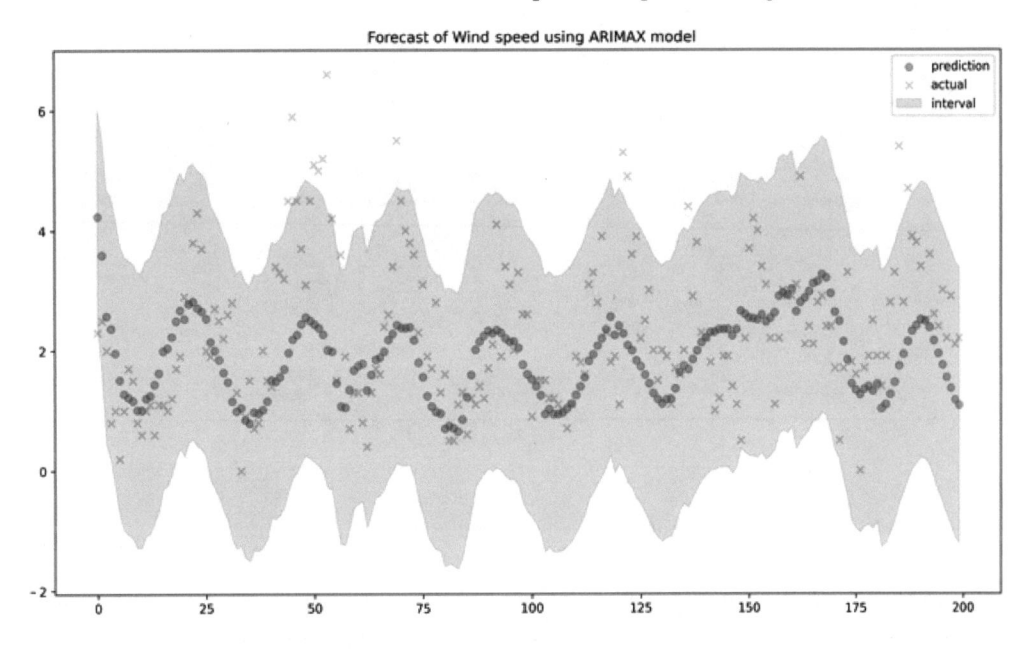

Figure 12.6 – Forecast of wind speed using an ARIMAX model

We can see the forecast for wind speed using the ARIMAX model we created in *Figure 12.6*. Two observations stand out from the plotted forecast:

- The forecast captures the oscillation of the `windspeed` variable
- The forecast fails to capture the high spikes in `windspeed`

To understand whether the exogenous variables provided material value in the model, we should compare the forecasts of the ARIMAX model, shown in *Figure 12.6*, to forecasts from a univariate model. The forecasts from a univariate model are shown i*n Figure 12.7*:

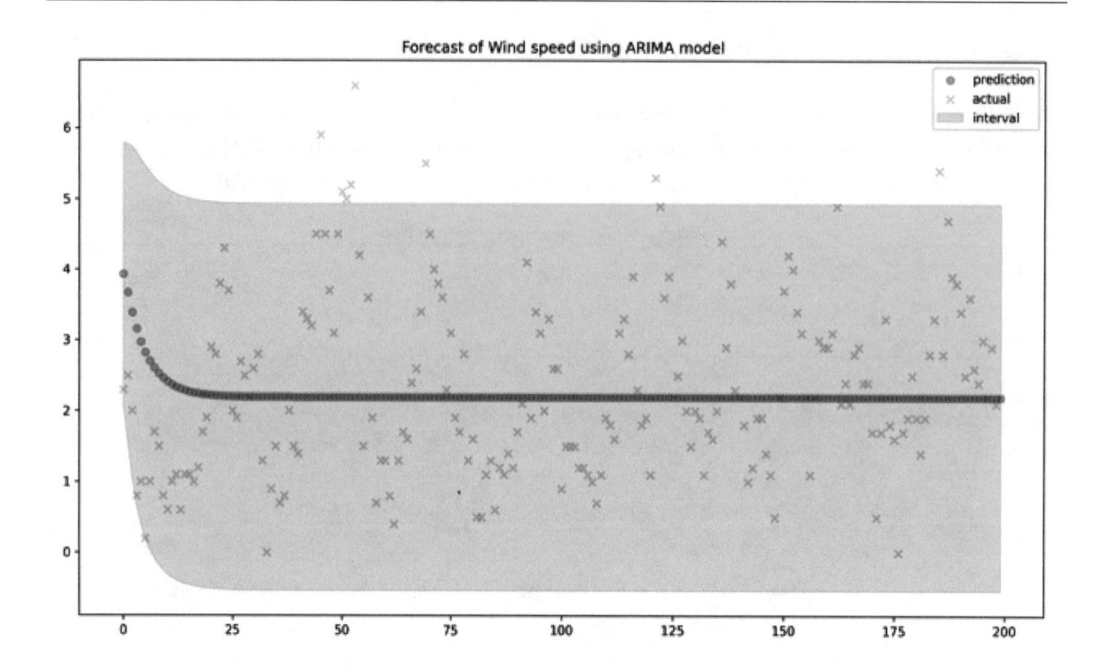

Figure 12.7 – Forecast of wind speed using an ARIMA model

The forecasts of the ARIMA model, shown in *Figure 12.7*, do not appear to capture much variation in the time series as the forecast length grows. This is actually a property of ARMA models. It should be evident from the two plots that the ARIMAX model provides a better forecast than the ARIMA model. Finally, let us compare the **mean squared error** (**MSE**) of the forecasts of the two models. The MSE from the ARIMAX model is approximately 1.5, while the MSE from the ARIMA model is about 1.7. The MSE value confirms what we saw in the plots—namely, the ARIMAX model provides a better forecast for this multivariate time series.

In this section, we discussed an extension to the ARIMA model that allows us to incorporate information from covariates in a multivariate time series. This is only one of many methods for multivariate time series. In the next section, we will discuss another method for working with multivariate time-series data.

VAR modeling

The AR(p), MA(q), ARMA(p,q), ARIMA(p,d,q)m, and SARIMA(p,d,q) models we looked at in the last chapter form the basis of multivariate VAR modeling. In this chapter, we have discussed ARIMA with exogenous variables (ARIMAX). We will now begin discussion on the VAR model. First, it is important to understand that while **ARIMAX requires leading (future) values of the exogenous variables, no future values of these variables are required for the VAR model** as they are all autoregressive to each other – hence the name vector autoregressive – and by definition not exogenous. To start, let us consider the two-variable, or bivariate, case. Consider a process y_t that is the output of two different input variables, y_{t1} and y_{t2}. Note that in matrix form, we are discussing the case of an *nxm* matrix ($y_{n,m}$) where *n* corresponds to the point in **time** and *m* corresponds to the **variables** involved (variables 1,2,...,m). We exclude the comma from notation going forward, but it is important to understand when we discuss multivariate autoregressive processes that we are discussing data within a multidimensional matrix rather than only the single-dimensional vector form we used for univariate time-series analysis. We can write our process, y_t, in vector notation as follows:

$$y_t = \begin{bmatrix} y_{t1} \\ y_{t2} \end{bmatrix}.$$

We can expand on the input variables' definitions as a VAR(1) process, here, for the two variables in y_t shown previously:

$$y_{t1} = (1 - \phi_{11})\mu_1 - \phi_{12}\mu_2 + \phi_{11}y_{t-11} + \phi_{12}y_{t-12} + \varepsilon_{t1}$$

$$y_{t2} = -\phi_{21}\mu_1 + (1 - \phi_{22})\mu_2 + \phi_{21}y_{t-11} + \phi_{22}y_{t-12} + \varepsilon_{t2}.$$

The terms $(1 - \phi_{11})\mu_1 - \phi_{12}\mu_2$ and $-\phi_{21}\mu_1 - (1 - \phi_{22})\mu_2$ are our **model constants** (β coefficients).

We can reduce the previous processes (using a zero-mean form) into the matrix-reduced form, like so:

$$y_t = \Phi_1 y_{t-1} + \varepsilon_t$$

Here, we have the following:

$$\Phi_1 = \begin{bmatrix} \phi_{11} & \phi_{12} \\ \phi_{21} & \phi_{22} \end{bmatrix}$$

Our forecast for a VAR(1) model is shown here:

$$\hat{y}_{t_0 1}(l) = (1 - \phi_{11})\bar{y}_1 - \phi_{12}\bar{y}_2 + \phi_{11}\hat{y}_{t_0 1}(l-1) + \phi_{12}\hat{y}_{t_0 2}(l-1)$$

$$\hat{y}_{t_0 2}(l) = -\phi_{21}\bar{y}_1 + (1 - \phi_{22})\bar{y}_2 + \phi_{21}\hat{y}_{t_0 1}(l-1) + \phi_{22}\hat{y}_{t_0 2}(l-1)$$

It also includes this:

$$\hat{y}_{t_0}(l) = \begin{bmatrix} \hat{y}_{t_0 1}(l) \\ \hat{y}_{t_0 2}(l) \end{bmatrix}$$

Here, l represents the forward lag $(t+1, t+2, \ldots, t+n)$, $\hat{y}_{t_0,m}$ represents the forecasted value at time t for variable m, and t_0 corresponds to the most recent point in time (often, the length of the data). It is also important to note the mxm covariance matrix is this:

$$\hat{\Gamma}(k) = \begin{bmatrix} \hat{y}_{11}(k) & \cdots & \hat{y}_{1m}(k) \\ \vdots & \ddots & \vdots \\ \hat{y}_{m1}(k) & \cdots & \hat{y}_{mm}(k) \end{bmatrix}$$

Here, each covariance, $\hat{y}_{ij}(k)$, with k corresponding to the lag, is calculated as follows:

$$\hat{y}_{ij}(k) = \frac{1}{n} \sum_{t=1}^{n-k} (y_{ti} - \bar{y}_i)(y_{t+k,j} - \bar{y}_j).$$

Our estimated cross-correlation between each variable y_{ti} and y_{tj} is therefore calculated as follows:

$$\hat{\rho}_{ij}(k) = \frac{\hat{y}_{ij}(k)}{\sqrt{\hat{y}_{ii}(0)\, \hat{y}_{jj}(0)}}$$

Understanding the calculations for cross-correlation and, especially, covariance helps with understanding the relationships between the variables and their impact on model fit. The method for combining the two models to generate one overall process forecast, $\hat{y}_t$, is through weighting the two input means and their correlations into one response. As we can see in our VAR(1) forecast, shown previously, the correlations between each given variable and the other variables across different points in time are able to be modeled using a VAR model. Notably, there are no truly independent and dependent variables in a VAR model; there is simply an interaction between each variable. It is important to note that all processes modeled with a VAR model should be **stationary**.

Using the *United States Macroeconomic* dataset found at https://www.statsmodels.org/dev/datasets/generated/macrodata.html, we can observe the relationship between real gross private domestic investment (realinv), real personal consumption expenditures (realcons), and real private disposable income (realdpi) with a zero-mean form in the context of VAR modeling, as follows:

$$realcons(t) = \phi_{11}\, realcons_{t-11} + \phi_{12}\, realinv_{t-12} + \phi_{13}\, realdpi_{t-13} + \varepsilon_{t1}$$

$$realinv(t) = \phi_{21}\, realcons_{t-11} + \phi_{22}\, realinv_{t-12} + \phi_{23}\, realdpi_{t-13} + \varepsilon_{t2}$$

$$realdpi(t) = \phi_{31}\, realcons_{t-11} + \phi_{32}\, realinv_{t-12} + \phi_{33}\, realdpi_{t-13} + \varepsilon_{t3}$$

Let us look at an example using this data in the VAR process in Python with the `statsmodels` VARMAX model. First, let's load the data:

```
import numpy as np
import pandas as pd
import statsmodels.api as sm
import matplotlib.pyplot as plt
from statsmodels.tsa.api import VAR
data=sm.datasets.macrodata.load_pandas().data
data.sort_values(by=['year', 'quarter'], inplace=True)
data['yr_qtr'] = data['year'].astype(str) + data['quarter'].astype(str)
print('{}% of yearly quarters are unique'.format(round(100*(data['yr_qtr'].nunique() / len(data)), 1)))
```

We can see here from the output that 100% of our yearly quarters are unique and thus, no timestamp duplication exists:

```
100.0% of yearly quarters are unique
```

One of the first things we always need to do when performing regression-based time-series modeling is to ensure stationarity. In VAR modeling, this means all variables must be stationary. The first step is to visualize the process realizations through line plotting by index. If data points are aggregated at an interval desired to be modeled, then we can leave the index as is, but if a different time indexing is desired, we need to specify this. Here, we are going to build a model to determine how **real consumption and real disposable income impact real investment**. In the following code sample, we change `quarter` to the length of the data since that is the smallest denomination of time and there is one yearly quarter per index, as we checked previously. We then set the index to `quarter` (this is why `quarter` appears twice; once as the index and once as the column name):

```
data['quarter'] = range(1, len(data)+1)
data.drop('yr_qtr', axis=1, inplace=True)
data.index = data['quarter']
```

Note the updated index and `quarter` values:

quarter	year	quarter	realcons	realinv	realdpi
1	1959	1	1707.4	286.898	1886.9
2	1959	2	1733.7	310.859	1919.7
3	1959	3	1751.8	289.226	1916.4
4	1959	4	1753.7	299.356	1931.3
5	1960	5	1770.5	331.722	1955.5

Figure 12.8 – First five rows of macrodata data set for VAR modeling

Here, we will walk through six steps to produce a VAR model using the dataset we imported and prepared, shown partially in *Figure 12.8*.

Step 1 – visual inspection

In the line plot of the data in *Figure 12.9*, we can see what appears to be a strong linear trend. We need to check the ACF plots to assess this. After, we will run a **Dickey-Fuller test** to confirm:

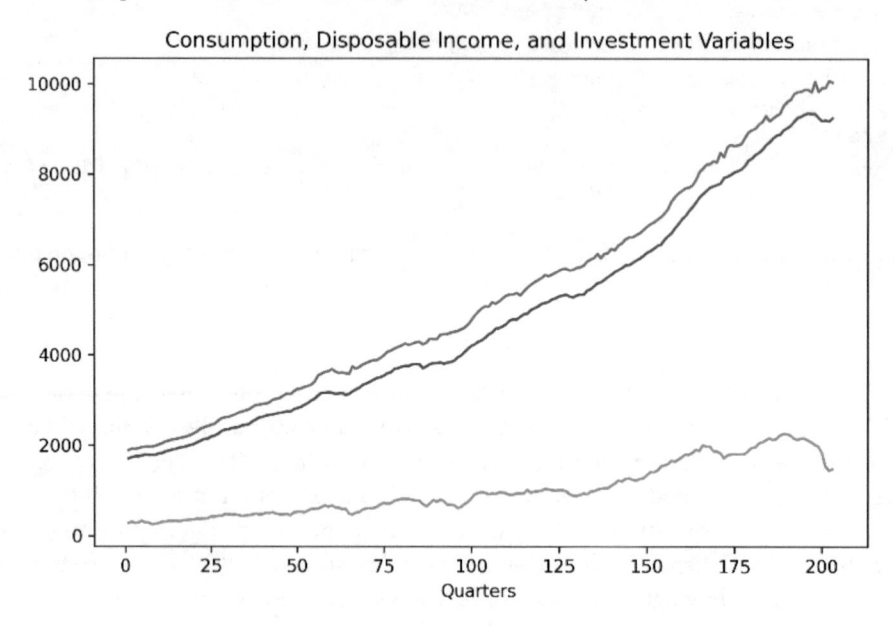

Figure 12.9 – Line plots of *realcons*, *realdpi*, and *realinv*

The ACF plots in *Figure 12.10* show a strong indication of a linear trend and at least one unit root, which likely gives us that trend:

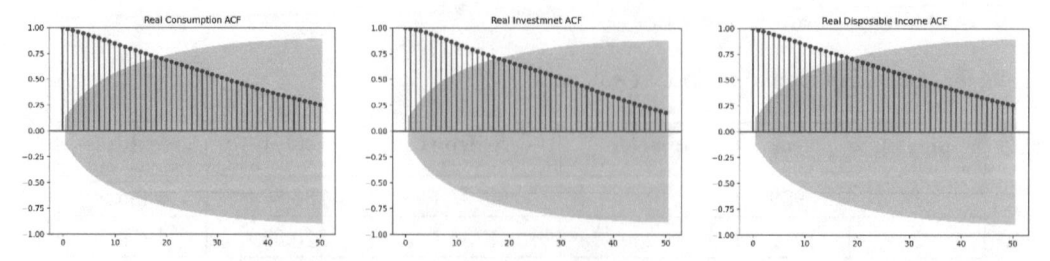

Figure 12.10 – ACF plots for VARMAX input variables

After running a Dickey-Fuller test, we can see that each variable contains a unit root and is thus trended. Therefore, we need to apply a first-order difference to all three variables so that we can work toward stationarizing the data. Recall that the **null hypothesis for the Dickey-Fuller is that a unit root exists**, and the alternative hypothesis is that one does not. We'll execute the following code:

```
from statsmodels.tsa.stattools import adfuller
for col in ['realcons','realinv','realdpi']:
    adfuller_test = adfuller(data[col], autolag='AIC')
    print('ADF p-value for {}: {}'.format(col, adfuller_test[1]))
```

Here we can see the Dickey-Fuller results for each variable. Recall the null hypothesis is the presence of a single unit root. Therefore, since the p-values are not low, we will not reject the null hypothesis. This gives us statistical evidence to assume the presence of trend:

```
ADF p-value for realcons: 0.9976992503412904

ADF p-value for realinv: 0.6484956579101141

ADF p-value for realdpi: 1.0
```

Let's difference the data here with a loop:

```
import numpy as np
data_1d = pd.DataFrame()
for col in ['realcons','realinv','realdpi']:
    data_1d[col] = np.diff(data[col], n=1)
```

Step 2 – selecting the order of AR(p)

Now, we can rerun the ACF plots to check whether we still see trending autocorrelation. Because the original line plots did not appear to have seasonality, we will go ahead and plot the **partial ACFs (PACFs)** to get an idea of what kind of AR(p) ordering might be useful. Had we suspected seasonality, it might have been useful to check the **spectral density of the sample autocorrelations to quantify seasonal periodicity**, a process we demonstrated in *Chapter 11, ARIMA Models*. When checking the PACF plots, we will use the **Yule-Walker method**.

We can see in the ACF plots there is no apparent seasonality. When using a VAR model, all terms are used to estimate all other terms as the estimates for a forecast of any term are developed. In other words, to forecast one, we must forecast all. The reason this matters is that when looking at the PACFs in *Figure 12.11*, we can see one variable could arguably be fit with an AR(1) (real investment) model, one with an AR(2) (real disposable income) model, and another with an AR(3) (real consumption) model. Using orders that are significant for each variable is one important aspect of making VAR modeling work.

Another important aspect of selecting ordering in a VAR model is cross-correlation. For example, we may have a model of variables whose highest order is order 3, but if the cross-correlation between the target variable and the input variable is at lag 5, we would most likely need to use lag 5. However, a model using lag 3 or 4 may also be useful. As with any regression model, however, we will always want to check the significance of coefficient terms following model training:

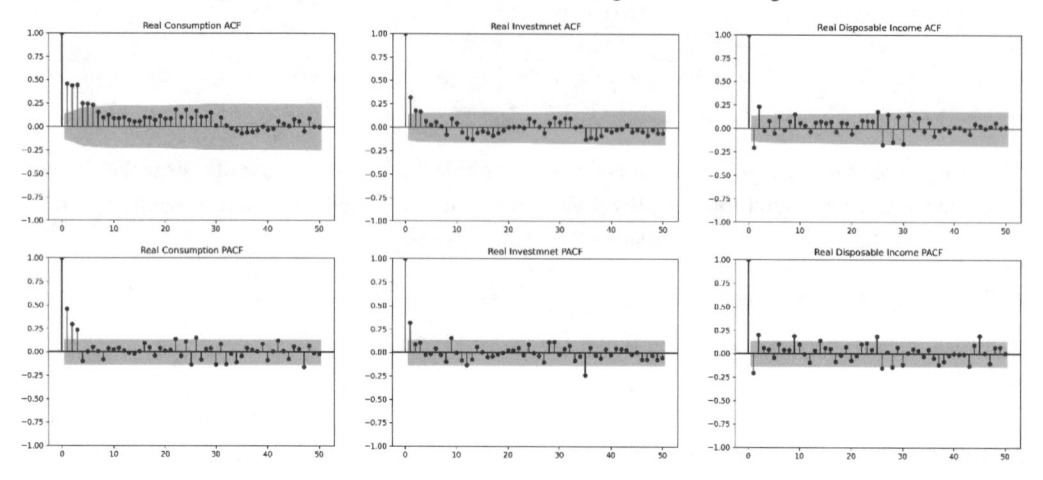

Figure 12.11 – Differenced VARMAX variables; ACFs and PACFs

Step 3 – assessing cross-correlation

For a VAR model, cross-correlation is most important between the variables we consider as inputs and the variable(s) we consider our target. This can be done using multiple lags or with the most significant lag if multiple iterations of the same variable using different shifts are a concern for overfitting. We must confirm cross-correlation analysis used in a forecasting model is done with respect to each input variable and the dependent variable. A cross-correlation that shows the input variable as a leading indicator for the target variable is not of major concern; it means historical values of the input are used to predict the target. However, if we have a scenario where the input is a lagging indicator—meaning the target predicts the input—and we want to use the input to predict the target, we will need to shift the input forward so that the correlation is at lag 0 and trim the data by the amount shifted.

In many cases, there is significant leading and lagging cross-correlation; this may indicate seasonality. In the case of seasonality with a VAR model, an indicator variable should be included with the data that provides a value of 1 associated with the peak in spectral density and a 0 value otherwise. Another special case with cross-correlation is that the input variable has prominent statistical significance but only at a lag far back in time. In this case, it may be appropriate to shift the data forward so that the model does not include terms (autocorrelations) that are not significant at the risk of overfitting. To demonstrate shifting based on the CCF, we will incorporate this into our process, next.

We observe in *Figure 12.12* that the strongest correlation between `realdpi` and `realinv` is at lag 0, but there are at least two other similarly correlated lags at *x=-27* and *x=-37* in the plot. As we saw in the *ARIMAX* section, we can include multiple shifted versions of the same variables as new variables to improve the predictive power of the model. This would enable us to take advantage of multiple correlations. However, for the purpose of model demonstration, we will not include that as the output from the VAR summary can become large. Also, as we mentioned, including too many lags of the same variable can result in overfitting. It is up to the practitioner to ultimately decide how many lags to use. This must be done by assessing model fit and bias/variance trade-offs. The function code given here is also in *Chapter 10, Introduction to Time Series*:

```
# code imported from chapter 10
def plot_ccf(data_a, data_b, lag_lookback, percentile, ax,
title=None):

    n = len(data_a)
    ccf = correlate(data_a - np.mean(data_a), data_b -
np.mean(data_b), method='direct') / (np.std(data_a) * np.std(data_b) *
n)

    _min = (len(ccf)-1)//2 - lag_lookback
    _max = (len(ccf)-1)//2 + (lag_lookback-1)

    zscore_vals={90:1.645,
                 95:1.96,
                 99:2.576}

    markers, stems, baseline = ax.stem(np.arange(-lag_lookback,(lag_
lookback-1)), ccf[_min:_max], markerfmt='o', use_line_collection =
True)
    z_score_95pct = zscore_vals.get(percentile)/np.sqrt(n) #1.645 for
90%, 1.96 for 95%, and 2.576 for 99%

    ax.set_title(title)
    ax.set_xlabel('Lag')
    ax.set_ylabel('Correlation')
    ax.axhline(y=-z_score_95pct, color='b', ls='--')# Z-statistic for
95% CL LL
    ax.axhline(y=z_score_95pct, color='b', ls='--')# Z-statistic for
95% CL UL
    ax.axvline(x=0, color='black', ls='-');

plot = plot_ccf(data_a=data_1d['realdpi'], data_b=data_1d['realinv'],
lag_lookback=50, percentile=95)
```

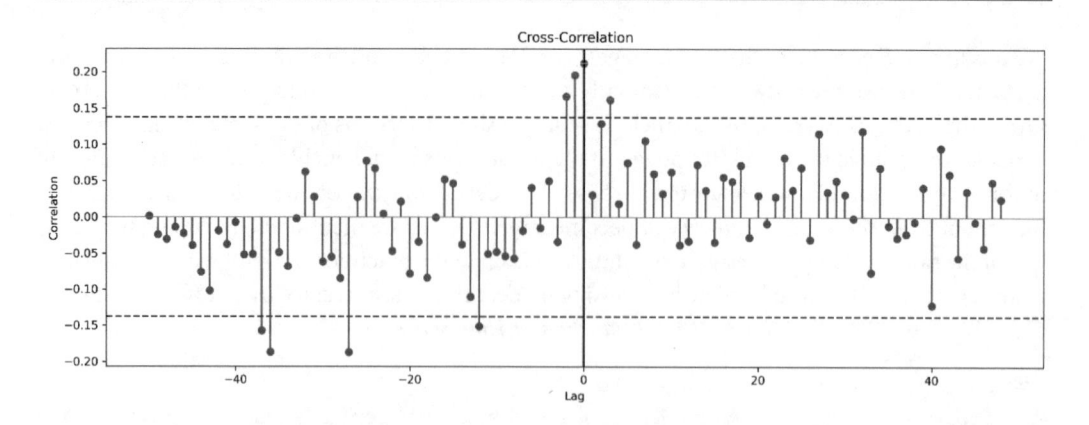

Figure 12.12 – Cross-correlation for *realdpi* and *realinv*

In *Figure 12.13*, we can see `realcons` is a leading indicator for `realinv` with a positive correlation. This means an increase in consumption results in an increase in investment. Nonetheless, we could still use a downshifted version of `realinv` to predict `realcons`—if this were our objective—since both variables are expected to continue indefinitely:

```
plot = plot_ccf(data_a=data_1d['realcons'], data_b=data_1d['realinv'],
lag_lookback=50, percentile=95)
```

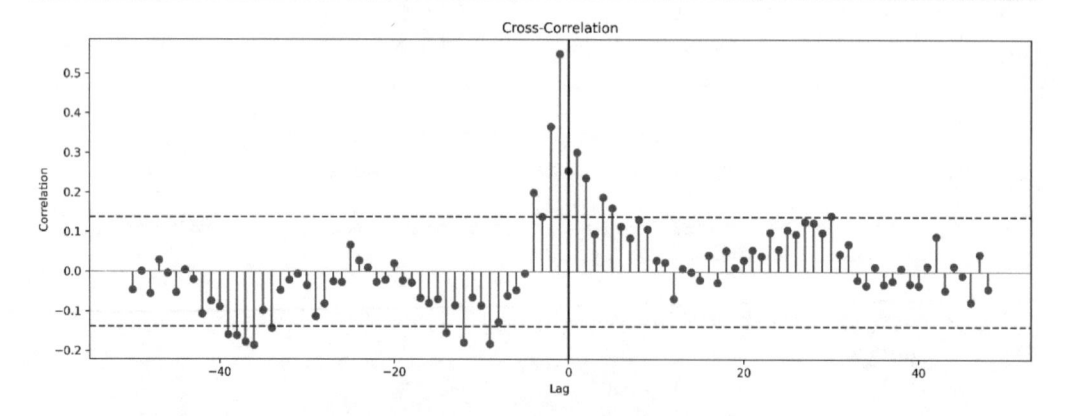

Figure 12.13 – Cross-correlation for *realcons* and *realinv*

To address the one-lag lead `realcons` has on `realinv`, let us shift `realcons` forward one index so that the strongest correlation is at lag 0. This is not required for a VAR model but can be useful in scenarios where the most significant lag is **much farther back** in time and could result in including an unreasonably high order for our model to include the necessary variable relationships.

> **Note on shifting data**
>
> When shifting data, null values will be included in the dataset. Consequently, the dataset must be trimmed by the length of the shift. This may come at the cost of model performance, so it is worth assessing the need to shift by comparing model errors.

After applying a forward shift to `realcons` in the following code snippet and rerunning the CCF, we can see in *Figure 12.14* the variables' strongest correlation is now at lag 0:

```
data_1d['realcons'] = data_1d['realcons'].shift(1)
data_1d = data_1d.iloc[1:]
plot = plot_ccf(data_a=data_1d['realcons'], data_b=data_1d['realinv'],
lag_lookback=50, percentile=95)
```

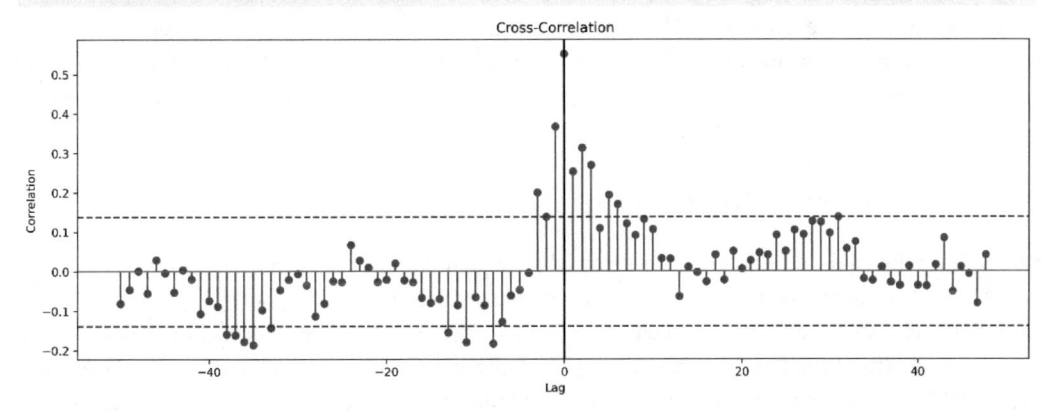

Figure 12.14 – Cross-correlation for *realinv* and forward-shifted *realcons*

Next, we want to run a function to append possible values of p and q (AR(p) and MA(q)):

```
from statsmodels.tsa.statespace.varmax import VARMAX
import warnings
warnings.simplefilter('error')
results_aic=[]
x_axis_list=[]
for p in range(1,6):
    for q in range(0,6):
        try:
            model = VARMAX(data_1d,
                            order=(p,q),
                            trend='c',
                            enforce_stationarity=True,
                            enforce_invertibility=True)
            results = model.fit()
            results_aic.append(results.aic)
```

```
            # x_axis_list.append(p,q,results.aic)
            print('(p,q): ({},{}), AIC: {}'.format(p,q,results.aic))
        except Exception as e:
            # print('(p,q): ({},{}), error: {}'.format(p,q,e))
            Pass
```

Here we can see the AIC error for VAR(1,0) and VAR(2,0) models:

```
(p,q): (1,0), AIC: 6092.740258665657
```

```
(p,q): (2,0), AIC: 6050.444583720871
```

We have two possible models through the AIC selection method. First, we built a model with p=2 and q=0, but after running, we observed that of seven coefficients predicting for `realinv`, only two were significant (even at the 0.10 level of significance), which is not good. After rerunning using a selection of p=1 and q=0, two of the four coefficients were found significant at the 0.05 level of significance and all at the 0.10 level of significance. Therefore, we can conclude the **VAR(1,0)** model is much better for predicting `realinv`.

Step 4 – building the VAR(p,q) model

We will skip the model validation steps we walked through in the last chapter as the process would be the same here. What we will do here is re-order the columns so that when we run the `get_prediction()` function, the results are for the first column. The model will stay the same since all variables are used to predict all other variables. After re-indexing, we will run the VARMAX model with **order (1,0)** using once-differenced data, with `realcons` shifted up one, and predict `realinv`. For our model, we used AR(1) and MA(0). Note that we use `trend='c'` for *constant* as we removed `trend` when performing differencing; the `trend` argument is for trend determinism. Trend differencing should be handled outside of the model since not all variables may have unit root trends. Handling trends outside of the model also helps us identify components such as seasonality or significance of the true autocorrelation structure following the removal of `trend`. It is worthwhile to note that seasonality is not addressed directly with the VAR (and VARMAX) model; indicator (dummy) variables for seasonality should be included.

Regarding model fit, we can refer to *Chapter 11, ARIMA Models,* for information on **Ljung-Box**, **Jarque-Bera**, **heteroskedasticity**, **skew**, and **kurtosis** values, so will not repeat that here. The order of values corresponds to the order of variables listed in the `Dep. Variable` list, shown next. Each model variable has listed the coefficients of the other variables' terms used for predicting each variable. For example, we can observe in `Results for equation realcons` that `realinv` does not contribute much predictive capability to `realcons`, but `realdpi` does. We can also observe that `realcons` is significant for predicting `realinv`. We can also see based on the p-value and confidence interval in the **error covariance matrix** (**ECM**) that there is not much cross-covariance between the residuals of `realinv` and `realdpi`. We can expect to see larger coefficient values if we have variance in one feature explained by variance in the other:

```
model = VARMAX(data_1d.
reindex(columns=['realinv','realcons','realdpi']), order=(1,0),
trend='c',
 enforce_stationarity=True,enforce_invertibility=True)
results = model.fit()
print(results.summary())
```

```
                        Statespace Model Results
==============================================================================
Dep. Variable:     ['realcons','realinv','realdpi']   No. Observations:           201
Model:                            VAR(1) + intercept   Log Likelihood        -3028.370
Date:                                Sat, 11 Feb 2023   AIC                    6092.740
Time:                                        12:10:38   BIC                    6152.200
Sample:                                             0   HQIC                   6116.800
                                                 -201
Covariance Type:                                  opg
==============================================================================
Ljung-Box (L1) (Q):       8.60, 0.10, 0.00   Jarque-Bera (JB):    9.82, 215.76, 60.32
Prob(Q):                  0.00, 0.76, 0.97   Prob(JB):            0.01, 0.00, 0.00
Heteroskedasticity (H):   3.14, 3.84, 4.74   Skew:                0.09, -1.07, 0.44
Prob(H) (two-sided):      0.00, 0.00, 0.00   Kurtosis:            4.07, 7.60, 5.54

                      Results for equation realcons
==============================================================================
                 coef    std err       z      P>|z|     [0.025     0.975]
------------------------------------------------------------------------------
intercept      14.5358    3.416      4.256    0.000      7.841     21.230
L1.realcons     0.4190    0.070      5.982    0.000      0.282      0.556
L1.realinv     -0.0308    0.052     -0.592    0.554     -0.133      0.071
L1.realdpi      0.1748    0.029      6.119    0.000      0.119      0.231

                      Results for equation realinv
==============================================================================
                 coef    std err       z      P>|z|     [0.025     0.975]
------------------------------------------------------------------------------
intercept     -11.0943    5.173     -2.145    0.032    -21.232     -0.956
L1.realcons     0.3425    0.093      3.699    0.000      0.161      0.524
L1.realinv      0.1540    0.087      1.778    0.075     -0.016      0.324
L1.realdpi      0.0789    0.041      1.902    0.057     -0.002      0.160

                      Results for equation realdpi
==============================================================================
                 coef    std err       z      P>|z|     [0.025     0.975]
------------------------------------------------------------------------------
intercept      38.0578    7.602      5.006    0.000     23.159     52.957
L1.realcons     0.3488    0.146      2.395    0.017      0.063      0.634
L1.realinv     -0.0404    0.086     -0.469    0.639     -0.210      0.129
L1.realdpi     -0.2576    0.059     -4.393    0.000     -0.372     -0.143

                         Error covariance matrix
==============================================================================
                          coef    std err     z    P>|z|   [0.025   0.975]
------------------------------------------------------------------------------
sqrt.var.realcons        28.396   1.252   22.688   0.000   25.943   30.849
sqrt.cov.realcons.realinv 18.901  2.919    6.476   0.000   13.181   24.621
sqrt.var.realinv         36.168   1.567   23.077   0.000   33.096    9.240
sqrt.cov.realcons.realdpi 17.584  3.612    4.869   0.000   10.506   24.663
sqrt.cov.realinv.realdpi  3.6078  3.744    0.964   0.335   -3.730   10.946
sqrt.var.realdpi         48.177   2.032   23.715   0.000   44.195   52.159
==============================================================================
```

Figure 12.15 – VAR(1) model output

Step 5 – testing the forecast

It is important to note that the prediction for the VAR model is actually quite stable compared to the long-running historical performance. The reason we see such extreme differences at the end of the forecast is that this time period corresponds to the **Great Recession**, which lasted from 2007 to 2009. The last five points in our test forecast correspond to the last two quarters of 2008 and the first three quarters of 2009:

```
df_pred = results.get_prediction(start=195, end=200).summary_
frame(alpha=0.05)
fig, ax = plt.subplots(1,1,figsize=(20,5))
ax.plot(data_1d['realinv'], marker='o', markersize=5)
ax.plot(df_pred['mean'], marker='o', markersize=4)
ax.plot(df_pred['mean_ci_lower'], color='g')
ax.plot(df_pred['mean_ci_upper'], color='g')
ax.fill_between(df_pred.index, df_pred['mean_ci_lower'], df_
pred['mean_ci_upper'], color='g', alpha=0.1)
ax.set_title('Test Forecast for VAR(1)')
```

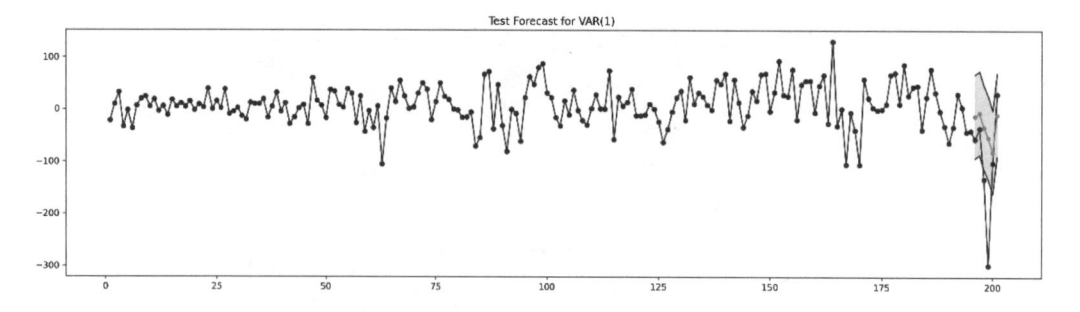

Figure 12.16 – Test forecast for VARMAX with p=1

We can compare our forecast (*mean*) to actuals (`realinv`) for the investment variable, as follows:

```
pd.concat([df_pred,data_1d['realinv'].iloc[195:]], axis=1)
```

Here in *Figure 12.17*, we can see the VAR (from the `VARMAX` function) model test forecast:

	mean	mean_se	mean_ci_lower	mean_ci_upper	realinv
196	-12.4348	40.79603	-92.3935	67.52398	-56.368
197	-6.26154	40.79603	-86.2203	73.69722	-35.825
198	-33.5406	40.79603	-113.499	46.41821	-133.032
199	-53.6481	40.79603	-133.607	26.31066	-299.167
200	-81.514	40.79603	-161.473	-1.55526	-101.816
201	-10.1039	40.79603	-90.0627	69.85483	29.72

Figure 12.17 – VAR(1) model test forecast

Step 6 – building the forecast

We can observe that because this is a model with an **autoregressive order p=1**, the forecast **tends very quickly toward the mean**. Including a higher order would likely result in more model variance, which we could see as a more confident, yet risky forecast. The multivariate AR(1) model produces more bias, which we can observe in *Figure 12.18* as the tendency to the mean we mentioned. We can run the following code to generate the data and plot for the forecast:

```
df_forecast = results.get_prediction(start=201, end=207).summary_
frame(alpha=0.05)
forecast = np.hstack([np.repeat(np.nan, len(data_1d)+1), df_
forecast['mean']])
  fig, ax = plt.subplots(1,1,figsize=(20,5))
ax.plot(data_1d['realinv'], marker='o', markersize=5)
ax.plot(forecast, marker='o', markersize=4)
ax.plot(df_forecast['mean_ci_lower'], color='g')
ax.plot(df_forecast['mean_ci_upper'], color='g')
ax.fill_between(df_forecast.index, df_forecast['mean_ci_lower'], df_
forecast['mean_ci_upper'], color='g', alpha=0.1)
ax.set_title('Forecast for VAR(1)');
```

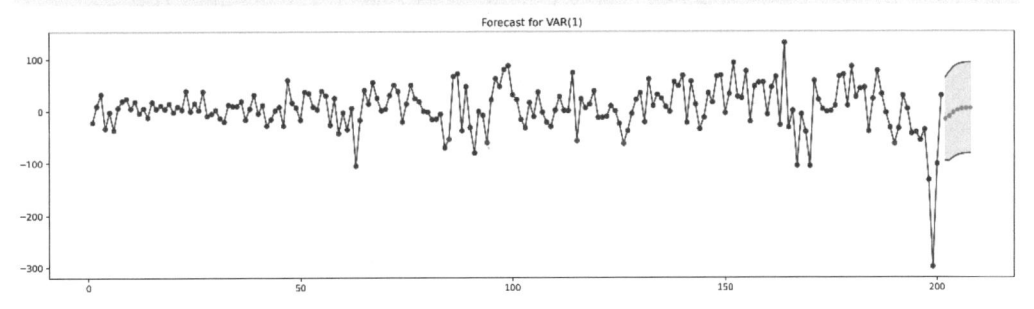

Figure 12.18 – VAR(1) model forecast, h=7

Here, we can see the forecasted points and the confidence interval for the forecast:

```
df_forecast
```

	mean	mean_se	mean_ci_lower	mean_ci_upper
202	-16.3442	40.79603	-96.3029	63.6146
203	-10.9571	43.69828	-96.6041	74.68998
204	-3.24668	44.4028	-90.2746	83.7812
205	1.301495	44.56735	-86.0489	88.65189
206	3.576619	44.60344	-83.8445	90.99775
207	4.650973	44.61108	-82.7851	92.08709
208	5.147085	44.61268	-82.2922	92.58633

Figure 12.19 – VAR(1) model output data

Note that our results are based on the differenced data. This model tells us the level of statistical significance between variables regarding their respective signals but does not give us forecast values in terms of the original data. To reverse differenced data, we need to estimate a constant of integration and then reverse-difference the data according to that constant or use a different transformation method that allows us to back-transform, such as a logarithmic transformation that can be exponentiated to reverse. Often, the VAR model is used for identifying potential causation between variables identified as highly correlated in a cross-correlation analysis as well as in economic shock analysis.

It can be argued that depending on the level of detail needed in the forecast, differencing and shifting of variables may not be required. However, because this is a fully endogenous model with no exogenous variables, it is important that if we apply differencing to any variables, we must also apply differencing to all variables containing non-stationary behavior. With respect to shifting, it may be more useful to simply apply a higher autoregressive lag order than to apply a shift; the drawback of shifting is we lose samples early on in the process. However, the drawback of using a higher lag order is the inclusion of more variables in the time dimension, which can increase the likelihood of overfitting. Logically, as more variables are included in the model, we must also increase sample size. As with any model, we must apply rigorous cross-validation to ensure performance stability and minimize risk.

Summary

In this chapter, we provided an overview of multivariate time-series and how they differ from the univariate case. We then covered the math and intuition behind two popular approaches to solving problems using multivariate time-series models—ARIMAX and the VAR model framework. We walked through examples for each model using a step-by-step approach. This chapter concludes our discussions on time-series analysis and forecasting. At this point, you should be able to identify and assess the statistical properties of time series, transform them as needed, and construct models that are useful for fitting and forecasting both univariate and multivariate cases.

In the next chapter, we will begin our discussion on survival analysis with an introduction to **time-to-event** (**TTE**) variables.

References

[1] *Liang, X., Zou, T., Guo, B., Li, S., Zhang, H., Zhang, S., Huang, H. and Chen, S. X. (2015). Assessing Beijing's PM2.5 pollution: severity, weather impact, APEC and winter heating. Proceedings of the Royal Society A, 471, 20150257.*

[2] *Dua, D. and Graff, C. (2019). UCI Machine Learning Repository* [`http://archive.ics.uci.edu/ml`]. *Irvine, CA: University of California, School of Information and Computer Science.*

Part 5: Survival Analysis

This part will cover another statistical approach named survival analysis by analyzing a time to event outcome variable. After an introduction of survival analysis and censored data, we will study models with survival responses.

It includes the following chapters:

- *Chapter 13, Time to Event variables - An introduction*
- *Chapter 14, Survival Models*

13
Time-to-Event Variables – An Introduction

In this short chapter, we will introduce another branch of statistics called **survival analysis**, which is related to survival and time-censoring studies. Survival analysis is also called **time-to-event variable analysis**, which is a particular statistical outcome type that requires other techniques than those used in the few last chapters that we have studied. A time-to-event variable analysis studies, for example, whether a participant has an event of interest during the study timeframe. In other words, we study the *proportion of a sample surviving after a specific time point* and the rate at which the survived sample proportion will fail or die, or whether there are survival differences in different treatment groups. The term *survival* in survival analysis is originally based on the time from treatment until death in the medical field. However, survival analysis is readily applicable to many fields including engineering (where it is referred to as **reliability theory** or **reliability analysis**), sociology (**event history analysis**), and economics (**duration analysis**). Survival studies might not be ambiguous in medical research when a death occurs, although some events such as organ failure can be poorly defined. Ambiguity does occur in other research fields, however: for instance, in engineering, a failure or a broken system can be partial or not strictly determined by time. We also examine survival data, the survival function, hazard, and hazard ratios in this chapter.

In this chapter, we're going to cover the following main topics:

- Censoring
- Survival data
- The survival function, hazard, and hazard ratios

What is censoring?

In the field of statistics, **Censoring** refers to a situation where the full extent or precise value of a measurement or observation is not entirely known. In Survival Analysis, this happens when we have information about sample observations and do not know when the given event happened, and is considered a key issue in survival analysis, distinguishing time-to-event analysis from the other statistical analyses mentioned in the previous chapters. There are several reasons why censoring happens; for example, a person withdraws from a study or exits prior to a follow-up, or the event in question has already happened before the study starts. The censored event is *non-informative*, that is, *censoring causes study failure due to some reason other than failure time. In other words, failure caused by censoring is not related to the probability of an event occurring*. **Informative censoring** happens when an observation is lost to follow-up because of research reasons. Three types of informative censoring are **right censoring**, **left censoring**, and **interval censoring**. In this book, we focus on **non-informative right censoring**.

Left censoring

When we know the *event has happened before the observation data is collected*, then the observation is **left-censored**. An example left-censored scenario is where we study the age when a middle-school student begins puberty. There may be some students reaching this developmental stage before they enter middle school. All of these students are considered left-censored. As another example, consider an HIV study for an at-risk group of patients where, by the time the study starts, some patients already have HIV.

Right censoring

When the *event happens after some time point* but we do not know the exact time, then it is considered **right-censored**. Applying the previous scenario here, imagine there are some students who withdraw from the puberty developmental stage study, or we lose them prior to following up (for example, they move to another school) or they have a delayed puberty (the developmental stage had still not taken place by the time they left middle school to begin high school). Another example is a cancer study where a group of patients is recruited for a clinical trial to see how a new treatment affects them, but a patient in the group dies because of a car accident after 2 years. In that case, we know this patient's survival is at least 2 years during the period of the study, but do not know how many years beyond 2 that this patient could have survived with the cancer diagnosis.

Interval censoring

Left censoring and right censoring are special cases of **interval censoring**. A clear example to illustrate interval censoring is a patient that tests negatively at time t_1 for HIV but positively at time t_2. However, we do not know the exact time this patient got infected between time interval $[t_1, t_2]$. This particular patient had an annual checkup last year and again this year. The lab results show he was HIV-negative last year but HIV-positive this year. Therefore, the infection occurred at some time point between last year's and this year's checkups, but we do not know exactly when.

Type I and Type II censoring

There are two types of right censoring: **Type I censoring** and **Type II censoring**. In **Type I censoring**, a study is designed to terminate after some time point t. Fixed censoring occurs when an object does not have an event after time t. This object is considered censored at time t. Random censoring occurs when censored objects do not have the same censoring time. In **Type II censoring**, a study terminates after some prespecified number of events occur. This type of censoring is frequently used in life testing experiments to save time and money. For example, we randomly select n objects as a sample to test whether a new innovation is really effective, but instead of testing all these n objects, we define that the study terminates if r failures occur to save time and costs. In this case, the first r objects have events at time $t_1, t_2, \ldots, t_r$ but the remaining $n - r$ objects do not have an event at time $T = max(t_1, t_2, \ldots, t_r)$ and they are therefore considered censored.

Let us visualize a cancer study for a period of 5 years testing a new treatment. The start time point is 0 when the study starts, and the end time point is 5 when the study ends.

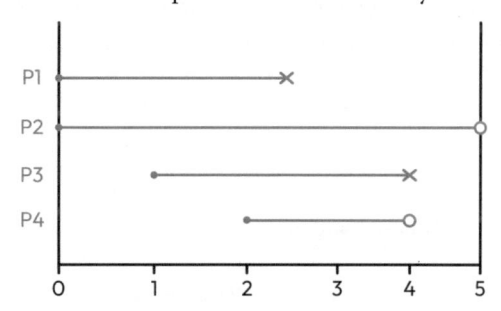

Figure 13.1 – Right censoring in survival analysis for a cancer study

The first patient was recruited at the beginning of the study, but he died in year 2-and-a-half. The second patient was also recruited at the beginning of the study but in year 5, he was still surviving. This patient has survived for more than 5 years, but we do not know for how many years he can ultimately survive, which means that right censoring occurred with this patient. The third patient was recruited in year 1 but she died in year 4. The fourth patient joined the study in year 2 but by year 4, we were unable to follow up with this patient. The reason could be that he moved to another city, for instance. Censoring occurs for patient 2 and patient 4.

The study seems like a regression problem but in this example, the second patient was considered to have survived when the study concluded. This makes survival analysis different from regular statistical analyses. We also do not remove the data of the second patient from the study data since the information is valuable: patients can survive for at least 5 years. In addition, patient follow-ups can be prevented because they may move to another place or are very sick but instead of removing the data, we record the reason why they dropped out of the study. Similarly, gender considerations can be biased because, for example, male patients may be more likely to withdraw from the study than female patients when they are sick, thus resulting in a potentially incorrect conclusion that females survive longer than males.

Survival data

Survival data focuses on whether or not an object in a study experiences an event. In addition, the follow-up time is also considered. *Time zero* or *time origin* is the time when the study starts. Depending on the purpose of a study, time zero or time origin can be different. For instance, in a prostate cancer study, researchers recruit 40-year-old and older male participants, but in a study of puberty developmental ages, male and female teenagers of 12 and older are recruited. If the research takes place over a period of time (which could be several months or years) then recording the time origin and follow-up time during the study is vitally important.

Lastly, we discuss the *relationship between survival and censoring times* and how we *record survival data with censoring*. Suppose that for each object in a sample, we know its true event time T and true censoring time C. Then, the *survival time of an object is the period of time until an event occurs* and the *censoring time of an object is the moment when censoring occurs*, when, for example, the study ends or we are unable to follow up on this object. Let δ_i denote the status indicator for the object i:

$$\delta_i = \begin{cases} 1 \text{ if the event was observed}, \\ 0 \text{ if censoring was observed}. \end{cases}$$

In other words, $\delta_i = 1$ when the event T_i occurs *before* censoring C_i and $\delta_i = 0$ when the event T_i occurs *after* censoring C_i. The observed outcome variable is

$$Y_i = min(T_i, C_i)$$

and the censored survival data for object i can be captured as (Y_i, δ_i).

The following example considers three patients in a cancer study. Time is considered here in years and the duration of the study is 5 years.

Patient	T	C	δ
1	3 (the patient died in year 3)	No censoring – patient died in year 3	1 (because the event was observed)
2	7 (the patient died in year 7 but the study ended in year 5)	5 (the study ends in year 5)	0 (because the censoring was observed)
3	No information	4 (the patient dropped out of the study)	0 (because the censoring was observed)

Figure 13.2 – Survival analysis for a cancer study

The following figure illustrates information about 3 patients in the cancer study.

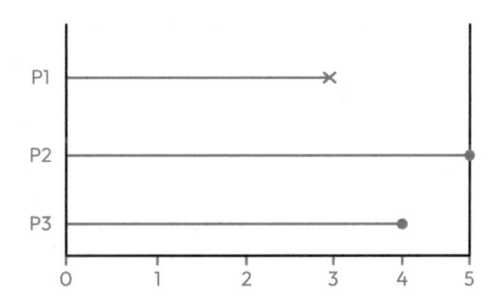

Figure 13.3 – Survival analysis for a cancer study

Survival Function, Hazard and Hazard Ratio

Let us first discuss the survival function. The formula of the function is defined as

$$S(t) = P(T > t)$$

and represents the probability that the object survives past time t. The survival function is a non-increasing function with t ranges from 0 to ∞. When $t = 0, S(t) = 1$ and when $t = \infty$, $S(t) = S(\infty) = 0$. It is a smooth function theoretically but practically, *events occur on a discrete time scale* (days, weeks, years).

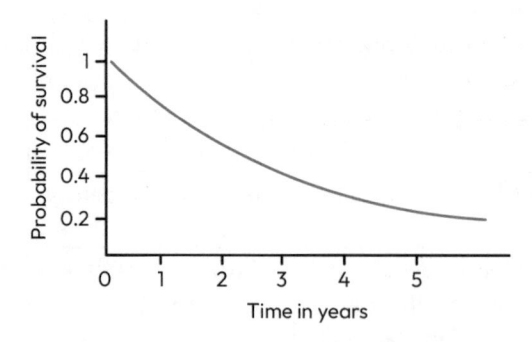

Figure 13.4 – Survival function illustrated

In this example, we go back to the cancer study that spanned 5 years. At time zero, when the study started, the survival probability was 1 or 100% but at year 5, the probability of survival was close to 0.2 or 20%.

Now we consider the *Stanford heart transplant dataset*. The dataset contains the information of 103 patients who participated in an experimental heart transplant program (see *Figure 13.5*). The patients were seriously ill and had the potential to obtain some benefit from being part of the program, as they could be waiting anywhere from a few weeks to several months to receive a new heart from a donor. The CSV dataset can be downloaded at https://www.openintro.org/data/csv/heart_transplant.csv.

	id	acceptyear	age	survived	survtime	prior	transplant	wait
0	15	68	53	dead	1	no	control	NaN
1	43	70	43	dead	2	no	control	NaN
2	61	71	52	dead	2	no	control	NaN
3	75	72	52	dead	2	no	control	NaN
4	6	68	54	dead	3	no	control	NaN

Figure 13.5 – First five rows of Stanford translant data

We are interested in two features here: `survived` (recording the survival status of dead versus alive) and `survtime` (the number of days the patients survived after they were determined to be a candidate for a heart transplant until the study ended). When we sort the data by `survtime`, among the first 15 observations, the patient in row 13 is still alive as shown in *Figure 13.6*. Therefore, the observation at row number 13 is time censored.

	survtime	survived	
0	1	dead	
1	2	dead	
2	2	dead	
3	2	dead	
4	3	dead	
5	3	dead	
6	3	dead	
7	5	dead	
8	5	dead	
9	6	dead	
10	6	dead	
11	8	dead	
12	9	dead	
13	11	alive	
14	12	dead	

Figure 13.6 – Status of first 15 transplant patients

In this example, to calculate the probability of survival beyond survtime = 3, we just need to calculate the **probability of survival** in each interval of time $[t_0, t_1], (t_1, t_2], (t_2, t_3]$, which is calculated by

$$P\big(Survival\ in\ (t_{i-1}, t_i]\big) = 1 - P\big(Dead\ in\ (t_{i-1}, t_i]\ |\ Alive\ in\ t_{i-1}\big)$$

where

$$P\big(Dead\ in\ (t_{i-1}, t_i]\ |\ Alive\ in\ t_{i-1}\big) = \frac{number\ of\ deaths\ in\ (t_{i-1}, t_i]}{number\ of\ alives\ at\ t_{i-1}},$$

then

$$S(t_3) = P(Survival > t_3) = \left(1 - \tfrac{0}{103}\right) * \left(1 - \tfrac{1}{103}\right) * \left(1 - \tfrac{3}{102}\right) = 96.12\%.$$

Next, we discuss hazard function or hazard rate. The **hazard function** is defined as

$$h(t) = \lim_{\Delta t \to 0} \frac{Pr(t < T \leq t + \Delta t | T > t)}{\Delta t},$$

the instantaneous rate at which events occur given survival up to that time. In other words, it is just the probability of *dying* in the next few seconds or the next day given being *alive* now. The hazard ratio is a ratio of death probabilities. For example, when a hazard ratio is 3, then the given group has 3 times the chance of dying compared to another group.

Summary

In this chapter, we provided an overview of time-to-event variable analysis and how it differs from other statistical analyses that we have studied in the previous chapters. We covered censoring intuition (left, right, and interval censoring) and discussed Type I and Type II censoring. We also discussed non-informative and informative events in this chapter. We then discussed survival data and the relationship between survival and censoring times and how we record survival data with censoring. The survival function, hazard, and hazard ratio were also mentioned in the last section of this chapter.

In the next chapter, we will consider the non-parametric Kaplan-Meier model, the parametric exponential model, and also the semiparametric Cox Proportional Hazards model. We will perform real data analysis in Python by applying these models for survival analysis.

14
Survival Models

In *Chapter 13, Time-to-Event Variables*, we introduced the topics of survival analysis, censoring, and **time-to-event** (**TTE**) variables. In this chapter, we will provide an in-depth overview and walkthrough of the implementation of these techniques with respect to three primary model frameworks:

- Kaplan-Meier model
- Exponential model
- Cox Proportional Hazards model

We will discuss how each approach provides probabilistic insight into the survival and hazard risk of study subjects using univariate Kaplan-Meier and exponential approaches as well as the multivariate Cox Proportional Hazards regression model. We'll walk through examples using real data and discuss the results so that the reader understands how to assess performance and translate test output into useful information. Finally, we will show how to use the trained models to provide forecast probabilities for unseen data.

Technical requirements

In this chapter, we use an additional Python library for survival analysis: `lifelines`. Please install the following versions of these libraries to run the provided code. Instructions for installing libraries can be found in *Chapter 1, Sampling and Generalization*:

- `lifelines== 0.27.4`

More information about `lifelines` can be found at this link:

`https://lifelines.readthedocs.io/en/latest/index.html`

Kaplan-Meier model

The first model for survival analysis we will discuss is the **Kaplan-Meier model** (also called the **Kaplan-Meier estimator**). We will start this section with a discussion model definition and learn how it is built. Then, we will close this section with an example of how to use this model in Python using the `lifelines` library. Let's get started.

Model definition

The Kaplan-Meier estimator is defined by the following formula:

$$\hat{S}(t) = \prod_{i:t_i \leq t} \frac{n_i - d_i}{n_i}$$

Here, n_i is the number of subjects at risk just before time t, d_i is the number of death events at time t, and $\hat{S}(t)$ (the survival function) is the probability that life is longer than t. The Π symbol used in the formula is like the symbol Σ; however, Π indicates multiplication. This means that the preceding formula will result in a multiplication of fractions for each time step t. This means that the Kaplan-Meier estimator is a series of declining horizontal steps. To make this clear, let's take a look at its application with some example data.

Let's say that we have the following data, where the `duration` column represents how long a subject survived and the `death` column indicates whether a death was observed in the study. Let's walk through the calculations for the Kaplan-Meier estimator for this data:

duration	death
5	1
7	1
3.5	1
8	0
2	1
5	1

Figure 14.1 – Death and duration example data

Initially (at $t = 0$), all subjects are at risk, but no subjects have died. That means the current survival rate is 1. We calculate the value of the survival function at $t = 0$ as follows:

$$\hat{S}(0) = \frac{n_0 - d_0}{n_0} = \frac{6 - 0}{6} = 1$$

The first event occurs at $t = 2$. At this point, all subjects (six subjects) are at risk just before the event, and one subject dies at this time. We calculate the current survival rate at this point, then multiply this value by all the previous values. The survival rate at $t = 2$ is given here:

$$\frac{n_2 - d_2}{n_2} = \frac{6-1}{6} \approx 0.83$$

Then, we calculate the survival function at $t = 2$:

$$\hat{S}(2) = \frac{\frac{n_0 - d_0}{n_0} * n_2 - d_2}{n_2} = \frac{\frac{6-0}{6} * 6 - 1}{6} \approx 0.83$$

The next event occurs at $t = 3.5$. We repeat the process as outlined previously. However, there are only five subjects at risk at this point since one has already died. Thus, the current survival rate at time $t = 3.5$ is this:

$$\frac{n_3 - d_3}{n_3} = \frac{5-1}{5} = 0.8$$

Then, we calculate the survival function at this point:

$$\hat{S}(3.5) = \frac{\frac{\frac{n_0 - d_0}{n_0} * n_2 - d_2}{n_2} * n_{3.5} - d_{3.5}}{n_{3.5}} = \frac{\frac{\frac{6-0}{6} * 6 - 1}{6} * 5 - 1}{5} \approx 0.67$$

The next event occurs at $t = 5$, where two subjects die. At $t = 5$, there are four remaining subjects at risk, and two subjects die. Thus, we have a survival rate of 0.5 at $t = 5$. The new value of the survival function is calculated as follows:

$$\hat{S}(5) = 1 * 0.83 * 0.67 * 0.5 \approx 0.33$$

The final event occurs at $t = 7$. At this point, the survival rate is 0.5 and the value of the survival function at this point is approximately 0.167:

$$\hat{S}(7) = 1 * 0.83 * 0.67 * 0.5 * 0.5 \approx 0.167$$

Since our study ends here, this is the final value of the survival function, and the function value remains constant through the end of the study.

As you can see, this survival model is simple to calculate. Notice also that we did not mention assumptions or underlying statistical distributions. The Kaplan-Meier estimator is a nonparametric survival model. As we discussed in *Chapter 5, Non-Parametric Tests*, this means it can be used as an alternative to parametric models when the data under analysis does not meet the assumptions of parametric models. The Kaplan-Meier estimator is a useful model to have in your tool belt for that reason.

Now that we have had a walkthrough of how this model works, let's look at how to calculate the Kaplan-Meier estimator survival function with the `lifelines` package in Python.

Model example

For this example, we will use the `larynx` dataset from `lifelines`. This dataset consists of six variables: `time`, `age`, `death`, `Stage_II`, `Stage_III`, and `Stage_IV`. The information for this dataset is limited, so we will make the following assumptions about the variables for the purposes of this example:

- `time` is the amount of time a patient lived after diagnosis of larynx cancer
- `age` is the age of the patient at the time of diagnosis
- `death` indicates whether the death was observed during the study
- `Stage_II` indicates that the cancer was at stage two at the time of diagnosis
- `Stage_III` indicates that the cancer was at stage three at the time of diagnosis
- `Stage_IV` indicates that the cancer was at stage four at the time of diagnosis
- When a patient is not indicated in stage two or above, they were in stage one at the time of diagnosis

Let's start by calculating the survival function of all the patients in the study. We will use `KaplanMeierFitter` from `lifelines`. This is an object that will be the Kaplan-Meier model we just discussed. To fit this model, we will use the `fit()` method, which takes two arguments: duration and event_observed. duration is how long the patient lived in the study. event_observed is whether the death was observed: `True` for observed and `False` for not observed. The data is already set up to work with these two arguments. We can fit the model as follows:

```
import matplotlib.pyplot as plt
from lifelines import datasets
from lifelines.fitters.kaplan_meier_fitter import KaplanMeierFitter
data = datasets.load_larynx()
kmf = KaplanMeierFitter()
kmf.fit(data.time, data.death)
# this prints the first 5 elements of the survival function
kmf.survival_function_.head()
```

The output of the preceding code is shown in *Figure 14.2*, which is the first five elements of the survival function:

timeline	KM_estimate
0.0	1.000000
0.1	0.988889
0.2	0.977778
0.3	0.944444
0.4	0.933333

Figure 14.2 – First five elements of the larynx survival function

The elements of the survival function shown in *Figure 14.2* are slowly decreasing, which is indicative of patients dying over time. However, it's difficult to get the full story from a series of numbers. Let's plot this series. We can also get the confidence intervals from the KaplanMeierFitter object, as shown in the next code block:

```
fig, ax = plt.subplots(1)
kmf.survival_function_.plot(ax=ax)
ci = kmf.confidence_interval_survival_function_
index = ci.index
ci_low, ci_high = ci.values[:, 0], ci.values[:, 1]
ax.fill_between(index, ci_low, ci_high, color='gray', alpha=0.3)
```

This code results in the plot shown in *Figure 14.3*, which is the survival function over time:

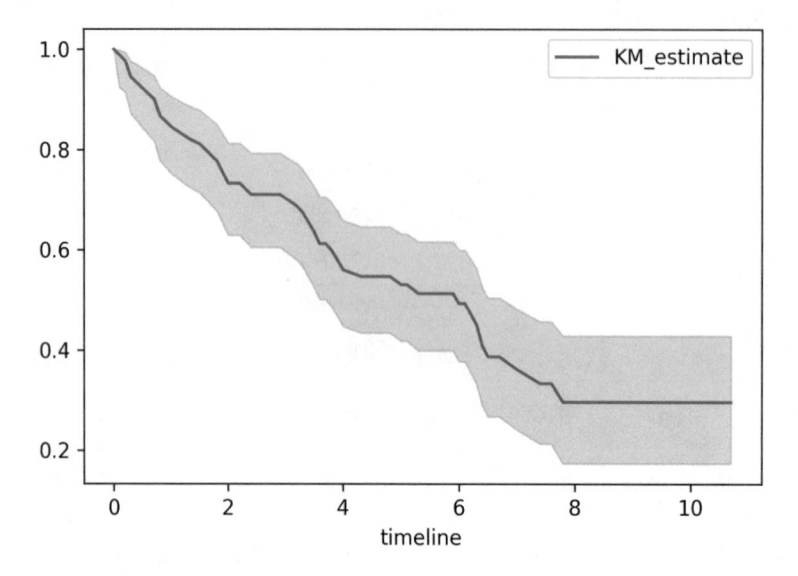

Figure 14.3 – Larynx survival function using Kaplan-Meier estimation

The Kaplan-Meier estimator survival function for the larynx data is shown in *Figure 14.3*. This function decreases over time until the last observed death at time step 8 where the function remains constant until the end of the study.

Now, consider we have a few features in this dataset that we may want to compare. For example, we have the stages of cancer at the initial diagnosis, and we have the ages of the individuals. Let us start by comparing the survival functions of two age groups. Then, let us compare the survival functions of groups based on the cancer stage.

The ages of individuals in the study range from about 40 to 85, with the bulk of the individuals in the range of 60 to 80 years old. Let's split the individuals into groups at the age of 65. Anyone younger than 65 will be in the younger group, and anyone 65 or older will be in the older group. The survival functions for the two groups are shown in *Figure 14.4*:

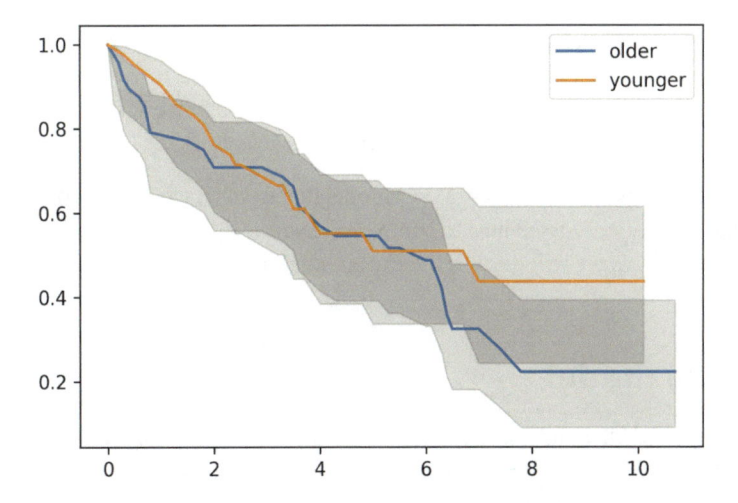

Figure 14.4 – Larynx survival functions for older and younger groups

The survival functions for the older and younger group shown in *Figure 14.4* appear to exhibit little difference until around time step 7. However, the confidence intervals never appear to diverge. It appears that age does not appear to be a strong factor for survival based on this data.

Now, let's do a similar survival analysis with the various stages. As mentioned previously, we will assume that any individual that is not indicated in stage two or above was diagnosed as stage one. The survival functions for these groups are shown in *Figure 14.5*:

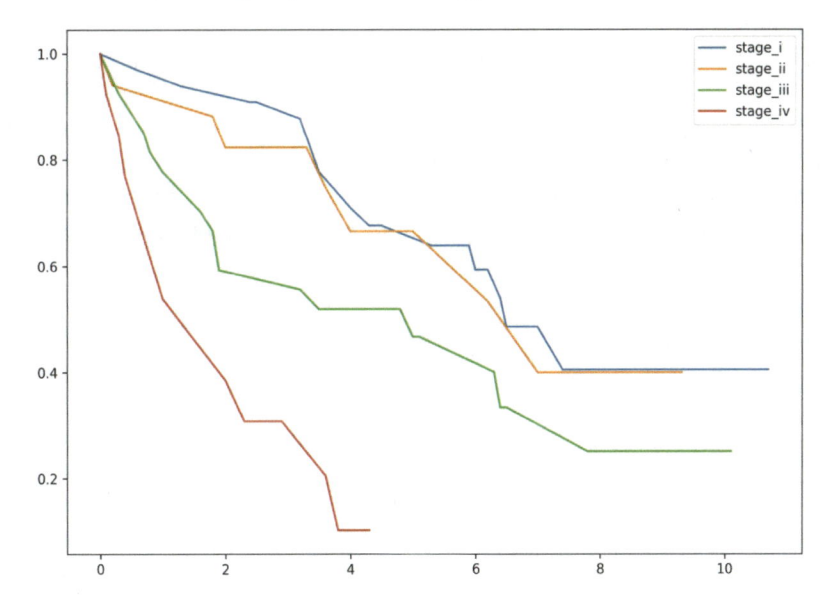

Figure 14.5 – Larynx survival functions based on stage

The survival functions for the groups based on stage are shown in *Figure 14.5*. This plot omits the confidence intervals for plot clarity. Based on the plot, there does not appear to be a significant difference between stage one and stage two. We start to see a difference at stage three. The difference between stage four and the other stages is distinctly noticeable.

In this section, we discussed the Kaplan-Meier model for survival, which is a nonparametric model for survival data. We also discussed how to compare the survival functions of multiple groups within a set. In the next section, we will discuss the exponential model, which is a parametric model for survival data.

Exponential model

In the last section, we studied the non-parametric **Kaplan-Meier survival model**. We will now bridge parametric modeling with the **exponential model** and then will discuss a semi-parametric model, the **Cox Proportional Hazards model**, in the next section. Before considering the exponential model, we will review what the exponential distribution is and why we mention it in this section. This distribution is based on the Poisson process. Here, events occur independently over time and the event rate, λ, is calculated by the number of occurrences per unit of time, as follows:

$$\lambda = \frac{Y}{t}$$

The Poisson distribution is a **statistical discrete distribution** concerning the number of events occurring in a specified time period. It is defined as follows. Let Y be the number of occurrences in time t. Y follows the Poisson distribution with parameter λ if a probability mass function is given by the following formula:

$$f(Y) = Pr(y = Y) = \frac{e^{-\lambda}\lambda^{Y}}{Y!}$$

The mean and the variance of the discrete variable y are equal to the event rate. Instead of focusing on how many events occur in a specific time period, we can think about how long the event occurs; that's why the exponential distribution is considered in this section. Let T be the time until the event occurs ($Y = 1$). The probability density function is calculated by the following formula:

$$f(t) = P(T = t) = \lambda e^{-\lambda t}$$

And the cumulative distribution function is calculated as follows:

$$F(t) = P(T \leq t) = 1 - e^{-\lambda t}$$

The function $F(t)$ represents the probability of not surviving past time t or the probability that survival time is less than t. We can see in *Figure 14.6* the following probability form:

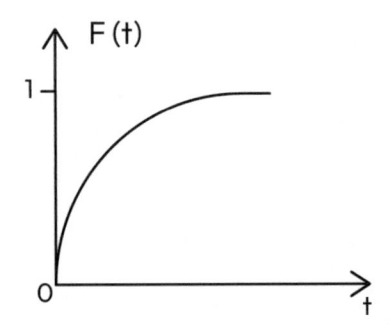

Figure 14.6 – Curve of the probability of not surviving past time t

In contrast, the survival function is related to the probability of surviving past a certain time t:

$$S(t) = P(T > t) = 1 - P(T \leq t) = e^{-\lambda t}$$

By visualization, the survival function will have the form seen in *Figure 14.7*:

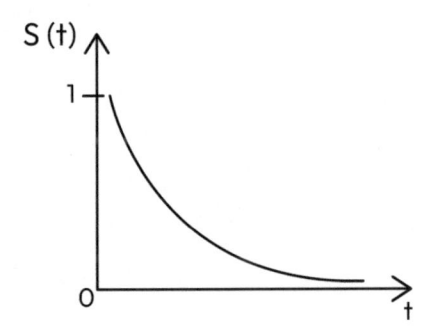

Figure 14.7 – Curve of the survival function

The **accumulated hazard function** is $H(t) = \lambda t$. We can then we can rewrite the survival function in terms of the hazard function as follows:

$$S(t) = e^{-H(t)}$$

Given $S(t)$ and $f(t)$, we can use the idea of the log-likelihood equation approach and estimate the value for λ via maximum likelihood estimation.

Model example

We will use the same `larynx` dataset from `lifeline` as in the last section. In this example, we will use `ExponentialFitter` from `lifelines`. Similarly, the `fit()` method takes two arguments: `duration` and `event_observed`. We can fit the model as follows:

```
from lifelines import ExponentialFitter
from lifelines import datasets
data = datasets.load_larynx()
exf = ExponentialFitter()
exf.fit(data.time, data.death)
# this print the first 5 elements of the survival function
exf.survival_function_.head()
```

In the Kaplan-Meier model, the curve looks like stair steps, but using the exponential approach, the curve is gradually decreasing continuously, as we can see here:

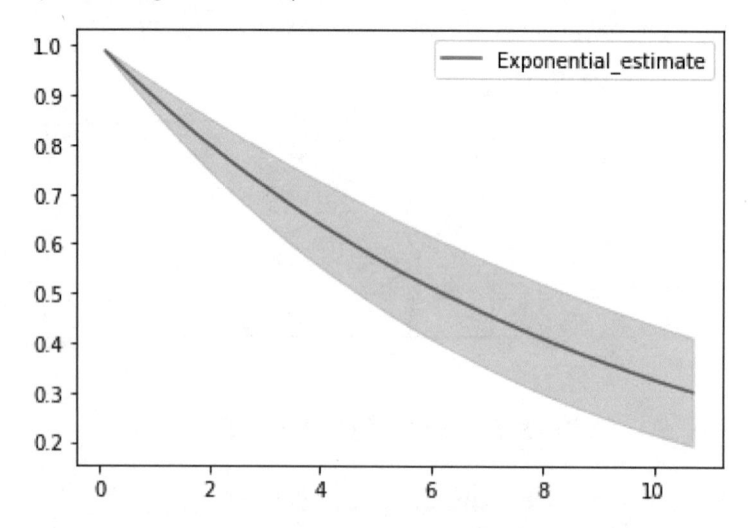

Figure 14.8 – Larynx survival function using exponential distribution

At time step 8, the function remains constant until the end of the study in the Kaplan-Meier model, but in the exponential model, the curve continues to decrease until the end of the study. We also perform similar studies between the younger group (less than 65 years old) and the older group (from 65 years old) (see *Figure 14.9*):

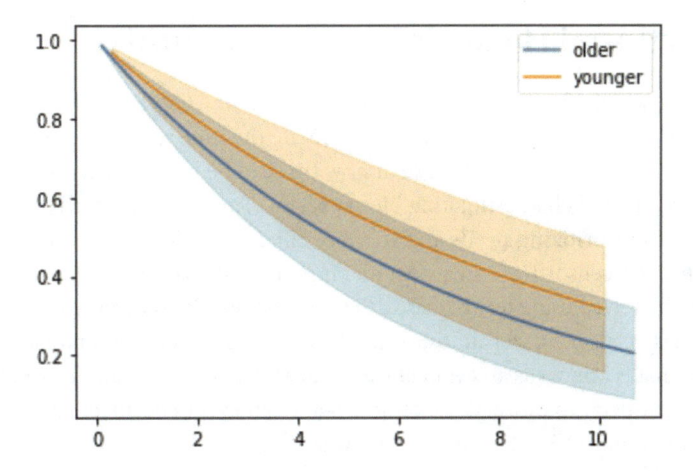

Figure 14.9 – Larynx survival functions for older and younger groups

We also do this on survival analysis with the various stages in the Larynx case (see *Figure 14.10*):

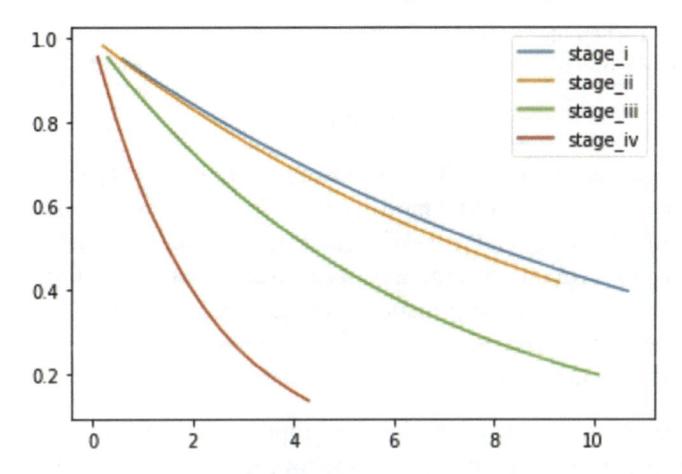

Figure 14.10 – Larynx survival functions based on stage

Both the exponential and Kaplan-Meier models give us similar information. There are advantages and disadvantages between the two models. In the Kaplan-Meier model, results interpretation is simpler; we can estimate the survival function value for both models, but one advantage of the exponential model is to be able to **estimate the hazard ratio**. Two disadvantages of exponential models are that results are not always realistic and that we must assume a constant hazard ratio, which is not always practical. In the next section, we will discuss another survival model, the Cox Proportional Hazards regression model, which is semi-parametric.

Cox Proportional Hazards regression model

Survival analysis, also called TTE analysis, as we discussed in *Chapter 13, Time-to-Event Variables*, is an analytical approach that uses probability to estimate the time remaining before an event occurs based on previous observations. We have seen how this can be helpful when including appropriate covariates in applications such as estimating life expectancy, mechanical failure, and customer churn, which can help with prioritizing needs and to more efficiently allocate resources. As we discussed in depth in *Chapter 13*, censoring is an aspect making survival analysis unique from other statistical questions that can be solved using techniques such as regression. Consequently—and because dropping an observation due to censoring will almost certainly mislead our model and provide results we cannot trust—we insert what is known as an **event status indicator** to help account for whether an event will occur or fail to occur prior to estimating the time before an event's occurrence. Recall that the event status indicator, δ, for object *i* is described as follows:

$$\delta_i = \begin{cases} 1 \text{ if the event was observed,} \\ 0 \text{ if censoring was observed.} \end{cases}$$

We noted that $\delta_i = 1$ when the true event T_i occurs *before* censoring time C_i and $\delta_i = 0$ when the true event T_i occurs *after* censoring time C_i and that thus, the outcome Y for object *i* is this:

$$Y_i = min(T_i, C_i)$$

Censored survival data for object *i* can be captured as (Y_i, δ_i). In this chapter, we have discussed the **non-parametric Kaplan-Meier survival model**, which uses a raw conditional probability of survival counts respective to the sample volume at each point in time without consideration of covariates. We then walked through the parametric **exponential survival model**, which, unlike the Kaplan-Meier model assumes the data fits specific distributions and additionally includes a hazard ratio to compare groups. We will now discuss the semi-parametric **Cox Proportional Hazards survival model**, also referred to as the **Cox Proportional Hazards regression model**.

The primary purpose and distinction of the Cox Proportional Hazards model is to use multiple covariates risk factors to model survival time. These covariates can be categorical or continuous. The Cox Proportional Hazards model is able to account for differences in survival between groups by leveraging the **hazard ratio**. As we have discussed, the hazard ratio is the ratio of the event rate, at any point in time, for one group with respect to another group provided that survival up to that point in time has been maintained among the two groups.

The hazard ratio can be calculated as follows:

$$\frac{\sum \text{Observed events in group A at time } t}{\sum \text{Expected events in group A at time } t} \bigg/ \frac{\sum \text{Observed events in group B at time } t}{\sum \text{Expected events in group B at time } t}$$

The issue with the hazard ratio is that when we fail to obtain a balance of values across covariates shared among the groups, we lose control of confounding variables. When we cannot control for confounding variables, we cannot reliably obtain inference. To account for any imbalance that may exist among covariate values in survival analysis, we can use the **Cox Proportional Hazards model** to introduce coefficients that help explain different levels of exogenous influence. The model follows this form:

$$h(t) = h_0(t) e^{(\beta_1 X_1 + \beta_2 X_2 + \ldots + \beta_n X_n)}$$

Here, $h(t)$ is the expected hazard value at the base time, t, $h_0(t)$ is the baseline hazard value at t when all covariates in the variable matrix X are equal to zero, and n is the count of variables included in the model. The coefficients β_n provide quantification into the impact (effect) of each columnar variable in X. Therefore, we are able to predict survival using the specific values of the input variables multiplied against their corresponding estimated coefficients.

Cox Proportional Hazards has three pertinent assumptions:

- It is assumed the hazard ratio *remains constant* through follow-up. If the nature of input for one group is subject to change, this may negatively influence the results of the model; for example, we compare the performance of two machines, but one is likely to improve when the weather is warmer and the other is not, so our model may not be able to properly model mechanical survival.

- Complete *survival independence*. This means the survival of one study participant or object has no dependence on the other. Consider two separate machines. If one's vibrations are likely to result in an early failure that would not otherwise occur, and that vibration is additionally felt by the other machine, also at that other machine's detriment, there may not be survival independence between the two machines. However, if the vibration of the first machine is not felt by the other, there is likely to be survival independence if all other confounders are controlled for.

- The model assumes *any censoring that takes place is non-informative*. As such, any censoring that takes place does not introduce confounding into the model. Losing a study subject due to censoring prior to follow-up must not be correlated to changes in survival risk. For example, when testing the mechanical survival of two types of machines in a specific area of a facility, if one machine is relocated to another area of the facility to minimize its own self-destructive vibrations, this is informative censoring. However, if the machine was moved to a different area simply to make space and the move does not impact the machine's survival risk or outcome, the censoring is non-informative. Therefore, it does not introduce model confounding. The Cox Proportional Hazards model is for **right-censored** data.

The steps to be taken when performing Cox Proportional Hazards regression modeling are outlined next. The sequence is important as the test should only be performed when practical. These steps follow the outline provided by the **National Institutes of Health (NIH)** (https://www.ncbi.nlm.nih.gov/pmc/articles/PMC7876211/).

Step 1

Set up the null hypothesis to be tested (hypotheses in the multivariate case, which we will look at). We will use the *Stanford heart transplant* dataset, which lists the survival of the patients waiting for transplants in the Stanford heart transplant program. We will use age and year with respect to the start date of the program, transplant status, and the existence of previous bypass surgery as predictors for survival, which ends in either death or transplant. Survival duration is the stop variable, which is the exit time of the study. event is the variable confirming right-censoring and corresponds to either death or transplant. The null hypothesis for each variable is that there is no statistically significant difference in survival between the groups. For example, age has no impact on survival, while awaiting transplant would be the null hypothesis. The code is shown in the following snippet:

```
import statsmodels.api as sm
data = sm.datasets.get_rdataset('heart', package='survival').data
train_data = data.iloc[0:171]
test_data = data.iloc[171:]
exog=train_data[['age','year', 'transplant','surgery']]#df_test[['age','year','surgery','transplant']]
endog=train_data['stop']
data.head(2)
```

Two additional aspects of our data to be aware of are (1) that age has negative values; the starting age value is 48 years old, so an age of -1 corresponds to a 47-year-old patient and (2) year starts November 1, 1967. id is unique to each patient's identification. We will not model this variable but can use it for reference. In *Figure 14.11* we can see, for reference, the first two rows of the data, including all variables we will use to model survival:

start	stop	Event	Age	year	surgery	transplant	id
0	50	1	-17.1554	0.123203	0	0	1
0	6	1	3.835729	0.25462	0	0	2

Figure 14.11 – First two rows of the Stanford Heart Transplant data set

Step 2

We estimate the survival function using a **Kaplan-Meier estimate**, which accounts for right-censoring in the data. In our example, right-censoring occurs when the event variable is equal to one. We can see 171 enter at point 0. On the first day (`stop=1`), corresponding to index column `event_at` equal to 1, we have right-censoring for three patients. We observed the event take place for one patient (`observed=1`) and the other two patients were censored. Here, we show the corresponding data. The patient with `id=45` stops but restarts the same day, then survives until day 45. We see the same for patient `id=3`, who survives until day 16 when they receive a transplant. The patient with `id=15` does not re-enter for follow-up; we assume since they did not receive a transplant, they did not survive.

In the following code sample, the `durations` variable corresponds to survival time and `event_observed` corresponds to the event (transplant or death) that results in right-censoring:

```
from lifelines import KaplanMeierFitter
import matplotlib.pyplot as plt
KaplanMeierFitter = KaplanMeierFitter()
KaplanMeierFitter.fit(durations=train_data['stop'], event_observed=train_data['event'])
KaplanMeierFitter.event_table.reset_index()
```

Note from the Kaplan-Meier results that `event=1` with `stop=1` results in `observed=1`, whereas `event=0` with `stop=1` results in `censored=1` and `stop=1` results in `removed=1`. All three conditions result in those corresponding samples being removed from the calculation for that day's calculation, but unlike the patient who experienced the event, those who were censored are considered to survive at least as long as the study continues. We can see this by considering the Kaplan-Meier survival estimation probability function, S_t:

$$S_t = \frac{n \text{ Patients at Start} - n \text{Patients with Event} = 1 \text{ at time } t}{n \text{ Patients at Start}}$$

Here, survival probability at any time, t, is denoted as S_t.

In looking at our event table from the `KaplanMeierFitter` function, we have the following:

$$S_t = \frac{n \text{ Patients at risk} - n \text{ Patients observed at time } t}{n \text{ Patients at risk}}$$

We can understand that a `removed` patient has limited impact on survival estimations unless they were also noted as `observed=1`, which we can ascribe to `event=1`.

Let's consider patients 3, 15, and 45 to understand this relationship better. We can see based on counts in *Figure 14.12* and *Figure 14.13* that these patients were removed from the denominator on day 1 (`event_at=1`). However, by looking at *Figure 14.14*, we can see patients 15 and 45 have later stops corresponding to events that get factored into the survival function. Let's look at *Figure 14.12* first so we can see the three removed patients on day 1:

	event_at	removed	observed	censored	entrance	at_risk
0	0	0	0	0	171	171
1	1	3	1	2	0	171
2	2	6	3	3	0	168
3	3	6	3	3	0	162
4	4	2	0	2	0	156
...	...	...	...	...	...	...
107	1401	1	0	1	0	5
108	1408	1	0	1	0	4
109	1572	1	0	1	0	3
110	1587	1	0	1	0	2
111	1800	1	0	1	0	1

Figure 14.12 – Summary of the events table for the Stanford Heart Transplant data set

Now, we can see in *Figure 14.13* all patients that had stops on day 1.

```
train_data.loc[train_data['stop']==1]
```

The patients with stops on day 1 are patients with `id` 3, 15, and 45:

start	Stop	event	Age	Year	surgery	transplant	id
0	1	0	6.297057	0.265572	0	0	3
0	1	1	5.815195	0.991102	1	0	15
0	1	0	-11.8166	3.263518	0	0	45

Figure 14.13 – Training data values for patients who exited the study on day 1

To get a better understanding of how this analysis works, we can look specifically at all records for the patients having `id` 3, 15, and 45:

```
train_data.loc[train_data['id'].isin([3, 15,45])]
```

These patients registered a stop on day 1, but patients 3 and 45 also registered stops on days 16 and 45, respectively:

start	stop	event	Age	year	surgery	transplant	id
0	1	0	6.297057	0.265572	0	0	3
1	16	1	6.297057	0.265572	0	1	3
0	1	1	5.815195	0.991102	1	0	15
0	1	0	-11.8166	3.263518	0	0	45
1	45	1	-11.8166	3.263518	0	1	45

Figure 14.14 – Training data values for patients 3, 15, and 45

We can also see patient 45 is the only patient with a `stop` value (corresponding to `event_at`) on day 45. Therefore, we know a transplant results in an observation and a removal. To identify the values at day 45, run the following code:

```
KaplanMeierFitter.event_table.loc[KaplanMeierFitter.event_table.
index==45.0]
```

The code outputs the table in *Figure 14.15*:

event_at	removed	observed	censored	entrance	at_risk
45	1	1	0	0	87

Figure 14.15 – Events on day 45

Now we can look at the events table filtered to events with a stop at day 45 by running the following code:

```
train_data.loc[train_data['stop']==45]
```

This gives us the tabular output seen in *Figure 14.16*:

start	stop	event	Age	year	surgery	transplant	id
1	45	1	-11.8166	3.263518	0	1	45

Figure 14.16 – Events having a stop on day 45

We can repeat this step to see if there is a difference in survival until the end of the study for those who receive a transplant and those who do not by splitting the two groups out in the following code:

```
from lifelines import KaplanMeierFitter
import matplotlib.pyplot as plt
KaplanMeierFitter_n = KaplanMeierFitter()
```

```
KaplanMeierFitter_y = KaplanMeierFitter()
KaplanMeierFitter_n.fit(durations=data.loc[data['transplant']==0]
['stop'], event_observed=data.loc[data['transplant']==0]['event'],
label='no transplant')
KaplanMeierFitter_y.fit(durations=data.loc[data['transplant']==1]
['stop'], event_observed=data.loc[data['transplant']==1]['event'],
label='transplant');
```

We observe in the **Kaplan-Meier survival function** in *Figure 14.17* a steady decrease in survival probability for both groups, but a steeper decrease for the group that does not receive a transplant. We can also see the group, overall, with no transplant does not survive if the group that does receive the transplant. By the time the program closed, there were no non-transplants surviving. The shading corresponds to the 95% confidence interval bands for each probability interval:

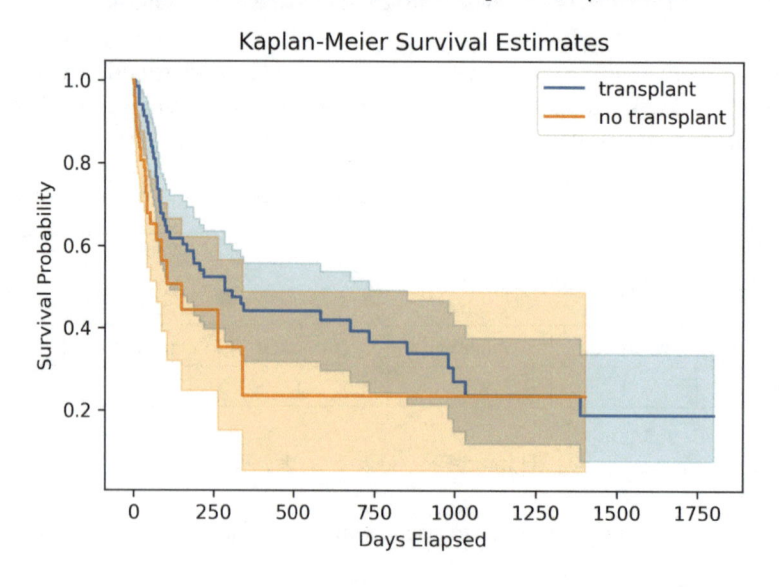

Figure 14.17 – Kaplan-Meier survival estimation probability function

To generate the exact probability of survival on any day (corresponding to the `stop` time), we can run the following code:

```
print('Non-Transplant Survival Probability at day 300: ',
KaplanMeierFitter_n.predict(300))
print('Transplant Survival Probability at day 300: ',
KaplanMeierFitter_y.predict(300))
```

Here we can see the probabilities of survival at day 300 for the transplant and non-transplant groups:

```
Non-Transplant Survival Probability at day 300:  0.3545098269168237
Transplant Survival Probability at day 300:  0.4911598783770023
```

Step 3

Now, we will use a **log-rank test** to determine if survivor curves are statistically different. At this step, we investigate if there is any statistical significance in survival between transplant and non-transplant groups. If there is no statistical significance, it would not be useful to understand the relationships between the covariates and their impacts on survival. We looked at **contingency table analysis** using the **Chi-Square test of independence** earlier in the book. This is essentially the log-rank test here, where we compare the survival distributions of the two groups (transplant versus non-transplant). We use this approach because the data is right-censored, and thus right-skewed; as such, we are working with non-parametric distributions. The null hypothesis is that the hazard ratio between the two groups is equal to 1, which would mean they are equal in terms of survival or hazard risk.

In the following code sample, we use the two different patient distributions (those with and those without transplants) as groups A and B in the `logrank_test` function:

```
from lifelines.statistics import logrank_test
lr_results = logrank_test(durations_A=data.loc[data['transplant']==0]['stop'],
                          durations_B=data.loc[data['transplant']==1]['stop'],
                          event_observed_A=data.loc[data['transplant']==0]['event'],
                          event_observed_B=data.loc[data['transplant']==1]['event'])
```

We can see the Chi-Square distribution is used:

```
print(lr_results.null_distribution)
```

This shows the default distribution is the chi-square distribution:

```
chi squared
```

We can run the following code to see the results of the test:

```
lr_results.summary
```

We can also see a statistically significant difference at a 5% level of significance between the two groups. Therefore, we can assume there is a purpose in proceeding to use the Cox Proportional Hazards model to understand the relationships between the covariates we outlined earlier and the hazard risk for survival:

test_statistic	P	-log2(p)
4.02651	0.044791	4.480663

Figure 14.18 – Log-rank test results

Step 4

Here, we run the Cox Proportional Hazards test and analyze the p-values and confidence intervals for the terms included to identify whether they have statistical significance with respect to influence on survival rate. We look at the hazard ratio to determine any effects they have provided:

```
from lifelines import CoxPHFitter
CoxPHFitter = CoxPHFitter()
CoxPHFitter.fit(df=train_
data[['age','year','surgery','transplant','stop','event']], duration_
col='stop', event_col='event')
CoxPHFitter.print_summary()
```

We can see three of our four terms are significant at the 5% level of significance for predicting survival; our `surgery` variable does not appear to be based on the p-value. The 95% confidence interval for surgery does not contain 0, however, which suggests statistical significance. The difference is because the p-value is calculated based on the test statistic with respect to the critical value, which is estimated using the *log of the hazard ratio*, and the confidence interval is based on the *hazard ratio* itself. This discrepancy could mean the `surgery` variable is significant, but also that it may not be. Since we are not statistically certain, we should err on the side of caution and not consider it useful when predicting future survival.

Note that `exp(coef)` corresponds to the hazard ratio and `coef` is the log of the hazard ratio. This follows the equation we provided earlier for the expected hazard value at the base time, *t*.

We can say, based on the hazard ratio (`exp(coef)`) of 1.03, that a single unit increase in age over age 48 results in an increase of the hazard by a factor of 1.03, or 3%. With respect to our `transplant` variable, we can say with a 95% level of confidence that having the heart transplant decreased the hazard by a factor of 0.54, which is equivalent to 46%.

Here in *Figure 14.19*, we can see the results of the Cox Proportional Hazards regression model:

Model	lifelines.CoxPHFitter
Duration column	stop
Event column	event
Baseline estimation	breslow
Number of observations	171
Number of events observed	74
Partial log likelihood	-298.47
Time fit was run	2023-03-18 16:23:26 UTC

	coef	exp (coef)	se (coef)	coef lower 95%	coef upper 95%	exp (coef) lower 95%	exp (coef) upper 95%	cmp to	z	p	-log2 (p)
Age	0.03	1.03	0.01	0.01	0.06	1.01	1.06	0	2.35	0.02	5.74
Year	-0.16	0.85	0.07	-0.3	-0.02	0.74	0.98	0	-2.25	0.02	5.35
Surgery	-0.64	0.53	0.37	-1.36	0.08	0.26	1.08	0	-1.74	0.08	3.62
Transplant	-0.62	0.54	0.27	-1.15	-0.08	0.32	0.92	0	-2.26	0.02	5.41

Concordance	0.67
Partial AIC	604.95
Log-likelihood ratio test	21.22 on 4 df
-log2(p) of ll-ratio test	11.77

Figure 14.19 – Cox Proportional Hazard model output

In *Figure 14.20*, we can see age, which applies only a 3% increase in hazard, does not explain much of the variance in the results. The age range spans roughly 9 years through 64 years of age. Without other covariates, we cannot say exactly why this might be, but including more covariates could prove helpful in understanding this variable more. Of the significant variables, `transplant` has notably the largest impact on survival outcomes:

```
plt.title('Coefficients within Confidence Intervals')
CoxPHFitter.plot()
```

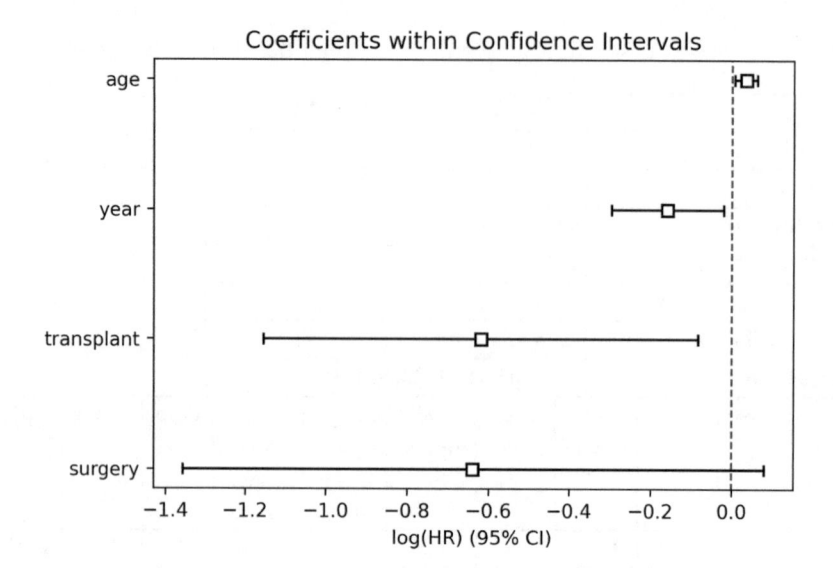

Figure 14.20 – Confidence intervals for hazard-ratio terms

We can see the assumption of a constant hazard ratio between all patients in the survival function plot in *Figure 14.21*. This is why it is important to prevent informative censoring. Let us assume this assumption is met.

Note that when predicting, we do not have start and stop times. Therefore, we can input only the quantifiable covariates for a given patient, as follows:

```
CoxPHFitter.predict_survival_function(train_
data[['age','year','surgery','transplant']]).plot()
plt.xlabel('Survival Time')
plt.ylabel('Survival Probability')
plt.title('Survival Function for All Patients')
plt.legend().set_visible(False)
```

Survival Function for All Patients

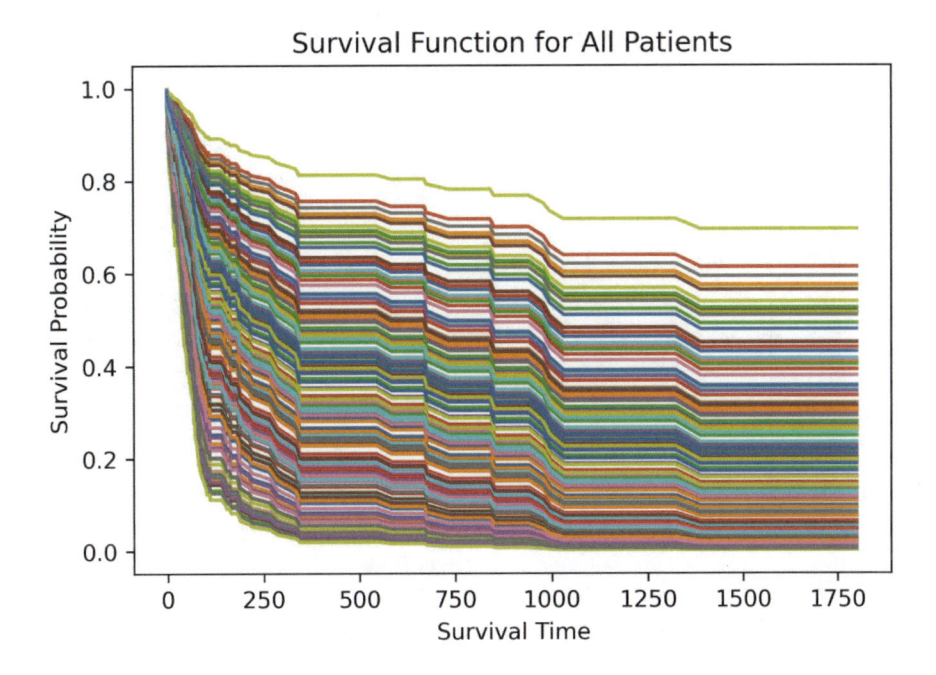

Figure 14.21 – Survival function for all patients

Step 5

In *Figure 14.22*, we can use the Cox Proportional Hazards model to predict the survival function for our held-out test patient:

```
CoxPHFitter.predict_survival_function(test_
data[['age','year','surgery','transplant']]).plot()
plt.xlabel('Survival Time')
plt.ylabel('Survival Probability')
plt.title('Survival Function for Holdout')
```

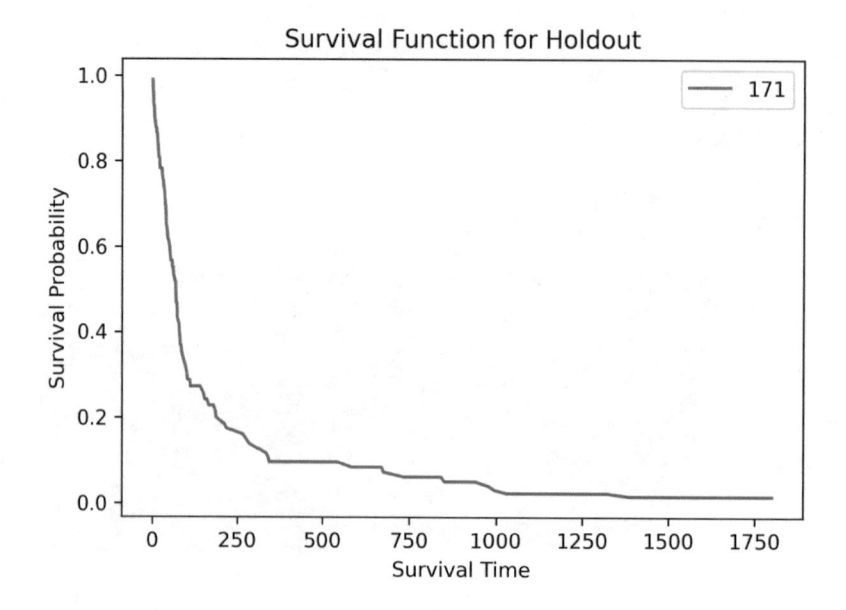

Figure 14.22 – Survival function for test patient, id=171

To generate the probability for this patient's survival, using their covariates, at each potential stop point, we can run the following code:

```
CoxPHFitter.predict_survival_function(test_data[['age','year','surgery','transplant','stop','event']])
```

Summary

In this chapter, we discussed three survival analysis models in depth; the Kaplan-Meier, the exponential, and the Cox Proportional Hazards regression models. Using these frameworks, we modeled survival functions and estimated survival probabilities and hazard ratios for various TTE, right-censored studies. For the multivariate case, we used Cox Proportional Hazards regression to model hazard ratios for covariate analysis on dependent variables. For all models, we demonstrated using the confidence intervals for assessing significance, as well as the corresponding p-values. At this point, the reader should be able to confidently identify the scenarios in which each model would outperform the others and appropriately fit and implement that model to obtain the necessary results for strategic success.

Index

A

accumulated hazard function 369
Aikake Information Criterion (AIC) 278
alternative hypothesis 62
Anaconda 4
　URL 4
Anderson-Darling test 91-96
ANOVA tests 117, 118
　versus pairwise tests 117
ARIMA models 296-299
　fitting 299-302
　forecasting with 302, 303
AR(p) end-to-end example 277
　building 280, 281
　forecast, building 283
　forecast, testing 281, 282
　order of AR(p), selecting 278, 279
　visual inspection 277
autocorrelation 253-257
　structure 253
autocorrelation function (ACF) 165, 322
autoregressive 98
autoregressive (AR) models
　AR(1) model 273-275
　AR(2) model 275, 276
　AR(p) end-to-end example 277
　AR(p) model 272
　order p, identifying using PACF 276
autoregressive integrated moving average (ARIMA) 252, 323
autoregressive integrated moving average with exogenous variables (ARIMAX) 321, 326-328
　exogenous variables, preprocessing 328, 329
　model, fitting 329-332
　model performance, assessing 333, 334
autoregressive moving average (ARMA) model 166, 256, 330, 287-289
　AR(p) model, building 291, 292
　forecast, building 294
　forecast, testing 293, 294
　order of ARMA(p,q), selecting 290, 291
　visual inspection 290
auxiliary OLS regression 222
Average Squared Error (ASE) 294

B

backshift operator notation 273
bagging 52
Bayesian Information Criterion (BIC) 278
Bayesian statistics 228

Index

Bayes' theorem 225
   conditional probability 227
   discussion 228, 229
   probability 225, 226
bias-variance trade-off 188
binary classification 205
Bonferroni correction 115-117
bootstrap aggregation 52
bootstrapping 45
   confidence intervals, creating 46-50
   correlation coefficients (Pearson's correlation) 51, 52
   standard error 51

C

categorical variable
   adding, to MLR 175, 176
   dummy variables 175
   levels 175
censoring
   informative censoring 354
   interval censoring 355
   left censoring 354
   non-informative right censoring 354
   right censoring 354
   Type I censoring 355, 356
   Type II censoring 355, 356
Central Limit Theorem 45
central tendency measurement 26
   mean 30-32
   median 29
   mode 26-28
characteristic equations 251
characteristic polynomial equations 257
chi-squared contingency test 137
chi-square distribution 133, 134

chi-square goodness-of-fit test 135, 136
   power analysis 138
chi-square test of independence 136-138, 379
cluster sampling 16
coefficient of determination 120
coefficients of correlation 148-151
coefficients of determination 151, 152
combinations 52-55
conditional probability 226, 227
confounding variables (confounders) 9
contingency table analysis 379
convenience sampling 17
Cook's distance 164
correlation of determination 158
covariate 321
Cox Proportional Hazards regression model 372, 373
   assumptions 373
   steps 374-384
Cox Proportional Hazards survival model 372
critical value 10
cross-correlation 257-263, 322
cross-correlation function (CCF) 322-324
cross-validation 186
cumulative density distribution 43
cumulative distribution function (CDF) 66
cyclical oscillation 157

D

data distributions
   central tendency measurement 26
   describing 26
   descriptive statistics 26
   measuring 26

shape measurement 38
variability measurement 33
data types 20
   interval data 21
   nominal data 20
   ordinal data 21
   ratio data 22
   visualization 22
data visualization 22
   qualitative data types, plotting 22-24
   quantitative data types, plotting 24, 25
dependent events 225
descriptive statistics 26
Dickey-Fuller test 278, 338
   null hypothesis 278
differencing operation 296
dimension reduction 196
   PCA 196-199
   PCR 199-202
discrete models 205
discrete regression 217
downsampling 312
duration analysis 353
Durbin-Watson test 98, 158
   statistic 158

E

Elastic Net 194-196
Empirical Rule 42
encoding 175
endogenous variable 326
environment setup 3-6
equal population variance 99
   Fisher's F-test 101, 102
   Levene's test 100
   robustness 99
   testing 100

error covariance matrix (ECM) 344
error rate
   selecting 81
event history analysis 353
event status indicator 372
exact testing 55
exogenous variables 326
exponential model 368, 369
   advantages 371
   disadvantages 371
   example 370, 371
exponential survival model 372

F

false negative error 64
false positive error 63
familywise error rate (FWER) 115
feature selection 178, 184
   performance-based methods 186
   recursive feature elimination (RFE) 187, 188
   statistical methods 184
first-order difference 259
first-order linear difference 259
Fisher's F-test 101, 102

G

Gaussian Discriminant Analysis (GDA) 238
Gaussian distribution 42
generalization 3
goodness-of-fit 120
Great Recession 346

H

hazard function 359
hazard ratio 359, 371, 372

heteroscedastic 154
heteroskedasticity tests 281
homoscedasticity 154, 163
hypothesis testing
  for mean 62
  goal 61
  steps 63

I

independent events 225
individual error rate (IER) 115
informative censoring 354
interquartile range (IQR) 34
interval censoring 355
interval data 21
invertibility 284
iris data 211
itertools
  reference link 54

J

Jarque-Bera test 281
Julia 3
Jupyter 4

K

Kaplan-Meier model 362
  definition 362, 363
  example 364-368
Kaplan-Meier survival function 378
Kolmogorov-Smirnov test 91-93
Kruskal-Wallis test 132
kurtosis 40-42, 281, 344

L

lag effect 257
lag-one autocorrelation 158
LASSO regression 189, 192-194
left censoring 354
left-tailed cumulative function 43
Levene's test
  for equality of variances 100
leverage 164
Linear Discriminant Analysis
  (LDA) 225, 229, 230, 232-236
  assumptions 229
  confusion matrix 235, 236
  supervised dimension reduction 236, 237
linear regression
  required model assumptions 152-155
  with OLS 145-148
linear relationship 162
Ljung-Box test 281
logistic distribution 207
logit model 205-210
log-linear model 215
log-linear negative binomial
  regression model 217
log-rank test 379
low-pass linear filter 250

M

machine learning (ML) 226
  machine learning (ML) models 122
MA(q) model invertibility
  identifying 284
math library 71
maximum likelihood estimation (MLE) 219
maximum likelihood method 207
mean 249, 250

mean absolute error (MAE) 169
mean absolute percentage error (MAPE) 187
mean squared error (MSE) 186, 334
median 29
  finding, when number of
    instances is even 29, 30
  finding, when number of instances is odd 29
model
  issues, addressing with residuals 164
  serial correlation, handling 165-169
model evaluation 311
  optimized persistence forecasting 314-317
  resampling 312, 313
  rolling window forecast 317, 318
  shifting 313
model fit evaluation 176
  homoscedasticity, of residuals 178-180
  independent samples 180
  linear relationships 177, 178
  multicollinearity 180
  residual value 178
model validation 169, 170
  diagnostic plots, for analyzing
    model errors 160-162
  Durbin-Watson test 158-160
  QQ plot, of residuals 163
  residuals influence, versus leverage
    influence 163, 164
  residuals vs. fitted 162, 163
  scale-location plot 163
moving average (MA) models
  MA(1) model 284, 285
  MA(2) model 286, 287
  MA(q) model 283, 284
multi-class regression 210
multicollinearity 176
multinomial logistic regression 210

multinomial logit model 210-212
multiple linear regression (MLR) 173, 174
  categorical variables, adding 175, 176
  model fit evaluation 176
  result interpretation 181
multiple tests
  for significance 114
multistage sampling 16
multivariate time series 321, 322

N

National Institutes of Health (NIH)
  reference link 374
natural logarithm 215
negative autocorrelation 158
negative binomial regression model 217
  negative binomial distribution 217-224
nominal data 20
non-informative censoring 354
non-parametric tests 125
non-probability sampling 16
  convenience sampling 17
  quota sampling 17
non-stationary time series
  ARIMA models 296
  differencing data 296
  models 295, 296
normal distribution
  cumulative density distribution 43
  probability density distribution 43
normal distribution of errors 163
Normal Distribution Theorem 42-44
normally distributed data, testing 89
  Anderson-Darling test 93-96
  Kolmogorov-Smirnov test 91-93
  Shapiro-Wilk test 96
  visual inspection 89-91

Index

normally distributed population data, parametric testing 88
  Durbin-Watson 98
  independent samples 97
  robustness 88, 89
  testing for 89
null hypothesis 62

O

observational study 9, 10
  benefits 9
odds ratio 206
one-proportion z-test 78, 79
one-sample t-test 104-108
one-sample z-test 72-76
one-sided tests 62
one-tailed hypothesis test 10
operator notation 273
optimized persistence forecasting 314-317
ordinal data 21
Ordinary Least Squares (OLS)
  using, for linear regression 145-148
outlier values 32
over-dispersion 217
overfit models 186

P

paired t-test 112-114
pairwise tests
  versus ANOVA tests 117
parametric distribution 42
parametric hypothesis test 102
parametric tests
  assumptions 87, 88
  assumptions, violating 125
  equal population variance 99
  normally distributed population data 88
parsimonious model 288
Partial Autocorrelation Function (PACF) 165, 339
PCA technique 196-199
PCR technique 196, 199-202
Pearson's chi-square test statistic 135
Pearson's correlation coefficient 51, 52, 118-122
penalty term 189
performance-based methods, feature selection 186
  models, comparing 186, 187
permutation calculations
  scaling 127
permutations 52-55
permutation testing 55, 126, 127
  performing 55-57
pip 4
  reference link 4
Poisson distribution 213-215, 368
Poisson model 213
  count data, modeling 215-217
pooled t-test 108
population 6
  versus sample 6-8
population inference, samples 8
  observational study 9-11
  randomized experiments 8
positive autocorrelation 158
posterior 228
power 81
  properties 81
power analysis 81, 82, 123
  examples 123
  for two-population pooled z-test 82-84
  one-sample t-test 123

PPMCC test 118
principal components (PCs) 196
prior 228
probability density distribution 43
probability of survival 359
probability sampling 11, 12
  cluster sampling 16
  simple random sampling 13
  stratified sampling 14-16
  systematic sampling 13, 14
probit model 205-210
probit regression 207
p-value 11
Python 3
  URL 3

Q

Quadratic Discriminant Analysis (QDA) 225, 238-243
qualitative data 20
  plotting 22-24
Quantile-Quantile (QQ) plots 89, 163
quantitative data 20
  plotting 24, 25
quota sampling 17

R

Random Forest 52
randomization testing 55
randomized experiments
  characteristics 8
  random assignment, of treatments 8, 9
  random sampling 8
Rank-Sum test 128
  example 129, 130
  normal approximation 129
  test statistic procedure 128
ratio data 22
recursive feature elimination (RFE) 187, 188
regularization 189
reliability analysis 353
reliability theory 353
re-randomization testing 55
resampling technique 312
residuals
  issues, addressing with 164
  versus fitted plot 162
  versus leverage influence plots 163, 164
result interpretation 181
  categorical variable coefficients, interpreting 183
  continuous variable coefficients, interpreting 183
  diagnostic tests 184
  high-level statistics and metrics 182
  model coefficient details 183
ridge regression 189
  example 190-192
ridge regression coefficient estimates 190
right censoring 354
  in survival analysis, for cancer study 355
rolling window forecasting 317, 318
root mean squared error (RMSE) 187
R-squared statistic 157

S

sample 6
sampling 3
sampling methods 11
  non-probability sampling 16, 17
  probability sampling 11, 12
SARIMA models 305

scale-location plots 163
scope of inference 62
seasonal ARIMA models 304
   fitting 306-308
   forecasting 309, 310
   seasonal ARIMA 305
   seasonal differencing 304
seasonal autoregressive integrated moving average with eXogenous factors (SARIMAX) 330
seasonality 296, 304
selection without replacement case 226
selection with replacement case 226
self-selection 8
serial correlation 158, 253
shape measurement 38
   kurtosis 40-42
   skewness 38-40
Shapiro-Wilk test 91, 96, 97
shrinkage methods 188
   Elastic Net 194-196
   LASSO regression 192-194
   ridge regression 189-192
shrinkage penalty 189
signal 250
signed-rank test 130-132
significance
   testing 156-158
significant p-values 157
simple linear regression (SLR) 173
simple random sampling 13
skew 344
skewed distributions 38-40
skew score 281
Spearman's rank correlation coefficient 139, 140
standard deviation 45

standardizing 96
standard normal distribution 42, 43
standard normal probability distribution 42
stationarity 265-270
stationary time series 250
   autoregressive (AR) models 272
   autoregressive moving average (ARMA) models 287-289
   moving average (MA) models 283
stationary white noise 252
statistical discrete distribution 368
statistical measurements 249
   autocorrelation 253-257
   cross-correlation 257-263
   mean 249, 250
   variance 251-253
statistical methods, feature selection 184
   correlation 184, 185
   statistical significance 185, 186
statistical power 82
strata 14
stratified random sampling 9
stratified sampling 14-16
survival analysis 353
survival data 356
survival function 357, 358
systematic sampling 13, 14

T

t-distribution 102, 103
test dataset 186
test statistic 10
time series cross-correlation 322
time-to-event variable analysis 353
training dataset 186

transformations 57, 58
   cube root transformation 57
   log transformation 57
   square root transformation 57
   transformed histograms 58
t-test 102, 103
   for means 103
   one-sample test 104-108
   paired t-test 112-114
   performing 103
   pooled t-test 108
   two-sample t-test 108-112
Tukey fences 37
tuning parameter 189
two-population pooled z-test
   power analysis 82-84
two-proportion z-test 79, 80
two-sample t-test 108
   pooled t-test 108-110
   Welch's t-test 111, 112
two-sample z-test 76-78
two-sided test 62
two-tailed hypothesis test 10
two-tailed probability density function 43
Type I censoring 355, 356
Type I error rate 116
Type I errors 63
Type II censoring 355
Type II error rate 116
Type II errors 63, 64

U

unit root 287
univariate time series 321
upsampling 312

V

variability measurement 33
   quartile ranges 33-36
   range 33
   Tukey fences 36, 37
   variance 37
variance 251-253
variance inflation factor (VIF) 180
vector autoregressive (VAR) 321, 335-337
   cross-correlation, assessing 340-344
   forecast, building 347, 348
   forecast, testing 346, 347
   order of AR(p) 339, 340
   VAR(p,q) model 344, 345
   visual inspection 338, 339
venv
   URL 4
visual inspection
   normally distributed data 89-91
Visual Studio Code 4

W

Wald approach 219
Welch-Satterthwaite adjustment 99
Welch's t-test 111, 112
white-noise model 263-265
white-noise variance 251
Wilcoxon signed-rank test 130
Wolfram Alpha
   URL 305

Y

Yates' continuity correction 138
Yule-Walker method 339

Z

z-ratio 219
z-score 65
  computing 65-67
z-statistic 68-71
z-table
  URL 66
z-test 65
  for means 72
  for proportions 78
  one-proportion z-test 78, 79
  one-sample z-test 72-76
  two-proportion z-test 79, 80
  two-sample z-test 76-78

Packt>

Packtpub.com

Subscribe to our online digital library for full access to over 7,000 books and videos, as well as industry leading tools to help you plan your personal development and advance your career. For more information, please visit our website.

Why subscribe?

- Spend less time learning and more time coding with practical eBooks and Videos from over 4,000 industry professionals
- Improve your learning with Skill Plans built especially for you
- Get a free eBook or video every month
- Fully searchable for easy access to vital information
- Copy and paste, print, and bookmark content

Did you know that Packt offers eBook versions of every book published, with PDF and ePub files available? You can upgrade to the eBook version at packtpub.com and as a print book customer, you are entitled to a discount on the eBook copy. Get in touch with us at customercare@packtpub.com for more details.

At www.packtpub.com, you can also read a collection of free technical articles, sign up for a range of free newsletters, and receive exclusive discounts and offers on Packt books and eBooks.

Other Books You May Enjoy

If you enjoyed this book, you may be interested in these other books by Packt:

Hands-On Simulation Modeling with Python - Second Edition

Giuseppe Ciaburro

ISBN: 9781804616888

- Get to grips with the concept of randomness and the data generation process
- Delve into resampling methods
- Discover how to work with Monte Carlo simulations
- Utilize simulations to improve or optimize systems
- Find out how to run efficient simulations to analyze real-world systems
- Understand how to simulate random walks using Markov chains

Modern Time Series Forecasting with Python

Manu Joseph

ISBN: 9781803246802

- Find out how to manipulate and visualize time series data like a pro
- Set strong baselines with popular models such as ARIMA
- Discover how time series forecasting can be cast as regression
- Engineer features for machine learning models for forecasting
- Explore the exciting world of ensembling and stacking models
- Get to grips with the global forecasting paradigm
- Understand and apply state-of-the-art DL models such as N-BEATS and Autoformer
- Explore multi-step forecasting and cross-validation strategies

Packt is searching for authors like you

If you're interested in becoming an author for Packt, please visit `authors.packtpub.com` and apply today. We have worked with thousands of developers and tech professionals, just like you, to help them share their insight with the global tech community. You can make a general application, apply for a specific hot topic that we are recruiting an author for, or submit your own idea.

Share Your Thoughts

Now you've finished *Building Statistical Models in Python*, we'd love to hear your thoughts! If you purchased the book from Amazon, please click here to go straight to the Amazon review page for this book and share your feedback or leave a review on the site that you purchased it from.

Your review is important to us and the tech community and will help us make sure we're delivering excellent quality content.

Download a free PDF copy of this book

Thanks for purchasing this book!

Do you like to read on the go but are unable to carry your print books everywhere? Is your eBook purchase not compatible with the device of your choice?

Don't worry, now with every Packt book you get a DRM-free PDF version of that book at no cost.

Read anywhere, any place, on any device. Search, copy, and paste code from your favorite technical books directly into your application.

The perks don't stop there, you can get exclusive access to discounts, newsletters, and great free content in your inbox daily

Follow these simple steps to get the benefits:

1. Scan the QR code or visit the link below

   `https://packt.link/free-ebook/978-1-80461-428-0`

2. Submit your proof of purchase
3. That's it! We'll send your free PDF and other benefits to your email directly

Made in United States
Cleveland, OH
09 July 2025